Each World Resources Institute study represents a significant and timely treatment of a subject of public concern. WRI takes responsibility for choosing the study topics and guaranteeing its authors and researchers freedom of inquiry. It also solicits and responds to the guidance of advisory panels and expert reviewers. Unless otherwise stated, however, all the interpretations and findings set forth in WRI publications are those of the authors.

The World Resources Institute (WRI) is a policy research center created in late 1982 to help governments, international organizations, the private sector, and others address a fundamental question: How can societies meet basic human needs and nurture economic growth without undermining the natural resources and environmental integrity on which life, economic vitality, and international security depend?

The Institute's current program areas include tropical forests, biological diversity, sustainable agriculture, global energy futures, climate change, pollution and health, economic incentives for sustainable development, and resource and environmental information. Within these broad areas, two dominant concerns influence WRI's choice of projects and other activities:

The destructive effects of poor resource management on economic development and the alleviation of poverty in developing countries.

The new generation of globally important environmental and resource problems that threaten the economic and environmental interests of the United States and other industrial countries and that have not been addressed with authority in their laws.

Independent and nonpartisan, the World Resources Institute approaches the development and analysis of resource policy options objectively, with a strong grounding in the sciences. Its research is aimed at providing accurate information about global resources and population, identifying emerging issues, and developing politically and economically workable proposals. WRI's work is carried out by an interdisciplinary staff of scientists and policy experts augmented by a network of formal advisors, collaborators, and affiliated institutions in 30 countries.

WRI is funded by private foundations, United Nations and governmental agencies, corporations, and concerned individuals.

Water and arid lands of the western United States

Water and arid lands of the western United States

A World Resources Institute Book

Edited by
Mohamed T. El-Ashry and Diana C. Gibbons
World Resources Institute

The right of the University of Cambridge to print and sell all manner of books was granted by Henry VIII in 1534. The University has printed and published continuously since 1584.

CAMBRIDGE UNIVERSITY PRESS
Cambridge
New York Port Chester Melbourne Sydney

CAMBRIDGE UNIVERSITY PRESS
Cambridge, New York, Melbourne, Madrid, Cape Town, Singapore, São Paulo, Delhi

Cambridge University Press
The Edinburgh Building, Cambridge CB2 8RU, UK

Published in the United States of America by Cambridge University Press, New York

www.cambridge.org
Information on this title: www.cambridge.org/9780521118224

© 1988 by World Resources Institute

This publication is in copyright. Subject to statutory exception
and to the provisions of relevant collective licensing agreements,
no reproduction of any part may take place without the written
permission of Cambridge University Press.

First published 1988
Reprinted 1991
This digitally printed version 2009

A catalogue record for this publication is available from the British Library

Library of Congress Cataloguing in Publication data

Water and arid lands of the western United States/[edited by]
Mohamed T. El-Ashry and Diana C. Gibbons.
p. cm.
"A World Resources Institute book."
Includes bibliographies.
1. Water-supply – West (U.S.) 2. Water resources development – West (U.S.) 3. Irrigation – West (U.S.) 4. Arid regions – West (U.S.) – Management. I. El-Ashry, Mohamed T. II. Gibbons, Diana C.
HD1695.A17W364 1988
333.91′00978 – dc19 888-3662 CIP

ISBN 978-0-521-35040-2 hardback
ISBN 978-0-521-11822-4 paperback

Contents

Contributors

Mohamed T. El-Ashry and Diana C. Gibbons
World Resources Institute

Charles W. Howe and W. Ashley Ahrens
University of Colorado, Boulder

Norris Hundley, jr.
University of California at Los Angeles

Ronald D. Lacewell and John G. Lee
Texas A&M University

William E. Martin, Helen M. Ingram, Dennis C. Cory, and Mary G. Wallace
University of Arizona, Tucson

J. Gordon Milliken
Milliken Chapman Research Group, Inc.
Littleton, Colorado

Charles V. Moore and Richard E. Howitt
University of California at Davis

Henry J. Vaux, Jr.
University of California at Riverside

Foreword

In the continental United States, the daily renewable supply of water totals about 1,400 billion gallons – 14 times what U.S. citizens consume per day. But national averages mask a central fact of American life: much of the western half of the country is arid or semiarid. Its rich greens would soon bleach to desert colors but for water pumped from aquifers or diverted from rivers by highly inventive, though sometimes extraordinarily expensive, means.

Geography may be destiny. Certainly, U.S. water riches are unevenly distributed. In the arid and semiarid West, annual water consumption averages 44 percent of renewable supplies. Everywhere else in the country, the average is 4 percent. This difference explains why rapid population and economic growth exerts particularly intense pressures in the West, why irrigation has become the lifeblood of western agriculture, and why water law and water allocation institutions are now being rattled to their foundations by the test of the times.

The American West is now living on borrowed water. Even discounting farfetched schemes to import water into the region from Canada, the West is using water faster than nature can replenish it. The borrowers are this generation, and the lenders the next.

Water-short or not, the West is still the place to go. Migration from other regions of the country and immigration from Mexico and other points south combine with rapid indigenous growth to make the western Sun Belt one of the fastest-growing regions in the United States. Tucson, where not long ago cowboy movies could be made with little fear that the telltale signs of big-city life would spoil the show, grew more than 50 percent in the 1970s. The entire state of Arizona grew an average of 4.4 percent per year.

Traditionally, over 90 percent of water consumption in the West has fed irrigated agriculture. As of the mid-1980s, irrigation still took the lion's share of water in every western state. But as cities expand and state economies diversify, municipal, commercial, and industrial demands for water multiply. Already the Colorado River Basin and southern California are mining groundwater and importing supplies from adjoining basins. Conventional irrigation practices have also degraded western lands and water. Soil erosion, salinization and waterlogging, high salinity levels in ground and surface water, and toxic elements in surface and subsurface return flows are the price that some water users and society as a whole are paying to make desert agriculture bloom.

Schemes to go farther afield in the search for water will certainly run afoul of growing concern in the West for environmental quality. Opposition to large water projects from environmentalists and from would-be water-exporting basins has grown more organized, vocal, and litigious. Competition among geographic regions with disparate water endowments is intensifying as surplus supplies dwindle, and the lines of the age-old battle in the West between those who would conquer nature and those who would preserve it are being redrawn, infusing water politics with irony.

Traditionally, new water sources were tapped whenever water demand increased. But conventional supply-side responses have now lost some of their luster in the West. The best sites for water development have already been taken; the rest will cost more to exploit. And federal funds may no longer flow as they once did, pooling magically around pork barrels.

As costs rise, water projects will come under closer scrutiny. So will today's water laws and institutions. States that depend mainly on waterworks – so-called structural solutions – will face crippling costs. Those that turn to conservation and such nonstructural means as water markets, marginal cost pricing, least-cost accounting, and interstate water-leasing arrangements will find the flexibility they need to prosper in the next water era.

Against this backdrop, Mohamed El-Ashry, Diana Gibbons, and the other contributors to this volume explore the nature of water demands in agricultural and municipal sectors in the U.S. West and outline policies for maximizing efficiency and minimizing the conflicts inherent in policy change. They show how the West can gradually shy away from expensive supply-side projects and water subsidies and how it can control demand and reallocate existing supplies within water markets, enhancing environmental quality in the process.

Water and Arid Lands of the Western United States also addresses the water management challenge other dry lands face. About one-third of the earth's surface is arid or semiarid. In China, the Indian subcontinent, and the Middle East – all long-settled areas – some lands are now being taken out of production because of irrigation-induced salinity. To transform the desert, water resources in many countries are being overexploited, and some groundwater aquifers have been depleted. Numerous countries are also experiencing conflicts as they try to reallocate limited supplies from agriculture to urban uses. Many of the recommendations included here, when adapted to prevailing social, cultural, and economic conditions, could help increase the productivity of water and land resources in other regions. Indeed, one of the most striking features of the case studies in this volume is the diversity of responses to aridity. Neighboring states have taken different, though perhaps equally successful, paths to sustaining agriculture and culture on desert lands.

Support for this volume and for a workshop on arid lands management attended by resource managers, academics, and government officials was provided by the John D. and Catherine T. MacArthur Foundation and the Joyce Foundation. The commitment and generosity of both are greatly appreciated.

James Gustave Speth
President
World Resources Institute

Acknowledgments

Many people have contributed to this book and the underlying research. We thank, in particular, the contributors and the distinguished advisory panel whose members are listed on page 396. In addition, we are indebted to Charles Meyers, Robert Hagan, John Leshy, Zach Willey, Robert Repetto, and Daniel Luecke for reviewing drafts of one or more chapters and providing valuable comments and suggestions.

We also thank Susan Schiffman, who assisted in the early planning of the project, and Kathleen Courrier and Sheila Mulvihill, who edited and sharpened the manuscript. We are grateful too to Cynthia Veney, our project secretary, Claudia Bedwell of WRI's publications department, and Sue Terry, our research librarian, for their help in finalizing the manuscript. Finally, our sincere thanks to Gus Speth and Jessica T. Mathews, who provided much-appreciated direction and support.

M. T. E.
D. C. G.

1 The West in profile

MOHAMED T. EL-ASHRY
AND DIANA C. GIBBONS

Water development in the American West has been shaped by the region's geography, legal and institutional arrangements, urban expansion, and the spread of irrigated agriculture.

Geography

From the grasslands of west Texas to the deserts of Arizona, the southwest quadrant of the United States is the most arid part of the country. The southern Great Plains states with their vast farmlands and grazing lands are semiarid and relatively flat. To the west, the Rocky Mountains and other ranges in Colorado and New Mexico rise out of the plains to form the massive peaks of the Continental Divide. These mountains shelter some well-watered valleys before giving way to arid basins and deserts in Nevada, Utah, and southern California. Terrain and local climate thus vary considerably throughout the arid and semiarid West. Of the 1.9 billion acres of land in the continental United States, almost half receive less than 20 inches of precipitation per year (Council on Environmental Quality, 1980). (See Figure 1.1.) Water supplies in this arid and semiarid region are both limited and variable. Precipitation in the form of rainfall and snowfall is unevenly distributed across and within the western states. For example, average rainfall in the mountainous area surrounding Flagstaff, Arizona, measures over 20 inches per year. But in central and southern Arizona, where most of the people live and most of the agricultural areas are located, average rainfall ranges from only 7 inches (in Phoenix) to 11 inches (in Tucson) per year (Ruffner and Bair, 1981).

Equally as important as the quantity of rainfall for western water users is the natural variation in supplies from season to season, year to year, and even decade to decade. For example, Tucson receives most of its rainfall during summer thunderstorms between July and September. Much of this precipitation is lost through evaporation and transpiration, so surface runoff is limited and unpredictable. Throughout the West,

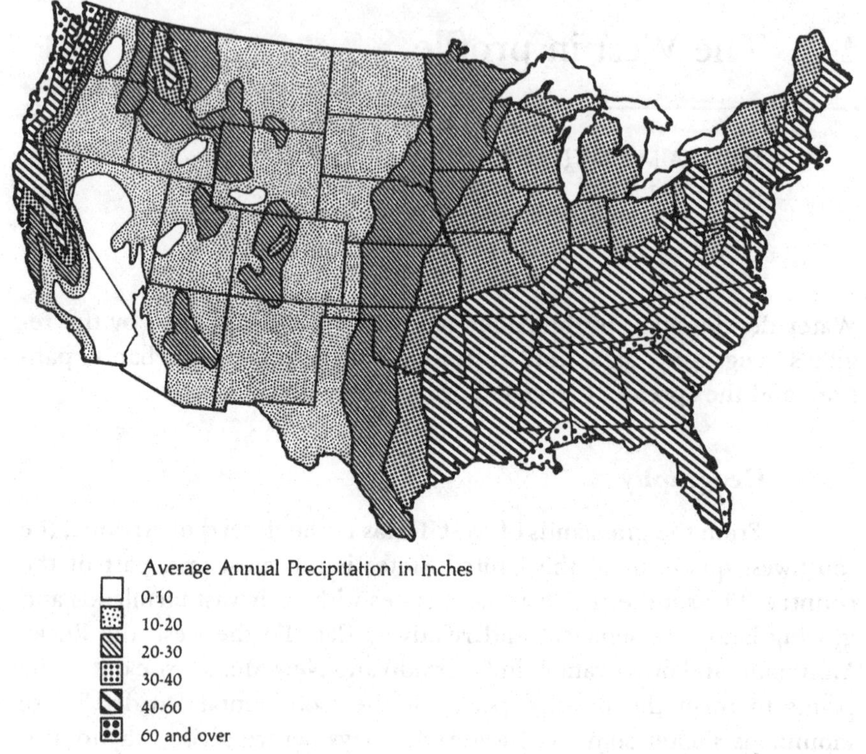

Figure 1.1. Average annual precipitation in the United States (Pye, Patrick, and Quarles, 1983, fig. 3–6).

seasonal variation in streamflows is also governed by the spring melt of the heavy snowpack in the mountains, forming the headwaters of all western rivers. Even long-term streamflows vary greatly. As Figure 1.2 illustrates, the Colorado River has persistent wet and dry cycles measurable in terms of decades.

The other source of water in the West is groundwater, which is a generic term for various subterranean supplies. In general, some groundwater supplies are interconnected with surface flows – tributary aquifers – and other supplies are held in place by impermeable rock layers. These less mobile supplies are often referred to as confined or artesian aquifers. In some cases, ages-old water that is recharged almost imperceptibly is referred to as fossil water. All these types of groundwater supplies are found in the West. For example, the portion of the Ogallala aquifer that underlies much of the High Plains of Texas is ancient and is rapidly being depleted. Similar aquifers are found in southern Arizona.

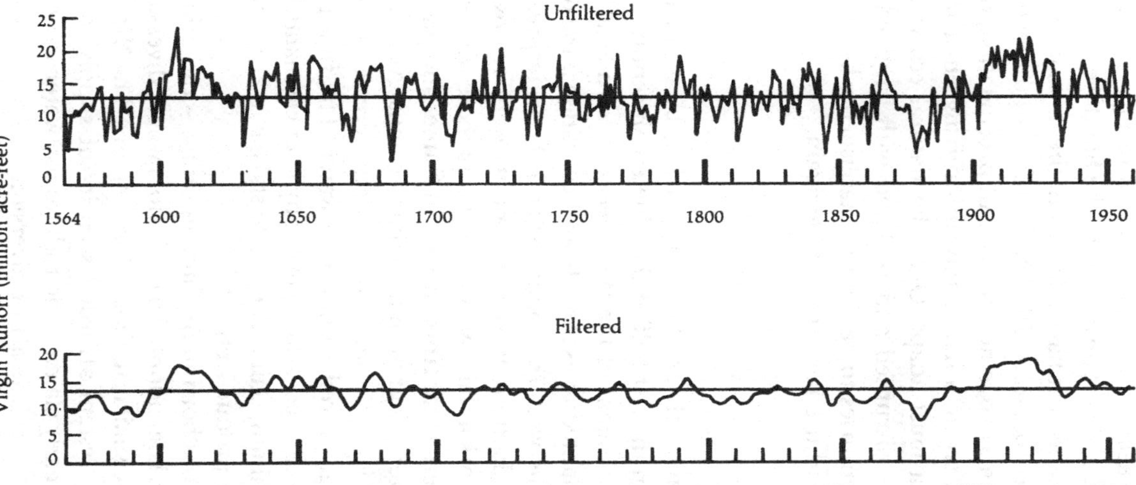

Figure 1.2. Annual runoff of the Upper Colorado River. This 400-year reconstruction of annual runoff at Lees Ferry, Arizona, is based on tree-ring chronologies within the Upper Colorado River Basin collected as part of the Lake Powell Research Project and on chronologies that were in the files of the Laboratory of Tree-Ring Research (Jacoby, 1975, fig. 4).

Hand in hand with the arid climate of many areas of the West are fragile natural ecosystems. The rangelands of the plains suffer frequent and prolonged dry spells that, along with wind erosion, threaten everything but the tumbleweeds. Soil erosion poses a serious environmental threat to the cultivated plains. The deserts of the Southwest are home to the cactus and other plants that have adapted to the dry soil and air and that can withstand a temperature differential of 50°–60°F between the hot days and chilly nights. Together, the lack of precipitation, thin soils, and climatic extremes create a harsh and unforgiving environment to which flora and fauna must adapt. Once the natural ecosystems and indigenous species are disrupted and replaced with irrigated agriculture, they may take centuries to become reestablished when farmlands are abandoned and land use patterns change again.

Water law*

The early history of the West was marked by tremendous inmigration beginning in the last half of the nineteenth century. Two developments during settlement of the West necessitated a legal doctrine for water allocation adapted to extremely arid conditions: the discovery of gold and silver in the western mountains and the widespread use of irrigation in crop production. Most ore deposits were removed from watercourses, and to work a claim all miners except those who panned in streambeds had to divert water from streams in flumes and pipes. Thus, a body of informal water law grew in the mining camps. As it happened, the procedures by which gold and silver claims were established and worked were easily transferred from ore to water. The first person to file a mining claim was allowed priority over any later claimants. To keep one's ownership of a mining claim, one had to stake it, take possession of it, and work the claim productively.

The water law that was born in the mining camps – the prior appropriation doctrine – was later adopted by irrigators whose diversion needs in the arid climate were similar to those of the miners. The "first in time, first in right" priority system afforded security of supply in times of drought to those with early claims. The right to use water was exclusive, absolute, and established by the act of diversion.

As the West grew and states codified their legal systems, some of the basic tenets of the prior appropriation doctrine gradually eroded. In practice, water use effects on third parties were rampant. Because diversion rights were defined in terms of withdrawal quantities, and because

*This section is based on Hundley, 1975, and Anderson, 1983, chap. III.

one user's return flows were another's source of supply, any change in point of diversion or type of use affected appropriators (third parties) downstream. Legislators attempted to correct these third-party effects and the incidence of inflated claims by adding restrictive clauses to the doctrine, such as beneficial and reasonable use requirements, land appurtenancy rules that tie water use to a particular parcel of land, and regulations for compensating parties injured by transfers. Often, these clauses were adopted from riparian water law, practiced predominantly in the East.

Although beneficial use became the basis for an appropriator's water right, efficiency criteria were not expressly defined. Rather, custom and tradition dictated the quantity of water considered reasonable. A water right could be lost if not put to beneficial use over a specified time – hence, the common admonition "use it or lose it" (Pring and Tomb, 1979).

The mixture of water laws adopted by each state often depended on how arid the land was. The Pacific coastal states with plentiful surface supplies adopted features of both doctrines; the dry interior states use prior appropriation water law exclusively. Laws governing the allocation of groundwater supplies likewise vary by state and range from full prior appropriation to absolute ownership to correlative and beneficial use doctrines.

Each western state developed different institutions for implementing the legal doctrines. In New Mexico, a state engineer makes decisions on water transfers and changes in point of diversion or type of use. In Colorado, water courts adjudicate conflicts over water rights. In most states, water rights holders themselves must prove that a proposed alteration of use will not damage the rights of downstream users.

Today, water law in the arid West does indeed protect senior users from supply interruptions and ensures that water entitlements will actually be employed, but efficiency is sacrificed. Prior appropriation doctrine and custom spell out a logical and orderly way to allocate water resources, but they compromise the potential benefits of the resource through cumbersome treatment of water rights transfers.

As western states developed legal definitions and allocation rules for water rights within their borders, it became apparent that the major rivers also needed to be apportioned among the states. The Colorado River Compact, the agreements hammered out among the states in both the Upper and Lower basins, congressional acts, and rulings by the U.S. Supreme Court together allocate the waters of the Colorado River to all the states through which they flow. If an Upper Basin state cannot use its share, that water continues downstream to the Lower Basin states, where it is further allocated to agricultural and urban interests through a priority system.

Significantly, the historic negotiations during 1922 for the Colorado River Compact took place near the end of a 20-year wet cycle for the Colorado River. The amount of water available over the long term now appears to be considerably less than was assumed a half-century ago, so the balance of entitlements is somewhat skewed in favor of the Lower Basin states. Although the compact allocates 7.5 million acre-feet (maf) to both the Upper and Lower basins, it obliges the Upper Basin to deliver 75 maf to the Lower Basin in any ten-year period. The data show that the long-run average yearly flow of the Colorado River may be as low as 13.5 maf (Weatherford and Jacoby, 1975).

Complicating the development of water law and allocation in the West were the economies of scale in water projects that came to light as agriculture grew. Farmers had to band together to afford large-scale works, such as storage dams and conveyance networks needed to move large quantities of water. State governments recognized the need for local umbrella organizations to organize these works and allowed farmers to form quasi-governmental entities with the power to require membership of all beneficiaries. These tax-exempt "districts" retain substantial powers of property taxation and regulation of water use within their jurisdictions; to this day they are central to the development, allocation, and management of water supplies in the West (Leshy, 1983).

Although large projects in the West were initially funded and constructed privately, the federal government soon stepped in to hasten and encourage the economic growth and settlement of the West. The Reclamation Act of 1902 is recognized as the beginning of many decades of federal involvement in constructing and heavily subsidizing water projects in the West. The Bureau of Reclamation was established to construct federally financed dams and other works and to contract water deliveries with the local beneficiaries (usually irrigation districts). The required reimbursement by the beneficiaries was based on their ability to pay, a loosely calculated rate that ensures the solvency of farms as well as government's ultimate failure to collect full repayment. Bureau subsidies added later included long-term contracts at little or no interest, with periods of suspended payments, extension of repayment periods, and below-cost electric power rates for water pumping. Eventually, even the 160-acre maximum farm size was abandoned. In California's Central Valley Project (CVP), only $50 million, or about 5 percent of the total $931 million spent on the project's irrigation facilities over the last 40 years, has been repaid. Because the current average water price is $6.15

per acre-foot, the subsidy amounts to more than 90 percent of the actual cost of $72.99 per acre-foot (LeVeen and King, 1985). These subsidies mean large economic rents, or profits, for a relatively small number of people who, predictably, lobby hard to keep them.

Despite recent attempts to reduce federal subsidies to irrigators, the federal presence in western agriculture – and, indeed, western water – is still strong. Almost 70 percent of the land in the West is publicly owned, much of it by the federal government. Because the West is blessed with many unusual and beautiful natural areas, including the Painted Desert and national parks like Grand Canyon, Bryce Canyon, and Zion, extensive acreage has been preserved in national parks, forests, and wilderness areas. Other lands that have been set aside include Indian reservations, military installations, wildlife preserves, and rangelands. In addition to the reach of land ownership, federal environmental protection laws and the commerce clause of the Constitution assure the federal government a permanent role in western water use and water development for the future (El-Ashry and Gibbons, 1986).

Municipal water demand

Although irrigated agriculture's future in the West was guaranteed by federal subsidies, other economic sectors grew rapidly, especially after World War I. Cities with a wide array of support industries sprang up in the desert as people moved and traveled to the Sun Belt to enjoy the warm, dry climate, the blue skies, and the scenic beauty. Many would argue that the West is still a frontier; the landscape is sparsely populated, with only 17 people per square mile, on average, in New Mexico, Colorado, Utah, Nevada, and Arizona. (The average figure for the entire United States is 64 people per square mile [U.S. Department of Commerce, 1984].) But the low average population density belies the number of people for whom the West is home: almost 30 million people live there. The biggest increases have been fairly recent, with the population up 27 percent in the 1960s and 29 percent in the 1970s (Council on Environmental Quality, 1980). (See Table 1.1.)

The western populace is concentrated in such large metropolitan centers as Denver, Phoenix, Salt Lake City, and Los Angeles. That part of the population living in metropolitan areas ranges from a high of 95 percent in California to a low of 42 percent in New Mexico, and it averages 76 percent throughout the region (U.S. Department of Commerce, 1984). In fact, common perceptions notwithstanding, the West has become the most urban region in the United States (Leshy, 1980).

Table 1.1. *Population growth in selected cities of the West*

City	Population (thousands)			Change (%)	
	1960	1970	1980	1960–1970	1970–1980
Albuquerque, New Mexico	276	333	454	20.7	36.3
Denver-Boulder, Colorado	935	1,240	1,620	32.6	30.6
El Paso, Texas	314	359	480	14.3	33.7
Las Vegas, Nevada	127	273	462	115.0	69.2
Los Angeles–Anaheim–Long Beach, California	7,752	9,981	11,496	28.8	15.2
Phoenix, Arizona	664	971	1,508	46.2	55.3
Salt Lake City, Utah	576	705	936	22.4	32.8
San Diego, California	1,033	1,358	1,862	31.5	37.1
Tucson, Arizona	266	352	531	32.3	50.9

Source: U.S. Department of Commerce, 1981, pp. 18–20.

Although the overall rate of population growth in the West has slowed in recent years, the number of residents continues to rise, and increases well beyond the year 2000 are forecast. The municipal sector in the West uses far less water than the agricultural sector; yet municipal consumption is not insignificant, and it sometimes reaches 10 percent of total water consumption (U.S. Geological Survey, 1984). (See Table 1.2.)

Municipal water use consists of numerous distinctly different demands that can be loosely categorized as residential, public, and commercial (Gibbons, 1986, pp. 7–21). Residential or household demand consists of such outdoor uses as lawn watering, car washing, and evaporative cooling and such indoor uses as bathing and cooking. Public use includes firefighting and maintaining public buildings and grounds. Commercial and light industrial demands are often met through a mu-

Table 1.2. *Municipal water use, 1980*

	Consumptive use (million gallons/day)	% of total consumptive use
Arizona	340	8
California	1,700	7
Colorado	160	4
Nevada	69	4
New Mexico	99	5
Texas	640	6
Utah	300	10

Source: U.S. Geological Survey, 1984.

nicipal delivery system and thus are included in most municipal data on water demand.

Much of the water that cities use is not actually used up. Bathing and cooking, for example, involve little consumption (evaporation) of water but result in municipal effluent of degraded quality. Outdoor uses, on the other hand, consume most of the water applied: the water evaporates, seeps into the ground, or is lost through plant evapotranspiration. Evaporative air conditioning units consume large amounts of water in climates where humidity is low, as it is in Arizona. Including all municipal water uses, consumption averages 25 percent of water withdrawal across the country (U.S. Geological Survey, 1984). In arid locales where lawn irrigation and air conditioning comprise a larger fraction of water demand, the percentage is higher: 61 percent for Arizona and 41 percent for California, for example, as compared to 6 percent for New Hampshire (calculated from U.S. Geological Survey, 1984). In Denver 51 percent of the municipal water supply is used to water lawns – a use that accounts for 94 percent of water consumption in the city (Getches, 1985, p. 11).

Municipal water demand is influenced by several factors. One is season: irrigation of lawns and urban parks and use of water-cooled air conditioners more than double the average monthly water consumption in the summer. Other factors include the number of residents per water meter and the household income level: the more people per household and the more water-intensive amenities (such as swimming pools) they can afford, the greater the household water demand. Communities with higher per capita income usually have more commercial establishments per resident and more golf courses and other public facilities requiring irrigation; thus, they have higher per capita water demand profiles. Population density is another factor: where households are clustered in apartment buildings, lawns and gardens are smaller, so per capita outdoor water use is lower. In general, per capita municipal use is higher in the arid West than anywhere else in the country. For example, monthly summer use by households in Raleigh, North Carolina, averaged 8.8 hundred cubic feet (cf) in 1973; the average in Tucson, Arizona, was 16.4 cf in 1979 (Gibbons, 1986, p. 18). This striking difference reflects the extensive irrigation of both residential and public landscaping in Arizona and the ubiquitous use of water-cooled air conditioning.

Price also determines water demand. Many economic studies confirm the general impression that municipal water use responds to changes in price, although it is termed inelastic because a 10 percent rise in price triggers less than a 10 percent decrease in overall demand. Compelling evidence from statistical studies of municipal water demand indicates that the so-called outdoor water uses are much more price sensitive (or elastic) than indoor uses. (For example, see Howe and Linaweaver, 1967, pp. 13–32; Danielson, 1977; Grima, 1972; Foster and Beattie, 1979.) Then, too, the price elasticity of water demand may be greater at higher water prices and is certainly greater over the long run as capital is invested in water conservation. In other words, individuals are more likely to respond to a price increase if they are already spending a significant portion of their budget on water bills. Similarly, price increases will spur short-term changes in habits, but reductions in household water demand are greater in the long term as consumers invest in such conservation devices as low-flow showerheads and water-saver toilets and replace thirsty gardens with desert flora and rocks.

Water for municipal use is most often supplied by an urban water department, public water utility, or water conservation district that acts as a broker for the consumers. There may be several organizational levels between the holder of the water right and the consumer, including water wholesalers, distributors, and retailers. The ultimate consumers, the households, are supplied with pressurized, treated water, delivered to the tap on demand, a service for which the provider charges according to a rate structure. Most towns have residential water meters and charge a flat rate plus a block-use charge that goes down as consumers use more water. In general, large-volume urban users are given price breaks. Laws prevent publicly owned utilities from making profits, and average cost pricing is one way to ensure that revenues exactly cover costs. Some cities have many unmetered residences: Denver has 88,000 residences without meters (Office of Technology Assessment, 1983). Such municipalities may base tolls on the assessed value of the property, or they may exact a flat fee for water and sewer service.

Most city pricing practices create no incentive to conserve water. To most consumers, the water bill is a minor line item in a monthly budget. At current prices and levels of use, the marginal value of household water is surprisingly small: in other words, with only a modest financial incentive, most households would willingly forego significant amounts of ordinary water use (especially outdoors in summer). With greater scarcity, however, the value of water for household use is almost infinite. As inessential uses are cut back, remaining water use is quite valuable to the consumer, confirming the general perception that water for human consumption has the highest legal priority and the highest requirement for security of supply.

Because the growth of population and associated support industries continues apace in the West, procuring water supplies for cities looms as a major political priority. Forecasts of population growth in the West imply that water demand will also grow, even if per capita use remains the same or falls slightly. By and large, water has to be brought to western cities, often from distant sources. Although most cities sprang up alongside rivers, their needs for water long ago outstripped local surface supplies. Two options once available to western cities – appropriating local instream flows and drilling new wells for groundwater – are less and less tenable (El-Ashry and Gibbons, 1986).

Sometimes a city, such as Tucson, can condemn or annex surrounding agricultural lands and take over the water rights previously used in irrigation or buy agricultural water rights outright. Leasing or exchange agreements can also increase the effective yield of a water system. Alternatively, the city may finance and construct new water storage and importation projects, most often with some financial backing from the state or the federal government. Facing shrinking budgets and rising construction costs, cities are finding large storage projects less attractive. Clearly, the options available to a specific urban area depend in large measure on the local and regional geography and on the water laws and institutions of the state (Getches, 1985, p. 29).

Most urban water management agencies are conservative – that is, their objective is to develop secure supplies to fulfill anticipated demand, whatever that might be. Few water service agencies try to control water demand through pricing, and few take a stand on urban growth. Clearly, however, land use planning, density zoning, and taxation are tools that a city council or other planning agency can use to control urban growth, and decisions on these matters affect future urban water demand.

Because they believe that their mandate is to meet rather than control demand, water service agencies act defensively, undertaking preemptive development to maximize supplies for urban growth far into the future. Acquisitions and projects are usually financed through municipal bonds or special taxes, because a dependable water supply is viewed as an asset to all citizens, regardless of how much water each personally uses. Through rate structures and financing methods, the entities that supply western cities with water have ensured that the household consumer never faces the marginal cost of use. In addition, citizens have few opportunities or incentives to become involved in decision making.

When droughts befall a city, the usual management strategies are to educate and cajole the public into cooperation and to restrict specific water uses through alternate-day lawn watering, a ban on washing cars, and similar measures. Because most utilities design water delivery capacity to meet peak summer afternoon demands, these measures ameliorate a drought's effects by shaving peak demand. Yet, in procuring reliable water supplies and planning for future demand, few cities consider adapting these and similar wise water-demand measures on a regular basis. However, at current rates of consumption, the value of the marginal uses foregone as a result of demand measures is almost invariably less than the cost of expanding the system's capacity.

Irrigation water demand

Although municipal and industrial water needs are growing, water use in the West remains dominated by irrigated agriculture. Everything from alfalfa to pecans can be grown in the desert if enough water is added to the soil. Even in the semiarid lands where dryland farming is possible, irrigation boosts yields and vanquishes anxieties over uncertain precipitation.

Consumptive use for irrigation is the largest single water use in the United States, often reaching 90 percent of total water consumption in western states (Folks-Williams, Fry, and Hilgendorf, 1985). (See Table 1.3.) Agricultural water use has the highest consumption-to-withdrawal ratio, which means that relatively more of the water diverted from streams or aquifers evaporates from the soil or transpires from crops instead of returning to the sources for reuse. This ratio averages about 60 percent, compared to 25 percent in municipal use and 0–25 percent in industrial use (U.S. Geological Survey, 1984).

Nationally, irrigation is a significant factor in the success and size of the agricultural economy. Although irrigated farms make up only one-seventh of all agricultural lands, they contribute more than one-fourth of the total value of crop production (U.S. Geological Survey, 1984). Irrigated acreage increased from about 4 million in 1890 to nearly 60 million in 1977, and about 50 million of these acres are located in the 17 western states (Frederick and Hanson, 1984, p. 1). Although the total acreage under irrigation in the United States is still growing, this growth is concentrated in southeastern states and elsewhere.

The most important determinant of the total number of irrigated acres is undoubtedly the overall crop price index. When national or international supply and demand cause the prices of food and fiber to fall, acreage decreases as well. Thus, unless crop prices rise substantially, water demand for agriculture in the West is not expected to grow. In fact, Bureau of Reclamation data show a sustained decline in irrigated acres in the West since 1979 (with the exception of 1982) and an increase since 1966 in land used for purposes other than crop production (for

Table 1.3. *Irrigated acreage and water use*

State	Irrigated land, 1982 (acres)	Irrigated land as % of total cropland, 1982	Consumptive water use, 1980 (million gallons/day)	% of total consumptive use, 1980
Arizona	1,153,478	74	4,000	89
California	8,460,508	75	23,000	92
Colorado	3,200,942	30	3,600	90
Nevada	829,761	96	1,500	88
New Mexico	807,206	36	1,700	89
Texas	5,575,553	14	8,000	80
Utah	1,082,328	56	2,400	83

Sources: For acreage and land use data, U.S. Department of Commerce, 1981; for water use data, U.S. Geological Survey, 1984.

example, residential, commercial, and industrial purposes) (Bureau of Reclamation, 1984, p. 6). In many areas, irrigated agriculture is threatened by urban encroachment. In metropolitan Phoenix the trend is evident, and on the front range of Colorado a string of cities from Colorado Springs to Fort Collins has advanced into formerly agricultural and grazing lands.

The water for western irrigation comes from several sources and through several institutional arrangements. Farms that use surface water may own appropriative rights, but they are more likely part of an irrigation district or a mutual water company that holds such rights or contracts for water from federal or state projects. The farmer owns shares in the ditch company, which are entitlements to water or contracts for water from the irrigation district; in neither case does the farmer hold the water right except through the umbrella organization.

Much of the water supplied to irrigators originates in state or federal (Bureau of Reclamation) water projects subsidized by taxpayers. When combined with use-it-or-lose-it provisions in state water law and restrictions on use and on the size and timing of return flow, these subsidies are a disincentive to conserving surface water. As long as water is cheap, it will be used inefficiently. And if farmers cannot consider the opportunity cost of retaining water for irrigation, perhaps because resale or leasing is prohibited, they have no reason to make sure that the economic return to the water used for growing crops approaches that in alternative uses (Gardner, 1983).

An exception to the norm of low water prices is the use of groundwater for irrigation. Farmers who rely on groundwater merely pump the water as needed, subject to state laws on pumping rates and well spacing. Because energy costs comprise the bulk of water-procurement costs for these farms, the rise in energy prices since 1974 has eroded profit margins and forced some farmers to conserve water. In many places, the conservation effect of higher energy prices is compounded by the need to go to ever-greater pumping depths as underlying aquifers are mined. As the costs of pumping groundwater have become prohibitive in places, most notably in parts of Texas and central California, the productivity of water use has increased. More efficient pumps and irrigation systems have become economical to install, and management practices have been adapted to the increased water scarcity. Rising water costs have also triggered shifts to higher-value crops and to crops that need less water to grow. When water costs make irrigated crop production less profitable, irrigated acreage finally reverts back to dryland farming, or it is abandoned. In the High Plains of Texas overlying the Ogallala aquifer and in other isolated areas of the West, this shift is already occurring (see Chapter 4).

The economic value of irrigation water depends on many factors, including which crops are grown and how efficiently water is used. Low-value crops – alfalfa and pasture, corn, sorghum, barley, wheat, and other small grains – account for about 74 percent of all irrigated acres in the West. In contrast, vegetable crops and fruit orchards have high average irrigation water values, but they comprise only 9 percent of total acreage (Frederick and Hanson, 1984, p. 25). In the middle range are cotton, soybeans, and other crops, which claim extensive acreage. (See Table 1.4.) Notably, as the efficiency of water use rises, the value of water for growing the crop also rises. According to one analysis, the marginal value of water for growing cotton rose from $61 to $94 per acre-foot as efficiency increased from average (50–70 percent) to high levels (60–80 percent) (Gibbons, 1986, p. 33).

Strictly defined, irrigation efficiency is a measure of how much of the water applied to a field is actually used by the plants. Water not used for crop evapotranspiration may evaporate, seep into the earth, or drain from the land as surface return flow. Some portion of the water seepage moves down through the soil and subsoil to the groundwater table, although some deep percolating water can be lost to the system for years. Systemwide efficiency is further reduced by the seepage and evaporation losses that inevitably occur as water is conveyed from the source. Irrigation efficiency is affected by the type of technology employed and the capital investment in the land and physical structures. Canals and ditches lined with concrete or plastic increase delivery efficiency by

Table 1.4. *Crop water values from three Arizona studies*[a] *(dollars per acre-foot)*

Crop	Willitt, Hathorn, and Robertson	Kelso, Martin, and Mack	Martin and Snyder
Grain sorghum	−1	3–28	23
Barley	5	27–35	32
Alfalfa	20	25–41	24
Wheat	18	30–32	40
Cotton	28–40	89–166	51–65
Vegetables	NA[b]	117	118

[a]All values were converted to 1980 values by use of individual crop price indices.
[b]NA = not applicable.
Sources: Willitt, Hathorn, and Robertson, 1975; Kelso, Martin, and Mack, 1974, pp. 122–126; Martin and Snyder, 1979.

lessening seepage, irrigation scheduling and water-saving equipment (such as trickle irrigation systems) lower evaporative losses, and laser-leveling of land reduces runoff (El-Ashry, van Schilfgaarde, and Schiffman, 1985).

Under present irrigation practices, crops use only about one-half of the water applied. And irrigation return flows, whether surface or subsurface, cause many water quality problems. Runoff from agricultural lands is often degraded by salts, sediments, pesticides, nutrients, and even such toxic trace elements as selenium.

The major water quality problem in the arid and semiarid western states is salinity, which affects nearly every river basin. From headwaters to the mouth, river waters become more saline as seepage and return flow from irrigated land empty into them. In the Texas Rio Grande, for example, the salinity concentration increases from 870 to 4,000 milligrams per liter in a stretch of 75 miles (El-Ashry, 1978).

The salinity levels in several western rivers result not only from irrigation return flows but also from natural saline springs and erodible salt-containing rock formations. By some estimates, as much as half the average salt load of the Colorado River may result from natural sources (U.S. Environmental Protection Agency, 1971).

In many areas, groundwater is severely polluted by deep percolation of irrigation water and seepage from irrigation conveyance systems. The groundwater system can act as a conduit for saline wastewater to enter rivers. In Colorado's Grand Valley, for example, about 145,800 acre-feet of irrigation water enter the groundwater system every year, contributing about 500,000–700,000 tons of salt annually to the Colorado River (El-Ashry, 1980).

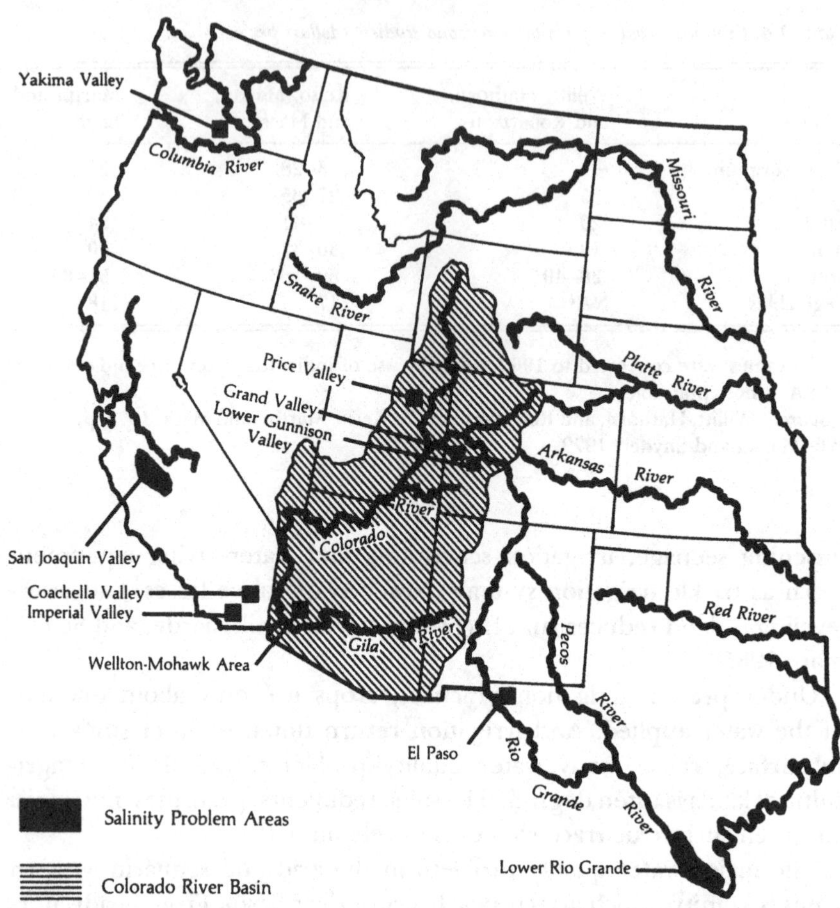

Figure 1.3. Areas with major salinity problems in the western United States (El-Ashry, van Schilfgaarde, and Schiffman, 1985, p. 49).

Salinity is not just a problem of instream water quality. A major threat to agriculture in the West, soil salinity can worsen if saline water is used for irrigation or poorly drained lands become waterlogged. The lack of drainage in many agricultural areas causes the water table to rise, subjecting the productive soil layer to severe salinization and reducing crop yields. An estimated 10 million acres in the West (about 20 percent of all irrigated land) suffer from salt-caused yield reductions (Hedlund, 1984). (See Figure 1.3.)

Each year, salinity causes millions of dollars in damages. In agriculture, the costs are reduced productivity and recourse to salt-tolerant but less profitable crops. Crop yield losses begin when the salt concentration in irrigation water reaches 700–850 milligrams per liter, depending on soil conditions and crop type. (For drinking water, the upper limit recommended by the U.S. Environmental Protection Agency is 500 milligrams per liter of total dissolved solids.) The corrosive effects of salinity on plumbing, industrial boilers, and household appliances plague municipal, commercial, and industrial water users, who also face increased costs for water treatment (U.S. Environmental Protection Agency, 1971).

Salinization's total cost is considerable. In the Colorado River Basin, the heavy salt load of 9–10 million tons annually costs all users more than $113 million a year (in 1982 dollars), and that estimate is expected to more than double by 2010 if controls are not instituted. In the San Joaquin Valley, crop yields have declined 10 percent with financial losses of $31.2 million annually since 1970 because of high saline water tables, and the losses are expected to mount to $321.3 million by 2000 if no action is taken (El-Ashry and Gibbons, 1986).

In all the affected rivers and river basins in the West, salinity problems have worsened progressively as water resources have been developed. This trend will continue with future water development unless the water users contributing to degradation face the damages and comprehensive basinwide water quality management schemes are instituted.

Competition, conflict, and change

The West now stands at a critical juncture. Water use in cities and industries grows each day, although irrigators hold the tickets to most of the West's limited renewable supplies. Except in small municipal projects, supply-side remedies are no longer the answer to water scarcity. Nevertheless, aggressive water development, which has a long and colorful history in the West, is entrenched despite the persuasive economic arguments supporting demand management. Making a smooth transition in the West from water development to water conservation and reallocation will require fundamental changes in long-held attitudes toward natural resources and man's dominion over them.

References

Anderson, Terry L. 1983. *Water Crisis: Ending the Policy Drought.* Baltimore: Johns Hopkins University Press.

Danielson, Leon E. 1977. "Estimation of Residential Water Demand." Economics Research Report 39. Raleigh: North Carolina State University.

El-Ashry, Mohamed T. 1978. "Salinity Problems Related to Irrigated Agriculture in Arid Regions." *Proceedings of Third Conference on Egypt and the Year 2000.* Cairo: Association of Egyptian-American Scholars, 55–74.

El-Ashry, Mohamed T. 1980. "Groundwater Salinity Problems Related to Irrigation in the Colorado River Basin." *Ground Water, 18*(1), 37–45.

El-Ashry, Mohamed T., van Schilfgaarde, Jan, and Schiffman, Susan. 1985. "Salinity Pollution from Irrigated Agriculture." *Journal of Soil and Water Conservation, 40*(1), 48–52.

El-Ashry, Mohamed T., and Gibbons, Diana C. 1986. *Troubled Waters: New Policies for Managing Water in the American West.* Washington, D.C.: World Resources Institute.

Folks-Williams, John A., Fry, Susan C., and Hilgendorf, Lucy. 1985. *Western Water Flows to the Cities.* Vol. 3 of *Water in the West.* Sante Fe, New Mexico: Western Network; and Corelo, California: Island Press.

Foster, Henry S., Jr., and Beattie, Bruce R. 1979. "Urban Residential Demand for Water in the United States." *Land Economics, 55*(1), 43–58.

Frederick, Kenneth D. and Hanson, James C. 1984. *Water for Western Agriculture.* Washington, D.C.: Resources for the Future.

Gardner, B. Delworth. 1983. "Water Pricing and Rent Seeking in California Agriculture." In Terry L. Anderson, ed., *Water Rights: Scarce Resource Allocation, Bureaucracy, and the Environment.* Cambridge, Massachusetts: Ballinger, 83–113.

Getches, David H. 1985. "Meeting Colorado's Water Requirements: An Overview of the Issues." Paper presented at a conference, Colorado Water Issues and Options: The 90's and Beyond. University of Colorado School of Law, Natural Resources Law Center. Oct. 8–9.

Gibbons, Diana C. 1986. *The Economic Value of Water.* Washington, D.C.: Resources for the Future.

Grima, Angelo P. 1972. *Residential Water Demand: Alternative Choices for Management.* Toronto: University of Toronto Press.

Hedlund, John D. 1984. "USDA Planning Process for Colorado River Basin Salinity Control." In Richard H. French, ed., *Salinity in Watercourses and Reservoirs.* Stoneham, Massachusetts: Butterworth Publishing Co., 63–77.

Howe, Charles W., and Linaweaver, F.P., Jr. 1967. "The Impact of Price on Residential Water Demand and Its Relation to System Design and Price Structure." *Water Resources Research, 3*(1), 13–32.

Hundley, Norris, Jr. 1975. *Water and the West: The Colorado River Compact and the Politics of Water in the American West.* Berkeley: University of California Press.

Jacoby, Gordon C., Jr. 1975. "An Overview of the Effect of Lake Powell on Colorado River Basin Water Supply and Environment." Lake Powell Research Project Bull. 14. University of California, Los Angeles, Institute of Geophysics and Planetary Physics.

Kelso, Maurice M., Martin, William E., and Mack, Lawrence E. 1974. *Water Supplies and Economic Growth in an Arid Environment.* Tucson: University of Arizona Press.

Leshy, John D. 1980. "Unveiling the Sagebrush Rebellion: Law, Politics, and Federal Lands." *U.C. Davis Law Review, 14*(2), 347.

Leshy, John D. 1983. "Special Water Districts – The Historical Background."

In James N. Corbridge, Jr., ed., *Special Water Districts: Challenge for the Future.* Proceedings of the Workshop on Special Water Districts at the University of Colorado, Sept. 12–13. Boulder, Colorado: University of Colorado School of Law, 11–30.

LeVeen, E. Phillip, and King, Laura B. 1985. *Turning Off the Tap on Federal Water Subsidies.* San Francisco: Natural Resources Defense Council and California Rural Legal Assistance Foundation.

Martin, William E., and Snyder, Gary B. 1979. "Valuation of Water and Forage from the Salt-Verde Basin of Arizona." Prepared for the U.S. Forest Service.

Pring, George W., and Tomb, Karen A. 1979. "License to Waste: Legal Barriers to Conservation and Efficient Use of Water in the West." *Proceedings of the Twenty-fifth Annual Rocky Mountain Mineral Law Institute, 25*(1), 25–67.

Pye, Veronica I., Patrick, Ruth, and Quarles, John. 1983. *Groundwater Contamination in the United States.* Philadelphia: University of Pennsylvania Press.

Ruffner, James A., and Bair, Frank E. 1981. *The Weather Almanac.* 3d ed. Detroit, Michigan: Gale Research Company.

U.S. Council of Environmental Quality. 1980. *Environmental Quality – 1980.* Washington, D.C.: U.S. Government Printing Office.

U.S. Department of Commerce. 1981. *Census of Agriculture.* Washington, D.C.

U.S. Department of Commerce. 1984. *Statistical Abstract – 1981.* "USA Statistics in Brief," supplement.

U.S. Department of the Interior, Bureau of Reclamation. 1984. *Summary Statistics.* Vol.1, *Water, Land, and Related Data.* Washington, D.C.

U.S. Department of the Interior, Geological Survey. 1984. *National Water Summary 1983 – Hydrologic Events and Issues.* Water Supply Paper 2250. Washington, D.C.

U.S. Environmental Protection Agency. 1971. *The Mineral Quality Problem in the Colorado River Basin.* Washington, D.C.: U.S. Government Printing Office.

U.S. Office of Technology Assessment. 1983. *Water-Related Technologies for Sustainable Agriculture in U.S. Arid/Semi-Arid Lands.* Washington, D.C.: U.S. Government Printing Office.

Weatherford, Gary D., and Jacoby, Gordon C. 1975. "Impact of Energy Development on the Law of the Colorado River." *Natural Resources Journal, 15*(1), 171–213.

Willitt, Gayle S., Hathorn, Scott, Jr., and Robertson, Charles E. 1975. "The Economic Value of Water Used to Irrigate Field Crops in Central and Southern Arizona, 1975." University of Arizona, Tucson, Department of Agricultural Economics Report 9.

2 The Great American Desert transformed: aridity, exploitation, and imperialism in the making of the modern American West

NORRIS HUNDLEY, JR.

The development of the modern American West dates from the mid-nineteenth century, when waves of frontiersmen, confronting what seemed a Great American Desert at the 98th meridian, leapfrogged across half a continent to the Pacific coast in response to the electrifying news of gold in California. From there, the newcomers quickly spread into other resource-rich areas of the region before gradually moving into the vast interior, where they joined with migrants from the East, Europe, and elsewhere to settle the intermountain plateaus and Great Plains. By 1890, there was no longer a discernible frontier line, and the stage was set for the phenomenal growth of the twentieth-century West. During the last century, the 17 contiguous western states and Alaska have moved from backwater status to world leaders in the development of large-scale mechanized agriculture and scientific stockraising, mineral extraction technologies, aerospace and electronic industries, massive multi-purpose public works projects, banking, motion picture and television industries, and tourism.* Growth in the American West, although uneven over the last century and a half, has been spectacular nearly everywhere. The greatest and most cosmopolitan population development has occurred in the urbanized areas along the Pacific coast and in the Sun Belt cities of the Southwest. By the mid-1980s, the West had some

*There is no book or essay that offers an interpretive overview of the American West since major settlement began in the mid-nineteenth century. Of value in formulating this appraisal have been findings gleaned from primary materials, the many writings dealing with aspects of the frontier years, and the fewer works on the twentieth century. Especially insightful are Webb, 1931, Gerald D. Nash, 1973, Worster, 1985, Athearn, 1986, and Limerick, 1987, but the present essay differs in some essentials and emphases from these writings. A useful – although now a decade-old – bibliography with a continental rather than a regional perspective is Paul and Etulain, 1977. More recent and analytical are Michael Malone, 1983, and Nichols, 1986. Special thanks are due to Lawrence Lee and Martin Ridge for commenting on a draft of this paper.

70 million residents, raising it to coequal status with the nation's other regions. At the same time, the West, while very urban, remained the most rural U.S. region, with 747 million acres (one-third of the nation's land area) available for agriculture and an additional 690 million acres in government ownership as national parks, forests, monuments, wilderness areas, mineral reserves, Indian reservations, fish and wildlife preserves, and grazing land (U.S. Department of Commerce, 1984a; 1984b, pp. 12, 204, 657).

Despite the remaining vast open spaces, the rapidity of western growth, characterized by burgeoning cities and massive industrial capacity, has produced uneasiness about what the region has and may become. The concern, especially pronounced during the last two decades, is shared not only by many westerners but also by discerning residents elsewhere who recognize the importance of the region economically, culturally, and aesthetically to the nation and the world. At the heart of this concern are institutions – customs, attitudes, laws, organizational structures – governing the West's natural resources. These institutions have promoted the phenomenal growth of the West, but they have failed to keep pace with the environmentally destructive by-products of that growth. By nature, they reflect a profound historical experience: an anxious people's struggle to survive in a water-shy but resource-rich area. Survival during that struggle became, in the popular imagination, not only dependent on the development of scarce water supplies and the exploitation of natural resources but also synonymous with the symbols of prosperity – great cities, industries, ranches, and farms. Such attitudes and habits dating from the years of earliest settlement are difficult to change.

The West since the mid-nineteenth century has gone through at least four major and interrelated stages of development. In the first, from the late 1840s to about 1900, the region's natural resources, especially those that could be extracted or used relatively inexpensively and without heavy outlays of capital, were vigorously exploited by private interests. Government at all levels favored these efforts, and although its role was important, it was largely indirect. Effective economic control of the region lay in the hands of eastern bankers and financiers who made their fortunes by taking the West's raw materials eastward for manufacturing and marketing. In the second stage, between 1900 and 1940, the West continued as an economic colony of private eastern interests, but government, especially the federal government, began developing and conserving the region's resources and encouraging transportation networks that fostered local manufacturing, attracted labor, and promoted the growth of the towns and cities that had always characterized western life. Scientific and technological advances allowed the exploitation of low-grade resources earlier thought worthless, promoted the more efficient use of raw materials in short or limited supply, and catapulted the West to the nation's leading agricultural area. In the third stage, from 1941 to the 1960s, the West retained its preeminence in agriculture, in stockraising, and as a source of raw material, and portions of the Pacific coast shook off their dependence on eastern financiers and emerged among the nation's leading manufacturing areas. An enormous infusion of government, especially federal, money dramatically increased most westerners' dependence on Washington and accelerated natural resource development, industrial growth, ground and air transportation, and scientific and technological achievements. While eastern capital continued to dominate much of the West, the region's great cities – Los Angeles–San Diego, San Francisco–Oakland, Seattle, Portland, Denver, Salt Lake, Phoenix–Tucson, and Albuquerque – increasingly played Rome to their own hinterlands and sometimes to one another. The fourth stage, from the 1960s to the present, has witnessed intensification of earlier trends, the persistence of imperialism, and continued spiraling growth. But with it has come an increasing awareness of the price paid: air and water pollution, congestion in cities and on highways, destruction of wildlife, loss of wilderness areas, degradation of the soil, depletion of natural resources. Yet most Americans seem less impressed with these negative features than with the economic abundance generated in the West. Any concern for environmental dislocations is piecemeal and fitful rather than systematic and full of conviction.

However distinct these several stages of western development, a persistent theme has been fascination with and commitment to rapid growth. The single most important consideration in explaining that growth has been the role of water. Except for isolated areas in the high mountains and a slice along the northwest coast, the area west of the 98th meridian, which bisects the tier of states from the Dakotas to Texas, is a land of few rivers and sparse rainfall. Consequently, those who have controlled the West's water have controlled the West. Another theme has been the vision of the West as a great storehouse of natural wealth, a cornucopia of virtually inexhaustible raw materials available for exploitation by the bold and adventurous – an idea emboldened by significant scientific and technological breakthroughs. Still other themes have been the potent roles of western urban centers and of outsiders – eastern capitalists, but especially the federal government – in shaping western development. Another theme has been the differential growth of the

West and the relationship of that growth to imperialism of eastern, western, and federal varieties. Variations in the influence of these themes account for the different stages in western growth.

Gold, grass, and grain: private exploitation in the ascendance, 1848–1902

Following the 1848 discovery of gold in California, the West developed rapidly, with attention focused primarily on precious metals in the mountains and stockgrowing and farming on the Great Plains and in the broad valleys of the far West. The natural resources being exploited – gold and silver for miners, grass for cattle and sheep, and fertile soil for growing wheat, corn, and other grains – either were in plentiful supply or promised lucrative rewards for what initially seemed relatively modest investments of capital and equipment. Great fortunes were made, but most investment capital came from the East, and most profits returned there. Westerners experienced both boom times and bust during the latter half of the nineteenth century, but they also left an indelible mark on western institutions and the environment.

Mining and the growth of towns and cities

When James Marshall discovered gold on January 24, 1848, on the headwaters of California's American River, he set off a rush that within three decades transformed the specie-poor United States into one of the world's greatest producers of precious metals. California's contribution alone during those years was a billion dollars, but other parts of the mountain West also contributed mightily (Paul, 1963; Greever, 1963; Trimble, 1914). In 1859, within a decade of Marshall's discovery, a rush nearly as large as that to California was on to the Pike's Peak area of Colorado and another to Virginia City, Nevada, where a lucky prospector discovered the Comstock lode, one of the greatest single concentrations of silver in North America. The following year, rich gold traces were found in Arizona in Tucson; two years later more traces were found in Prescott. As fortune seekers headed for Arizona, others set out for Idaho when gold was discovered along the upper Snake River. Similar findings in Montana at Bannock City (1862) and Butte (1864) and in Wyoming in the South Pass area (1867) prompted stampedes that helped boost the West's non-Indian population from a few thousand in the 1840s to more than 2.3 million by 1870 (U.S. Department of Commerce, 1975, vol. I, pp. 24–37).

Among the immediate spinoffs of the mining discoveries was the rapid growth of towns and cities. In a region of vast distances and sparse water supplies, towns and cities, rather than states and territories, became the goals of migrants and, increasingly, their permanent homes as well. Many of the mining camps, such as Virginia City, Sonora, and Grass Valley, evolved into permanent settlements. Other towns and instant cities, including San Francisco and Denver, mushroomed as commercial centers to supply the needs of miners in nearby areas. Those who did not work in the towns and more urbanized areas still frequently visited them, found their lives dominated by the cultural and economic influences of such communities, and gave them their allegiance. Later development, such as the railroad, intensified town development and expansion (Barth, 1975; Smith, 1967; Wells, 1974; Pomeroy, 1965, pp. 120–121, *passim*; Quiett, 1934).

The rapid population increases in city and countryside following the mining discoveries had immediate political impacts. California, the fastest growing area, entered the Union in 1850 without first becoming a territory; Nevada, within two years of the Comstock discovery, was a territory and, within five years, had achieved statehood; and territorial status quickly followed the mining rushes in Colorado, Arizona, Idaho, Montana, and Wyoming.

The discovery of precious metals had other far-reaching political and legal implications for western development. Virtually all the discoveries occurred on public land, and the miners were technically trespassers. The federal government, however, had neither the inclination nor the ability to eject the miners. Experience and practicality largely explain Washington's posture. During the early years of the new nation, Congress had conveyed a special status on mineral lands and withheld them from sale. Mounting pressures soon led the government to lease some lands and, when that proved unsatisfactory, to begin selling selected acreage in 1829 (IV Stat. L. 364, 1829; Gates, 1968, pp. 701–707). Following the California gold discovery, Congress considered selling the ore-bearing land but quickly dropped the idea when the stampede of hundreds of thousands of miners made it impractical. To turn back the hordes seemed equally out of the question, the more so because the infusion of precious metals into the economy would strengthen the nation. "This is public land," stated Richard B. Mason, the senior U.S. military officer in California, to a group of miners, "and the gold is the property of the United States; all of you here are trespassers, but as the Government is benefited by your getting out the gold, I do not intend to interfere" (Wiel, 1911, vol. I, p. 72).

Mining and the evolution of western water law

Left on their own, the miners in California and elsewhere established regulations to govern resource use and created local associations to enforce them. The most important legal institutions dealt understandably with mining claims, but also of critical significance were the principles they evolved concerning water. Without water, the gold-bearing materials could not be adequately worked, and because water was in short supply nearly everywhere, it took on an importance second only to the ore-rich soil itself. The miners responded by devising rules that treated land and water in similar fashion. The details varied throughout the mining districts, but invariably there was recognition of the same basic principle: first in time, first in right. Gold and water on public lands were free for the taking, and the person with the prime right was the person who first exploited them. "Priority of discovery, location and appropriation" constituted the basis of all mining rights (Wiel, 1911, vol. I, p. 73; Shinn, 1884, pp. 105–258; Meyer, 1984, pp. 145–167). The only condition was that the person working the claim or using the water had to continue doing both to maintain the right to both. The lust for gold and the need for water made any other arrangement seem inconceivable. Others along a stream could appropriate water so long as the supply remained adequate. In times of shortage, however, the latecomers were the first to forego water. There was no *pro rata* sharing of the flow. These principles, as they applied to water, came to be known as the doctrine of prior appropriation. The doctrine was eventually adopted in some form by all western states and as the only system of water law in Colorado, Nevada, Idaho, Utah, Wyoming, New Mexico, Arizona, Montana, and Alaska. Because Colorado in 1876 was the first state to adopt the principle unequivocally, it also became known as the Colorado doctrine. Eastern states had long before begun qualifying their riparian law tradition in favor of some appropriation practices, but those western states following Colorado's lead rejected riparianism outright (Colo. Const., art. 16, secs. 5–7, 1876; Kinney, 1912, vol. 2, pp. 1098–1124; Dunbar, 1983, pp. 78–85; Horwitz, 1977, pp. 32–42, *passim*).

The priority principles as applied to public land changed as the more accessible deposits of placer gold played out and production costs increased. Quartz mining promised rich dividends but required heavy machinery, the digging of numerous shafts, and often massive pumping operations to keep the shafts free from underground water. Expensive equipment also became mandatory for hydraulic mining, in which enormous hoses were used to wash down entire mountainsides and expose the ore buried in prehistoric streambeds. The need for capital investments prompted entrepreneurs to demand ownership of the land they sought to work. Congress responded with the Mineral Land Act of 1866 and important amendments in 1870 and 1872 (XIV Stat. L. 251, 1866; XVI Stat. L. 217, 1870; XVII Stat L. 91, 1872). Together, these measures essentially gave the mining interests what they wanted. Just as some congressmen had advocated at the time of the California discoveries, lands containing gold, silver, and other valuable minerals could now be purchased. In addition, the earlier tacit approval given prospectors to range over public lands now received express statutory assent. Equally important were the provisions concerning water use. Congress officially abandoned its authority (thus far unexercised) in this vital area to "local customs, laws, and the decisions of courts." In effect, Congress with this action formally approved the doctrine of prior appropriation – not only to "rights to the use of water for mining," but also "agricultural, manufacturing, or other purposes" (XIV Stat. L. 253, 1866).

Washington's ratification of past means of exploiting the West's water and precious mineral resources and its endorsement of local institutions encouraging similar practice in a wide range of economic activities proved a boon to continued development of the area. Gone was the fear that investments of capital and labor in water systems could be lost because latecomers upstream diverted needed water. The emphasis remained on the swift commandeering of scarce water supplies.

Cattle industry: from principles of priority to principles of management

While the mining boom was at its height, other entrepreneurs began exploiting the rich grasslands of the Great Plains. This vast expanse of flat and rolling countryside stretching from the Mississippi Valley to the Rocky Mountains and from the Gulf of Mexico on the south into Canada on the north had supported millions of buffalo and other game animals for thousands of years. The mid-nineteenth century witnessed dramatic ecological change. The destruction of buffalo by U.S. soldiers as part of the army's efforts to break Indian resistance and by hungry railroad construction crews, the wholesale slaughter of buffalo and other game by hunters supplying hides to eastern tanners, and Civil War demands for beef helped usher in the era of the cattle industry. From 1860 to about 1890, stockgrowing went through two major phases – the period of the open-range cattle kingdom, lasting until the late 1880s, and then the range-ranching era, which characterized the last decade of the nineteenth century and set the basic pattern for the twentieth.

The open-range cattle industry began in south Texas. The generally mild winters, the comparative lack of predatory animals, the general availability of water, the year-round abundance of lush grasses, and a liberal Texas land policy made the region ideal for breeding cattle (Fugate, 1961, pp. 155–158; Osgood, 1929, pp. 24–58). Cattle drives began even before the Civil War, but dramatic increases accompanied the outbreak of hostilities and the construction of railroads onto the plains, where major towns appeared at the end of well-traveled trails (at Wichita, Dodge City, Abilene, Caldwell, and elsewhere). The long drives also took advantage of the succulent grasses of the northern plains, which in summer were more nutritious than those of the southern plains and enabled drivers to fatten cattle made lean by months on the trail (Dale, 1930, pp. 15–76; Pelzer, 1936, pp. 37–69; Gard, 1954; Dykstra, 1968; Hayley, 1953). The rich grasses, the railheads' steady movement westward, and the emergence of the towns encouraged ranches to proliferate on the plains north of Texas beginning in the 1870s. By 1880, nearly 5 million head of cattle stocked the central and northern range (Bureau of Agricultural Economics, 1938, pp. 77, 79, 81, 83, 113, 117, 121, 123).

The cattlemen's initial success, like that of the miners, was not always within the law. Most lands on the plains north of Texas belonged to the federal government, which in the 1862 Homestead Act offered them for settlement in small 160-acre tracts. Because such allocations proved incompatible with large-scale cattle-growing operations, ranchers ignored the law. Like the miners, they established institutions based on the principle of first in time, first in right. The first cattleman to begin using a stretch of land obtained a so-called range right superior to those who came later. The extent of the right depended on a number of variables, especially the availability of water and the size of the herd. Unlike miners (and later the irrigation farmers), cattlemen were less interested in diverting water from creeks, streams, and rivers than in making accessible to their livestock the watercourses that crisscrossed the open grasslands. Hence, they found little use for (and often vigorously opposed) the appropriation doctrine and concentrated instead on amassing large claims embracing as much as several hundred thousand acres and a stream's entire watershed. Seasonal roundups to separate the inevitable intermingling of herds on the open (unfenced) range and to prepare for drives to railheads were usually organized over drainage basins. The range right was enforced much as mining codes were. Cattlemen formed local associations to resolve disputes among themselves and to protect their enterprises from Indians, thieves, farmers, and unwanted ranchers. As threats mounted, local associations banded into regional and then statewide organizations and finally, in 1884, into the National Cattle Growers' Association, which lobbied Congress in their behalf (Osgood, 1929, pp. 181–185, *passim*; Dale, 1930, pp. 82–89; Treadwell, 1931, pp. 78–94, 344–349; Burroughs, 1971).

Despite the cattlemen's vigorous efforts to promote their interests and the prosperous years in which the return on investments reached 30 percent and higher (and attracted many eastern and British financiers), the open-range cattle industry died young. By the late 1880s it was in serious trouble, and by the early 1890s in ruin. Heady success contributed significantly to the demise. Rising prices led to overproduction, market glut, and then falling prices. Cattlemen fought declining prices by further increasing production, thereby worsening their position in the marketplace. The inevitable overgrazing of the range and the eventual deterioration of the carrying capacity of the land, where native grasses often gave way to scrubby brushwood and other non-native vegetation, only made matters worse, as did the growing number of cattle thieves, quarantines against the free passage of tick-infested Texas cattle, range wars intensified by rivalry over water and ever-decreasing tracts of good grazing land, and farmers pushing onto the plains (Frink, Jackson, and Spring, 1956, pp. 59–92, *passim*; Gressley, 1966, *passim*; Osgood, 1929, pp. 216–258; Dale, 1930, pp. 81–98). The presence of farmers forced many ranchers onto drier areas of the plains; other cattlemen responded with violence and fenced their illegal claims. In the often bloody confrontations, Congress and local authorities increasingly sided with the more numerous farmers. Unlike the miners, who had Congress's blessing, the cattlemen never received congressional approval of their range rights; in 1885 their attempts to fence their lands were officially outlawed in the federal Enclosure Act (XXIII Stat. L. 321, 1885). The coup de grace came with the severe winter of 1886–1887, when stockgrowers on the northern plains lost 40–80 percent or more of their cattle (Mattison, 1951, pp. 5–21).

The catastrophes of the late 1880s led to range ranching, which was more efficient and paid more attention to scientific methods of range management. Such improvements came slowly, but as early as the 1890s significant achievements could be discerned. Rather than allowing the stock animals to fend for themselves during severe winters, cattlemen followed the lead of sheepmen and purchased a section or half section of land on which they raised hay or fodder and provided shelter for the winter months, dug wells, and constructed reservoirs to hold spring freshets. Their stock now grazed on public lands only during summer and fall. Cattlemen also benefited from scientific breakthroughs that controlled the tick and ended the quarantines. As they began breeding better-quality beef for a more discriminating market, they established a pattern that came to dominate twentieth-century ranching (Dale, 1930, pp. 171–185; Schlebecker, 1963, pp. 1–43).

Agriculture: boom, bust, and irrigation

Farmers were not far behind the miners and stockgrowers in pushing beyond the 98th meridian. Like their predecessors, they experienced boom and bust cycles, and their methods changed significantly in response to regional and outside pressures. But their impact on the land was in many respects far more dramatic than that of their predecessors. In the vast areas they put to the plow, they wrought abrupt ecological change by substituting eastern U.S. and European crops for natural vegetation. Except for the relatively small areas cultivated earlier by Indians, Hispanic settlers, and Mormons, these changes occurred first in the wake of the mining rush when pioneer farmers flocked to the well-watered and fertile interior valleys of California, Oregon, Washington, and other western locales to supply the food needs of the hundreds of thousands of fortune seekers. In the 1880s, farmers advanced onto the plains from the eastern prairie provinces. By 1890, they occupied the heartland of the plains, and by the close of the century, they were approaching the front range of the Rockies.

The farmers pouring onto the plains and into the far West had many reasons to come. The mining camps and then the towns and cities that sprang up to supply the commercial needs of the miners and cattlemen provided a ready and growing market for food supplies. Fodder and grain were also needed for the horses and other animals that accompanied the waves of migrants. Additional attractions included the Homestead Act with its promise of virtually free land, the inexpensive state lands put up for sale to raise the funds for schools and colleges, the railroad lands advertised at bargain rates to generate freight traffic, easy credit from rich eastern industrialists who now sought out borrowers and other investment opportunities in the rapidly developing West, and the growing realization that the era of good free or cheap land was rapidly drawing to a close. Another important but insidious attraction was the weather. Climate cycles of wet and dry years on the plains are unpredictable. Agriculture arrived in the region during a spell of wet years beginning in 1882, when annual rainfall averaged nearly 22 inches, half of it falling during the growing season (Fite, 1966, pp. 15–

174; Shannon, 1945, pp. 51–75, 148–172, 215–220; Meinig, 1955, pp. 221–232; Olsen, 1970; Jelinek, 1982, pp. 23–46).

The five years following 1882 witnessed bountiful crops. Harvests of wheat and corn were phenomenal. Soon wheat became the principal crop throughout the West, produced mainly on so-called bonanza farms that were huge and highly mechanized. Kansas, Nebraska, the Dakotas, and the Red River Valley of Minnesota became particularly noted for their wheat harvests. In the far West, the Central Valley of California and parts of eastern Washington and Oregon achieved similar fame. Corn also emerged as a major crop, especially in Kansas, Nebraska, and Iowa, where much of it went for fattening steers and hogs before they were shipped to meat-packing houses in Omaha and Kansas City (Drache, 1964; Fite, 1966, pp. 75–174; Miner, 1986, pp. 172–229; Shannon, 1945, pp. 154–161; Olsen, 1970, *passim*; Jelinek, 1982, pp. 39–46).

The boom in agricultural production did not mean untroubled times for farmers. Increases in grain harvests elsewhere in the world – Canada, Australia, and South America – led to glutted markets, declining demand, and falling prices. For plains farmers, however, the severest blow came when the weather changed in 1887. For the next ten years, there was considerably less rainfall than in other recent cycles. Crop failures were common. More than a third of all farmers went bankrupt, and many left the area. Making the situation worse was continued overproduction in foreign countries and the trade barriers raised against U.S. grain in many parts of the world. When the devastating dry cycle ended in 1897, the farmers who returned to the plains in large numbers had a fuller knowledge of the fickleness of nature and a willingness to adopt new methods and experiment with newer crops (Shannon, 1945, pp. 307–309, *passim*; Hicks, 1931, pp. 30–34, 54–95; Miller, 1951).

Less affected by the severe drought were farmers engaged in irrigation along the West's relatively few streams and rivers. Indians in parts of the Southwest had irrigated land from prehistoric times, and the Spaniards had brought their own irrigation institutions to the pueblos, presidios, and missions they began establishing in the seventeenth century. As the U.S. frontier pushed westward, the Mormons were in the forefront of irrigation. They no sooner arrived in the valley of the Great Salt Lake in July 1847 than they began to build small earthen dams in local streams and to divert water onto the nearby dry but fertile fields. Irrigation spread quickly as the Mormons established colonies elsewhere in Utah and in neighboring Idaho, Arizona, Wyoming, and southern

California (Hewett, 1930; Hollon, 1966, pp. 21–107; Arrington, 1958, pp. 52–54, 84–88). Non-Mormon irrigation colonies also soon dotted the West, the first in 1857 in Anaheim, California, and the second and perhaps best known of these ventures in 1870 in Greeley, Colorado, where the Union Colony attracted national attention. In addition, farmers individually and in groups, as well as investors in the West and East, initiated irrigation enterprises. These efforts received a major boost when state governments authorized individuals to create irrigation districts that could raise taxes to build needed canals and diversion structures. Approved first in California's Wright Act of 1887, the idea found favor in the other 16 western states where it was tested and improved in the following years (Raup, 1932, pp. 123–146; Willard, 1918; Thomas E. Malone, 1965; Dunbar, 1983, pp. 20–35).

Although irrigation spread to many corners of the West, the land cultivated during these years remained modest – 1 million acres in 1880 and only some 3.6 million acres by 1890 (U.S. Department of Commerce, 1975, vol. I, p. 433; Lee, 1977, p. 1000). Nonirrigation farming was comparatively cheap, and heavy capital investments were needed to construct dams, aqueducts, and laterals. Congress encouraged irrigation with the Desert Land Act (1874) and the Carey Act (1894) (XIX Stat. L. 377, 1874; XXVIII Stat. L. 422, 1894). The former conveyed title to 640 acres for a modest price to any individual who would irrigate a portion of the land, and the latter offered a million acres to any state in the semiarid West that would reclaim the land with irrigation. Neither measure had the desired effect. The Desert Land Act promoted speculation, not irrigation, and the Carey Act was undercut by flaws in design and by the depression following the panic of 1893 (Ganol, 1937, pp. 142–157; Hibbard, 1924, pp. 434–439). By the turn of the century, most irrigable lands that could be farmed without heavy capital investments had been developed.

A new hybrid water law and the reaffirmation of priority

Modest but significant advances in irrigation accompanied important developments in western water law, especially in states containing both arid and humid sections. These were the states located on the eastern and western periphery of the arid West, where irrigation, dryland agriculture, and stockgrowing were key elements in the economy. In the arid western heartland, the basic law remained the appropriation doctrine, but the states on the periphery adopted a hybrid principle that combined elements of both the appropriation doctrine and riparian rights.

With the federal government continuing to remain aloof, California pioneered the new system just as it had the earlier appropriation doctrine. The first state legislature had adopted the common law of England in 1850, thereby automatically settling on the state the common law doctrine of riparian rights (Calif. Stat., Apr. 13, 1850, p. 219). The riparian doctrine guaranteed an owner of land abutting a stream the full flow, less only a reasonable amount for domestic needs and for watering cattle and other livestock. A riparian's right was coequal with other riparians along the stream regardless of whether it was exercised. In contrast, the appropriation doctrine being developed in California's gold fields, gradually making its way into accepted legal practice and endorsed by the state supreme court in 1855 (*Irwin* v. *Phillips*, 5 Cal. 140, 1855) and by Congress in the 1866 Mineral Land Act, conveyed the prime water right to the first person using the flow of a stream for as long as the water was used beneficially (XIV Stat. L. 251, 1866). Moreover, as noted earlier, the right did not require ownership of land bordering a stream, and it permitted the diversion of a river and use of the water consumptively for mining, irrigation, or any other reasonable purpose even if the flow of a stream diminished or changed as a result. In addition, the right constituted personal property and could be sold without selling the land. Thus, within a few years of statehood, California had adopted two contradictory systems of water law that reflected different economic interests and environmental realities.

The inevitable confrontation between the conflicting systems occurred when the California Supreme Court agreed to hear the case of *Lux* v. *Haggin* (69 Cal. 225, 1886). The landmark decision of 1886 left the state with a new water law consisting of elements of both competing systems. The court reaffirmed the riparian principle adopted by the first legislature by holding that riparian rights inhered in all public lands when those lands passed into private hands. At the same time, the court held that an appropriator could possess a right superior to a riparian if he began using water *before* the riparian had acquired his property. Put simply, the court approved both systems and made timing the criterion for resolving conflict. The decision reflected the divergent climates and needs not only of California but also of the eight other western states (Washington, Oregon, Nebraska, Oklahoma, Texas, Kansas, and North and South Dakota) that eventually adopted similar principles and christened them the California doctrine (Hutchins, 1971–1977, vol. III, pp. 286–300, 333–358, 408–418, 423–435, 441–467, 478–498, 505–527, 571–601). More important, the decision reaffirmed the advantage given to the earliest settlers in a water-shy area. In this key respect, the

California and the appropriation doctrines were more similar than different.

As this brief overview shows, westerners from the outset looked on water as most crucial to their survival. With the exception of the pioneer Mormons, they would fight more legal battles over it than over any other resource. Moreover, unlike earlier generations of Indians, Spaniards, and Mexicans, who saw water as a common resource belonging to all, they viewed water as a commodity to be used for personal profit. Under the appropriation doctrine, the firmest right went to the person first using the water, not the person with the most economically productive or socially beneficial or politically popular project. The use had merely to be beneficial and continuous – tests that proved easy to meet for newcomers eager to survive and prosper in a land short on water but rich in other resources. The national government encouraged this laissez-faire spirit and promoted rapid economic and urban growth by first allowing and then officially endorsing the water law principles and mining customs devised by westerners. Further encouragement came in the government's liberal land policy, its initial toleration of cattlemen's range rights, its frequent reliance on local initiative to break Indian resistance in order to open additional lands to settlement, and its encouragement of major transportation networks through land grants and loans to railroads. These largely indirect means of stimulating western growth gave way in the twentieth century to increasingly direct governmental activity.

Washington, war, and depression: catalysts for economic development and environmental change, 1900–1940

The close of the nineteenth century marked the first time since Jamestown that no discernible line separated the continent's settled and unsettled portions. The end of the frontier meant that the era of good and relatively cheap or free land was all but over. Absence of easily exploited acreage for mining, stockraising, farming, or lumbering prompted the federal government, especially during the Progressive-era administrations of Theodore Roosevelt and William Howard Taft, to set aside valuable resources for future use. At the same time, the government lent its support to exploiting those resources not set aside and encouraged efforts to develop marginal lands and resources. Scientific and technological advances boosted these efforts by accelerating the creation of new job opportunities and population growth. Playing an especially significant role were the national and local governments, which after 1900 responded to western demands with direct assistance in a broad category of activities that widened dramatically during the Great Depression. By the eve of World War II much of the West, while retaining distinctive characteristics, was beginning to resemble the East.

Agriculture and stockraising: science, technology, and Washington

Western agriculture's hard times in the 1880s and 1890s were followed by unprecedented prosperity during the early twentieth century. By the end of World War I, western agricultural output had nearly doubled, and the region was producing more than half the country's food supplies. California was on its way to becoming the nation's leading agricultural state (a distinction it first achieved in cash income in 1929 and then consistently maintained after 1949). More than half of all the wheat grown in the United States came from the plains. Every part of the West registered significant agricultural gains during the decade before 1918 (U.S. Department of Commerce, 1919, pp. 156–157; 1922, pp. 147–151; 1935, p. 589; Schlebecker, 1975, pp. 206–211; Benedict, 1953, pp. 138–168). The demands of war were a major reason, but so too were advances in the two linchpins of western agriculture: dry farming and irrigation.

Dry farming became synonymous with plains agriculture in the wake of the devastating drought of 1887–1897. It had been practiced earlier, but because wet years coincided with settlement of the plains, most newcomers followed the traditional cropping patterns of the humid East and Europe. The drought abruptly changed such practices. Successful dry farming required some rainfall, but not the 20 or more inches needed for agriculture in humid areas. Ordinarily, plains precipitation was adequate so long as farmers followed methods calculated to conserve the moisture from one growing season to another. By plowing deeply after rains, farmers could provide a reservoir for water. By pulverizing the surface and creating a shallow dust blanket, they could keep enough moisture from evaporating to produce significant yields, especially of wheat, corn, and barley. By eradicating water-consuming weeds and fallowing lands in alternate years, they could build up adequate moisture in the subsoil (Hargreaves, 1957, pp. 21–22, *passim*). Of course, dryland farming required planting drought-resistant crops that use water economically. All wheats possess this quality to some extent, but farmers were soon experimenting with hybrids and importing new strains: Turkey Red from the Crimean areas of Russia; Red Kharkov from southeastern Russia; Kanred, a hybrid produced by agricultural scientists in Kansas; Marquis, another hybrid typical of the many varieties that spread over the plains. A similar pattern emerged with other grains – for example, more than 80 varieties of sorghum corn were introduced – as well as with special strains of alfalfa, cotton, soybeans, and even vegetables. Many plant varieties could be successfully grown with only 9 or 10 inches of rain (Klose, 1950, pp. 109–119, *passim*; Ball, 1930; Malin, 1944).

Besides special farming methods and new crop varieties, mechanical innovations helped make dryland farming popular. The wheats grown on the plains provided the nation's best bread, though millers found it difficult to grind these hard grains until the introduction of stub rollers, which overcame this problem just as the perfection of the disc plow and rotating rod weeder increased yields. So, too, introduction of the gasoline-powered tractor replaced the slow, cumbersome, and expensive-to-operate steam tractor. Especially after 1909, technological advances made possible a machine that was not only highly maneuverable but was also more economical to maintain than a horse. By 1930, there were more than 90,000 gasoline tractors in Kansas and Nebraska alone, and gasoline-driven combines that reaped grain and threshed it simultaneously were also appearing. The relative flatness of the land encouraged mechanization and facilitated the development of large holdings. Flat land also required fewer workers, an important consideration during the labor shortages of World War I. Another popular crop that mechanization significantly aided following its introduction on the plains in the early twentieth century was cotton. Stripper harvesters and mechanical pickers greatly reduced costs for a crop that growers also favored because it required little water and resisted many of the insects and diseases that often ravaged cotton in the humid South. The highly destructive boll weevil, for example, found the plains too cold in winter and too dry in summer (McKibben and Griffin, 1938, pp. 2–7, 33–34, 84; Cooper, Barton, and Brodell, 1947, pp. 30–55).

No less important to the initial success of dry farming than science and technology was the federal government. Indeed, the government developed and disseminated important technical knowledge. Through the Office of Foreign Seed and Plant Introduction, created within the Department of Agriculture by Congress in 1897, experts introduced to the arid West plants brought from all parts of the world: special strains of alfalfa from Turkey, soybeans from Japan, crested wheat grass from Siberia, and a host of other crops. Agricultural experiment stations, established following creation in 1906 of an Office of Dry Land Agriculture in the Department of Agriculture, provided scientific knowledge as well as practical instruction in farming techniques (Klose, 1950, pp. 120–121, *passim*; Rasmussen and Baker, 1972, *passim*; Schlebecker, 1975, pp. 265–266, *passim*). Help of a different sort came in the Enlarged Homestead Act of 1909, which authorized 320-acre homesteads in much of the dry West. But the biggest boost came with World War I. The demands of the military, the destruction of European fields, and the need to feed those drawn to war-created jobs in the cities sent prices and production soaring. So, too, did the subsidies paid by the U.S. Food Administration headed by Herbert Hoover, whose battle cry was "Food Will Win the War!" (Gates, 1968, pp. 503–509; Schlebecker, 1975, pp. 208–211; Lyons, 1964, p. 102).

The rapid expansion in dry farming before and during the war accompanied significant, if less dramatic, increases in irrigation farming. Here the federal government's role was also important. Because most of the major dams that private investors could afford to build were completed by 1900, westerners turned increasingly to Congress for help and mounted an aggressive lobbying campaign. Their demands resulted in the landmark Reclamation Act of 1902, which put the federal government into the business of building irrigation works (XXXII Stat. L. 388, 1902).

To prevent the wealthy from taking advantage of government-constructed dams and reservoirs to create large estates, the new law limited the maximum permissible homestead to 160 acres. The land was free, but farmers had to pay for the construction and operation of the works to provide the water. Only when the debt had been removed (payback was initially set at ten years) would the water user receive title to the land, although storage dams were to remain in government control. Adjacent lands in private ownership before the project was built could also obtain water, but for only 160 acres. Lands in excess of that amount remained dry or could be sold (but not at a price inflated as a result of the government project) to others who could then obtain water for 160 acres. An advance occurred in 1911, when, to check speculation, Congress withdrew from entry all remaining public land susceptible of irrigation until the Secretary of the Interior had determined when water would be available. Government was to build the first works with proceeds from western land sales; later works would be constructed with funds repaid by settlers on the earlier reclaimed lands. Soon the Reclamation Service discovered that hydroelectricity generated at its dams represented a potential source of considerable revenue. In 1906 it persuaded Congress to earmark those funds for irrigation projects. Because preferential pricing for power was given to municipalities, reclamation law proved a boon for urban as well as agricultural development. Congress subsequently authorized additional revenue sources for the reclamation fund, such as the receipts from federal power licenses and oil and mineral leases. These arrangements made cost recovery a key feature of reclamation projects, but not all who benefited bore the financial burden equally. Indeed, in the case of flood control costs, all U.S. taxpayers shouldered the expense because Congress did not require repayment. Those who reaped the most benefit were western irrigation farmers. Their financial liability decreased dramatically when Congress exempted them from interest charges; when the Reclamation Fund swelled as a result of handsome revenues from oil and mineral leases; and especially when the enormous returns from the purchasers of hydroelectricity, most of whom were urban, began pouring in. In its drive to bring people onto the land, the federal government adopted a policy (still in effect) under which nonagricultural interests in the West and taxpayers throughout the nation heavily subsidize the cost of water to farmers (Gates, 1968, pp. 28–29, 667–668; Golzé, 1961, pp. 70–71, *passim*).

Reclamation Service engineers fanned out across the West after 1902, identifying reservoir and diversion sites, measuring streamflow, and constructing dams, first on the Truckee and Carson rivers in Nevada, then on the Salt River in Arizona, later on the Milk River in Montana, and eventually at 664 locales within 75 years (Robinson, 1979, pp. 19–23, 108; Smith, 1986; Glass, 1968; Fahey, 1986, pp. 87–109). Despite such feverish activity, the Reclamation Service's first three decades were marked by major disappointments generated by unrealistic cost estimates, poor choices of location for some projects, political opposition, and settlers' difficulties in paying off construction costs. (The payoff period was continually extended.) In part to overcome such difficulties and to attract and keep settlers, the Reclamation Service officials winked at the 160-acre limitation requirement. When in 1916 they interpreted the law to allow a husband and wife to hold 320 acres, they only did in a small way what they later allowed on a grand scale for agribusiness interests in California's Imperial and Central valleys and elsewhere. Despite the initial problems, the land under cultivation gradually increased. In 1910, some 400,000 acres were being cultivated with Reclamation Service water; by 1920, under the impetus of the First World War and federal subsidies, the figure reached 1,255,000 acres (Robinson, 1979, p. 100; U.S. Department of Commerce, 1922, vol. VII, p. 46). This acreage was small compared to the amount of land being irrigated by other agencies and private enterprise (nearly 18 million acres) and especially miniscule in relation to the vast acreage being dry-farmed. Nonetheless, it marked the federal government's entry into an area where its ultimate impact – through heavily subsidized, large-scale projects – would be extremely significant.

The major advances (nearly three-fourths of all irrigated lands by 1910) occurred initially in Colorado, California, Montana, Idaho, Wyoming, and Utah. California witnessed especially dramatic gains as irrigation acreage increased nearly 60 percent between 1909 and 1919, making the state first in the nation in irrigated lands. Cotton emerged there and in Arizona and New Mexico as a major new crop in response to the demand for the fiber for military uniforms and for clothes for the general public. Expensive irrigation systems, whether built with public or private money, encouraged profitable cash crops and a cheap, mobile labor force to harvest them. In the late nineteenth century, such work was done largely by Chinese, until federal exclusion laws forbade their immigration and they were replaced by Japanese and then by East Indians, Filipinos, and Mexicans, with most newcomers finding employment in the fields as a result of U.S. labor shortages during World War I (U.S. Department of Commerce, 1922, vol. VII, p. 46; Pisani, 1984; Holley and Arnold, 1938, pp. 7–8).

Livestock production also expanded vastly during the war years. Before the war, cattlemen found it extremely difficult to compete in the world market despite advances in ranching techniques and attention to improving beef strains. The war opened formerly closed markets and created others, causing U.S. beef exports to jump from 86 million pounds in 1914 to more than 700 million pounds four years later (U.S. Department of Commerce, 1919, p. 488). The increase reflected not only military demand and overseas needs but also the higher wages of urban workers, who added more meat to their diets, and the federal government's decision to open more public lands (including the national forests) to grazing. Additional federal support came with the Stockraising Homestead Act of 1916, which authorized 640-acre homesteads in many semiarid areas (XXXIX Stat. L. 862, 1916).

Mining production: science, technology, and war

Mining enjoyed similar prosperity during the war years, and again governmental expenditures, along with advances in science and technology, were decisive. Activity focused primarily on low-grade deposits of copper, gold, lead, and other minerals earlier thought not worth the expense. Froth flotation – a water-dependent process in which traces of minerals are pulverized, concentrated, and removed from surrounding dirt and rock – proved to be so cheap and effective that many old mines were reopened and new ones discovered (Rickard, 1921; 1932, pp. 397–414). Applied first to lead-zinc ores and then to copper, this process revolutionized the western mining industry.

Copper production grew particularly quickly, especially in Arizona, Montana, and Utah. In 1907, Arizona, followed closely by Montana, became the nation's leading producer (which it remains). The success of western producers rested on a combination of the froth-flotation process and open pit mining, in which immense deposits of low-grade ores were literally gouged from the earth (Geological Survey, 1909, p. 195; Bureau of Mines, 1982, vol. I, p. 280; Rickard, 1932, pp. 277–300, *passim*; Arrington and Hansen, 1963). The demand for copper accompanied the introduction of electricity and recognition of the metal as an excellent conductor whose qualities were further enhanced by the British-developed electrolytic method of refining. War demands and high prices stimulated additional copper production and also increased production of lead, zinc, and other minerals (Paul, 1977, p. 737).

New technologies and wartime government demands also accompanied sharp advances in the still relatively new petroleum industry. As early as 1911, when the Supreme Court dissolved the Standard Oil Trust, the West accounted for more than 70 percent of national production. That share soon increased with the discovery of the great fields, especially in California, Oklahoma, and Texas, that helped usher in a golden age of expansion by the 1920s (White, 1977, p. 860; Gerald D. Nash, 1968, pp. 1–48; Williamson et al., 1963, pp. 261–338, *passim*). Development of the internal combustion engine and its application to automobiles, airplanes, war and agricultural machinery, and myriad other purposes had created an insatiable market for fossil fuels. Technology served as catalyst, especially with the achievement in 1912 of a commercially successful cracking method that allowed for the extraction of more gasoline from crude petroleum and the improvement of quality through higher octane and increased anti-knock fuel. Offshore drilling, inaugurated as early as 1894 at Summerland on the Santa Barbara Channel in California and perfected during subsequent decades, opened up additional vast new fields that fed the western oil boom. Further stimulating the industry were government wartime expenditures and the Mineral Leasing Act of 1920, which opened federal lands (except national parks) to prospectors and private developers of such resources as oil, natural gas, and coal (Rae, 1965, pp. 48–104; Cleland and Hardy, 1929, pp.

185–186; Gerald D. Nash, 1968, pp. 23–48; Williamson, 1963, vol. 2, pp. 136–150, 261–295; XLI Stat. L. 437, 1920).

New industries and urbanization

Just as the war drove to new heights the exploitative capacity of the West's extractive enterprises, it also spurred new industries, including a host of service occupations that catered to the needs of those attracted to manufacturing jobs. Most of the increases occurred on the Pacific coast and in parts of the Southwest. Plants for manufacturing airplane parts and fuselages appeared in Los Angeles and San Diego, where the favorable climate offered good flying conditions nearly year-round. Shipbuilding expanded significantly on the northwest coast, and many existing western industries, such as canning, meat packing, and lumber milling, increased production. Small local manufacturers nearly everywhere registered gains as they sought to take up the slack created when overtaxed eastern firms found it impossible to fill all their orders (Rae, 1968, pp. 8–10, *passim*; Cleland and Hardy, 1929, pp. 133–164; Elliott, 1970, pp. 1–32). The creation of thousands of new jobs accelerated population growth, especially in the towns and cities where the new industries concentrated. In 1900 the West's population had stood at 11.2 million. Ten years later it was 16.5 million, and by 1920 there were more than 20 million westerners (U.S. Department of Commerce, 1975, vol. I, pp. 24–37).

Many of the newcomers found employment in the fast-growing service industries. Tourism, transportation, banking, real estate, law, and entertainment assumed greater importance in the West than in other regional economies. Tourism had long been a major source of western income as visitors poured west to enjoy the climate, observe Indian societies firsthand, and see the region's many spectacular natural wonders (Pomeroy, 1957, *passim*; Runte, 1979, pp. 82–105; Jakle, 1985, *passim*). The unique beauty, as well as the money-making tourism potential of the scenic outdoors, had years earlier prompted Congress to begin setting aside unusual sites as national parks: Yellowstone (the first in 1872), Yosemite (1890), Mt. Rainier (1899), Crater Lake (1902), Wind Cave (1903), Mesa Verde (1906), Glacier (1910), Rocky Mountain (1915), and others in subsequent years. In 1916, Congress created the National Park Service "to conserve the scenery and the natural and historic objects and the wildlife" of the parks "by such means as will leave them unimpaired for the enjoyment of future generations" (XXXIX Stat. L. 535, 1916). It quickly became clear that "unimpaired" did not prevent construction of trails, roads, shops, parking lots, or hotels (Ise, 1916, pp. 202–205, 326–328, 606–618, *passim*; Pomeroy, 1957, pp. 150–152; Runte, 1979, pp. 104–189; Jakle, 1985, pp. 53, 81, *passim*). In the public mind, the parks, like other resources, existed to be *used*.

Railroad advertising increased the stream of visitors to the West, but nothing had a more dramatic effect than the automobile and highway construction. Cheap and mass-produced, the automobile not only enabled tourism to boom as never before (more than 400,000 automobiles entered the national parks annually by 1926) but also ended the isolation of rural settlements and farms not easily accessible to railroad transportation. Automobile registrations nationally reached nearly 1.26 million by 1913, the first year of substantially reliable data. During the next six years the number jumped to more than 7.5 million, and by 1925 it was 20 million. Westerners adapted rapidly to automobiles, acquiring them in greater numbers than any other region and embarking on a love affair that has not yet ended. The repercussions were widespread: 169,000 miles of federal highways by 1925; urban sprawl; scores of new gasoline stations, restaurants, motor camps, repair shops; and almost countless new jobs (Paxson, 1946, pp. 236–253; U.S. Department of the Interior, 1926, p. 16; U.S. Department of Transportation, 1965, p. 120).

Underwriting much of this growth were the West's rapidly growing financial institutions. Most spectacular was California's Bank of Italy (later Bank of America), which, through the imaginative use of branch banking, emerged in the 1920s as the region's largest and the nation's third largest bank (Giovinco, 1968, pp. 195–218; James and James, 1954, p. 198; Butt, 1960). Also contributing significantly to economic growth and tourism, especially in California, was the motion picture industry. By the mid-1920s, it represented an investment of $1.5 billion and had a payroll of 300,000 people earning $75 million annually. By 1929, five studios alone – Paramount, Fox, Loew's, Warner Brothers, and RKO – had a combined total net income of $60 million (Giannini, 1926, p. 48; Hampton, 1970, p. 408).

Heightened competition for water: urban and state imperialism

The West's rapid growth aroused uneasiness about the adequacy of water supplies. Among those most concerned were business and political leaders in the fastest-growing towns and cities. Because of the expense involved, government, at first local and then federal, played the decisive role in providing additional water – most dramatically, in California.

Los Angeles, which had grown from a small community of 11,000 in 1880 to a large city of 300,000 by 1910, early established control of local supplies by invoking the so-called pueblo water right, a legacy (according to some scholars, an alleged legacy) of the community's Hispanic founding. When these sources could no longer support the growth favored by the city fathers, Los Angeles officials then talked the federal government out of building a reclamation project on the nearby Owens River so that the waters of that stream could be diverted to the ambitious city. By 1913 they had completed a 233-mile-long aqueduct, had left in their wake a bitterness among valley residents that still endures, and had acquired most of the land in the valley as a hedge against threats to the city's water supply (Ostrom, 1953, pp. 3–37, 121–127; Kahrl, 1982, pp. 7–202; Hoffman, 1981, pp. 18–172). But the continued rapid growth of Los Angeles – by 1920 the population was about 600,000 – soon had city officials looking elsewhere for electricity and additional water. When they learned of the Reclamation Bureau's plans to build a high dam on the Colorado River, they endorsed those efforts as a first step in obtaining Colorado River water and hydropower. In so doing, they joined forces with nearby communities and with farmers in California's Imperial Valley who had been agitating for flood protection and for a canal, an All-American Canal they called it, that would free them from the need to divert water through Mexico, where revolutionary conditions and international complications threatened the supply. The resulting package of interests emerged in Congress as the Boulder Canyon Project Bill (Hundley, 1975, pp. 17–52, 116–117, 251; Moeller, 1971, p. 14, *passim*).

Enactment of the bill proved difficult because six other states relied on the Colorado River. They feared that fast-growing California would establish rights to the lion's share of the water, thus sharply curtailing their own future growth. Their fears intensified in 1922, when the Supreme Court in *Wyoming* v. *Colorado* (259 U.S. 419, 1922) affirmed that the appropriation principle of "first in time, first in right" could be applied to states as well as to residents within a state. Now greatly alarmed, the six states bottled up the pending legislation, forcing California in November 1922 to agree to the Colorado River Compact, which divided the river's waters between the states of the Upper Basin (Wyoming, Colorado, Utah, New Mexico) and the Lower Basin (Arizona, Nevada, California) at Lees Ferry, a historic river-crossing station in northern Arizona just south of the Utah border (U.S. House of Representatives, 1923). The agreement, which came only after months of wrangling, foresaw delivery of 7.5 million acre-feet (maf) annually to each basin. In addition to this basic allocation, the Lower Basin states could increase their apportionment by 1 maf, a provision added because Arizona insisted that the Lower Basin (virtually all of which was in Arizona) receive compensation for the water in its tributaries. In allocating the water, the negotiators accepted the Reclamation Bureau's assumption that the Colorado River's average annual flow at Lees Ferry was 16.4 maf, although there was no gauging station there (the estimate was derived from measurements made far downstream near the Mexican border) and years of low flow before 1905 were not considered.

This resort to a compact or treaty to resolve interstate disputes over water use was a pioneering effort. It was a precedent soon imitated on other western rivers – the Rio Grande, Pecos, La Plata, South Platte, Arkansas, Red, Republican, Belle Fourche, Costilla, and Snake – but none of the subsequent agreements would involve as many states, as much land, or so complex a set of issues. Even with the compact, however, fear of California's ambitions led to prolonged debate and further restrictions on the state before the Boulder Canyon Project Act was finally passed in 1928 (XLV Stat. L. 1057, 1928). The measure gave congressional approval to the Colorado River Compact and authorized the construction of Hoover Dam, the creation of Lake Mead (still the nation's largest man-made reservoir), the generation and sale of hydroelectricity (contracted for by southern Californians) to meet building costs, and the construction of the All-American Canal. By controlling the river's flow, the dam would enable Los Angeles and other interested nearby communities, which had banded together in 1924 as the Metropolitan Water District of Southern California, to begin building an aqueduct to the river that would set the stage for greater urban, industrial, and agricultural growth (Metropolitan Water District, 1947).

Other western cities emulated Los Angeles's example. In 1902, San Francisco announced plans to dam spectacular Hetch Hetchy Valley in the Sierra Nevada and divert waters of the Tuolumne River to the city. The news sparked a national campaign led by naturalist John Muir to defeat the project, which required federal approval because the dam would be on public land. Congress temporized but finally gave its permission, which President Woodrow Wilson endorsed in 1913. Other western cities, such as Phoenix, Tucson, and Denver, attracted less attention than San Francisco and Los Angeles, but they too took steps to strengthen their control over local supplies and to plan their own aqueducts (*Congressional Record*, 1913, p. 1189; Clements, 1979, pp. 185–216; Richardson, 1959, pp. 249–258; Dunbar, 1983, p. 214).

Depression and increased dependence on the federal government

Western prosperity in the wake of World War I continued into the 1920s, although the end of hostilities meant a precipitous decline in the fortunes of the region's farmers, miners, and stockgrowers. Agriculture in particular experienced a severe depression after 1919 that lasted until World War II restored prosperity. Population growth slowed and in some areas, especially the plains states, even declined as people seeking work migrated to the cities (U.S. Department of Commerce, 1975, vol. I, pp. 24–37).

The sharp reduction in government purchases was a major reason for the depression in the rural United States. But the situation for farmers and livestock growers worsened following recovery in Europe and increased production in many other parts of the world – wheat in Canada, Argentina, and Australia; cotton in India, Egypt, and Brazil; cattle in Australia and South America. While the world market set the price for U.S. agricultural products and livestock, U.S. growers purchased what they needed at domestic prices. They fought back by increasing production, further saturating an already glutted market. Nor was the domestic grain market helped when horses and mules were replaced by automobiles, trucks, and tractors. The market also softened when the U.S. population growth rate dropped (Shideler, 1957, pp. 46–57, *passim*; Fite, 1964, pp. 67–102). Nature then worsened the situation. Prairie dogs scourged the southern plains, consuming grass in enormous quantities, and on the northern plains, grasshopper plagues in 1921 and 1922 became so severe, especially in Wyoming and Montana, that livestock had to be moved to other ranges. These disasters were followed in 1924 and 1925 by hoof-and-mouth disease epidemics that hit Texas particularly hard (Schlebecker, 1963, pp. 72–89; Nelson, 1946, pp. 300–301).

After 1929, the Great Depression moved from the rural states into the cities and workshops across the nation, sparing almost no person or economic activity. Tourism declined, and the rippling effect led to heavy unemployment in the West's numerous service industries. Only the motion picture business seemed unscathed as millions nationally sought escape from their misery in Hollywood's celluloid make-believe worlds. For the West's already distressed extractive industries, misery gave way to despair. The oil market became almost hopelessly glutted by the combination of declining markets and recent major discoveries, especially in California, Texas, and Oklahoma. The decline in manufacturing led to vastly reduced demands for the West's minerals; along with agriculture and stockgrowing, mining was among the most depressed sectors of the economy (Lowitt, 1984, pp. 8–32, 100–121, *passim*; Gerald D. Nash, 1968, pp. 98–127).

For farmers and stockmen on the plains, the situation deteriorated spectacularly with the worst weather to hit the area since major settlement began. For nine years between 1929 and 1939, rainfall was below average, and 1934 and 1936 were the driest years ever recorded. In 1936, the plains also experienced the hottest (121°F) and coldest (−60°F) temperatures on record. Grasshopper plagues descended frequently on the area during the 1930s, clogging automatic machinery and devouring clothing on the washlines as well as crops, until some semblance of control was established through the wide use of poisons purchased with state and federal funds. The sparse rainfall, the grasshoppers, and the overcropping by farmers and overgrazing by stockmen in a futile attempt to offset declining prices with increased production proved too much for the soft soils of the plains. Especially hard hit was the Texas panhandle and the adjacent area of western Oklahoma, eastern Colorado, parts of New Mexico, and western Kansas. Hot winds whipped up the pulverized surface dirt, creating severe dust storms that often blocked out the sun for days and removed as much as 2 inches of topsoil in many areas. Clogged rivers, harbors, and reservoirs all lost valuable storage area to the sedimentation. Many people fell ill to dust pneumonia, animals had difficulty breathing, and crops and trees were destroyed. The 1930s witnessed perhaps the greatest ecological calamity in U.S. history. Some 300,000 "dusters" headed for Arizona and California, where they inherited the field jobs of Mexicans – some of them U.S. citizens – frightened into returning home by a depression-embittered people ("Drought and Insects," 1934; Hurt, 1981, pp. 33–66; Worster, 1979, pp. 10–25; Schlebecker, 1953, p. 91; Lowitt, 1984, pp. 8–46; Hoffman, 1974; Balderrama, 1982).

In the West, the depression intensified a trend already underway: the region's increasing dependence on the federal government. The inability of local governments and the private sector to provide relief caused westerners as well as residents everywhere in the nation to look to Washington for help. For the West's extractive industries, where the major problem was lack of markets, the government sought to restrict production. To help alleviate unemployment and to provide water and electricity for farms and growing cities, federal agencies created public works projects, including gigantic reclamation ventures. Recognizing the devastating effects of people and climate on the Great Plains, the government also made soil preservation a national responsibility.

Washington's efforts to help the extractive industries raised hopes but produced mixed results. For farmers, the most significant program was the Agricultural Adjustment Administration (AAA), established in 1933 as part of President Franklin Roosevelt's New Deal. It paid farmers to reduce production, a practice that reappeared as a key element in subsequent soil bank plans. Farmers often undercut the program by planting only their best land and fertilizing it richly. Nonetheless, the results frequently proved beneficial for the largest farmers, who had the most land that could be withdrawn from production (Hoover, 1934, pp. 581–583; Lowitt, 1984, pp. 34, 36, 39, 176; Gerald D. Nash, 1973, pp. 161–162; Saloutos, 1969, pp. 345–355).

Among the more successful programs on the plains were those calculated to restore the physical environment. AAA efforts to remove acreage from production contributed measurably to that goal. But the most significant direct action was the Soil Conservation Act of 1935, which established the Soil Conservation Service (SCS) within the Agriculture Department, marking for the first time Congress's official acknowledgment of the federal government's responsibility for soil protection. SCS's mandate, primarily educational, consisted of showing farmers how to avoid future dust bowls by rotating crops, fallowing, contour plowing, strip cropping, liming pastures, taking marginal lands out of production and returning them to grass, and adopting other practices to avoid soil erosion. An important institutional by-product was the soil conservation district, authorized by Congress in 1937. The state-operated districts consisted of the farmers who drew up erosion control plans for their areas and the individual farmers within them. During the next two decades, 90 percent of the nation's farmland came within such districts (Held and Clawson, 1965, pp. 41–56; Hargreaves, 1976, pp. 561–582; Simms, 1970, *passim*). A major experiment of the Soil Conservation Service that at first provoked skepticism and then admiration was the shelterbelt project, which resulted in the planting of millions of trees as windbreaks on the plains (Droze, 1977; Hurt, 1981, pp. 84–86, 100–101, 123–137).

Stockgrowers also benefited from federal action. The AAA purchased great numbers of cattle, sheep, and goats, which were either shipped to the East or slaughtered so the meat could be distributed through other New Deal welfare programs. By the end of January 1935, when the buying program ceased, the federal government had purchased 8.3 million head of cattle. Other federal measures helped stockgrowers maintain foundation herds, fight livestock diseases, dig wells, and produce forage (U.S. Department of Agriculture, 1935, p. 18; FitzGerald, 1935, pp. 200–201).

Clearly, the most important development for stockgrowers was the Taylor Grazing Act of 1934 (XLVIII Stat. L. 1269, 1934). This measure sought to "stop injury" caused by overgrazing of cattle and sheep on the remaining public lands still open to homesteading. Those lands amounted to approximately 170 million acres concentrated in the semiarid plains and intermountain plateaus. Stockgrowers did not look eagerly on federal management of their activities, but they realized that the chaotic abuse of the range could not continue. The Taylor Act immediately withdrew 80 million acres from private entry, and most of the remainder was withdrawn a short time later. This acreage was placed under the supervision of the Grazing Service (subsequently consolidated with the General Land Office to form the Bureau of Land Management), a new agency within the Interior Department, and was then organized into grazing districts. A stockgrower could obtain a grazing permit for livestock by paying a fee. Within two decades, some 9 million head of cattle, sheep, horses, and goats ranged over federal grazing lands, accompanied by 900,000 antelope, deer, elk, moose, and mountain sheep (Foss, 1960, pp. 4, 39–116).

The Taylor Act and the Soil Conservation Act marked the end of an era; no longer was the public domain to be distributed through homesteading. Rather, the remaining lands were to be retained for planned use in the national interest. The need for federal land use planning was a major new concept that emerged during these years of severe economic collapse and land deterioration. State governments also recognized the need to plan more carefully for the lands in their ownership, and all but two states created planning boards that often worked with each other and with federal agencies (Lowitt, 1984, pp. 218–228; Hargreaves, 1976, pp. 561–582; Merk, 1978, pp. 551–554, 606–611).

Federal aid to the mineral industry, except for the silver producers, was more indirect than the assistance to farmers and stockgrowers. Led by western Congressmen, a powerful silver bloc obtained passage in 1934 of the Silver Purchase Act, transformed the Treasury Department into the principal market for western silver, and saved a failing industry (Brennan, 1969; Lowitt, 1984, pp. 112–121). For most other mineral producers, concentrated primarily in the Rocky Mountain states and the Southwest, the 1930s were a severely depressed period characterized by mass unemployment and the failure of small firms. Petroleum interests faced a glutted market at home and abroad until the mid-1930s, when a slow recovery began as a result of production quotas established by the federal government. In the 1935 Interstate Oil Compact, the oil-producing states received Congress's blessing to invoke their own powers and reduce production (Gerald D. Nash, 1985; 1968, pp. 128–156; Lowitt, 1984, pp. 100–112).

Major assistance to the West's extractive industries, towns, cities, and manufacturing came with the release of funds for labor to build Hoover Dam (completed in 1935), the Colorado River Aqueduct to the southern California coastal plain (begun in 1933), and the All-American Canal to California's Imperial Valley (begun in 1934). In 1937, Colorado won congressional approval of the Colorado–Big Thompson Project to generate electricity and transfer water from the western slope of the Rockies to parched lands on the eastern slope. Four years earlier, the Bureau of Reclamation had received funds to begin construction on the Columbia River in the Pacific Northwest of what would become the largest irrigation development ever underwritten by the federal government. The linchpin was Grand Coulee Dam, which when completed was a mile long and three times larger than Hoover Dam. California leaders dusted off decades-old plans to transfer water from the northern part of the state, where it was in greater abundance, to the arid and semiarid south. When state resources proved unequal to the task, the federal government was asked to take over in 1935. Four years later, the Bureau of Reclamation began building the $440 million Central Valley Project (Cole, 1948, pp. 49–65, 118–133; Knight, 1956, pp. 157–169; Golzé, 1961, pp. 176–187; McKinley, 1952, *passim*; Kahrl, 1979, pp. 46–49; Pisani, 1984, *passim*). Additional major federal assistance came to the West as it did to the rest of the nation in the many New Deal laws that sought to restore credit, create jobs, and reform the economy.

Thus, between 1900 and 1940 government, especially the federal government, played a major role in shaping western urban and economic development and the impact of that development on the physical environment. The impact differed over the years and by locale. Between 1900 and 1918, Washington set aside for future use some of the West's natural resources and designed subsidies, educational programs, reclamation projects, and other schemes to help westerners overcome the area's climatic restrictions and meet the demands generated by World War I. The end of the war, the immediate collapse of agricultural and livestock markets followed in 1929 by the collapse of industrial markets, and the environmental abuses resulting from attempts to meet earlier demands and to overcome falling prices prompted the federal government to reverse many of its former policies. Now the emphasis was on restricting production and curbing land abuse, a decision that had the indirect result of conserving natural resources in the West. Although the new policies did not end the environmental despoliation or the depression, they underscored how closely intertwined the government had become with the regional economy: western states received $7.5 billion from Washington, three times more in federal funds than they paid in taxes and more federal money on a per capita basis than any other region (Arrington, 1969, pp. 311–316). To westerners, such aid seemed altogether appropriate for a region where the federal government's landholdings were enormous – more than 40 percent of Arizona, California, and Oregon; 64 percent of Idaho and Utah; 81 percent of Alaska; and 86 percent of Nevada.

War, Washington, and renewed environmental exploitation, 1940–1960s

World War II had an impact on the American West greater than World War I, the Great Depression, or any other single event. It ended the depression and transformed the West into one of the nation's major urban and industrial areas while allowing it to retain its hegemony as the preeminent agricultural, stockgrowing, and mineral-producing region of the United States. The impact was not a change so much as an acceleration of trends evident since the turn of the century. Moreover, the transformation was fueled primarily by an exceptional outpouring of federal dollars that dwarfed earlier expenditures and left the West more dependent on Washington than ever before.

Selective urban and industrial takeoff

The huge flow of federal funds into the West's wartime economy meant jobs and extraordinary, if uneven, population growth. Cities grew at a particularly fast pace as waves of immigrants from outside the area as well as rural westerners flocked to the new urban manufacturing jobs in steel, shipbuilding, aircraft, aluminum, machine tools, textiles, and atomic energy. The growth of both population and manufactures was greatest in California, which gained 2 million newcomers during the war and an additional 6 million by 1960. Three years later, with a total population of nearly 18 million, California became the most populous state in the nation and was behind only New York in the value of its manufactured products (U.S. Department of Commerce, 1975, vol. I, pp. 24–37; 1965, p. 11; *California Magazine*, June 1964). The Pacific Northwest also registered significant though somewhat less spectacular population and industrial gains. Newcomers there were attracted especially to the aircraft plants and shipyards along the coast. Population and industry in the Denver, Phoenix, Tucson, and Albuquerque areas grew impressively as well, but elsewhere in the West population and manufacturing gains were modest at best. In the plains states, the number of people declined even faster than in the prewar years, a trend that continued into the postwar period. Young people went into the military or to jobs in the booming cities, while older citizens in growing numbers retired in the more pleasant climates of the far West and Southwest, including the hotter desert areas, as the rapid spread of air conditioning revolutionized living conditions there (U.S. Department of Commerce, 1975, vol. I, pp. 24–27; Gerald D. Nash, 1985, pp. 14–17, *passim*; Gerald D. Nash, 1973, pp. 200–201, 204–210, 217–218).

Many of the new jobs were with "instant industries" created by the federal government through the Reconstruction Finance Corporation. Times had changed since World War I, when private capital financed all but 10 percent of the war-related plants. Now, the federal government often put up most of the funds for new aircraft, magnesium, aluminum, and rubber plants. Except for rubber production, these factories were nearly all in the West (Gerald D. Nash, 1985, pp. 19–20; White, 1980, *passim*).

Among the most powerful stimulants to western industrial development during and after the war was the cheap electricity made available by the gigantic public multipurpose projects. The power plants at Grand Coulee and Bonneville dams in the Northwest, Hoover Dam in the Southwest, Shasta Dam in California, and similar projects not only promoted the development of local manufactures and the processing of regional raw materials but also significantly raised the quality of life on the farm and in the city. The profound impact – both immediate and potential – of electricity led to complex and often monumental struggles over power distribution, especially between private energy companies and the advocates of public power. On the Pacific Northwest's Columbia Basin Project, the federal government kept control over the generation and distribution of power, but elsewhere compromise was more often the norm. At Hoover Dam, two of the three major contractors for energy were public agencies – the city of Los Angeles and the Metropolitan Water District of Southern California. The third was a private firm, the Southern California Edison Company. On California's Central Valley Project, the powerful Pacific Gas and Electric Company (PG&E) battled for years to limit the Bureau of Reclamation to the generation of electricity and to obtain for itself control of distribution. In a compromise in 1951, the Bureau received authorization to transmit power to drive the project's water-pumping stations, and PG&E to distribute energy to municipalities and other public-preferred users. Controversies notwithstanding, the abundance of publicly generated energy fed the surging western economy, which received another major boost in 1964 with the Pacific Northwest–Pacific Southwest Power Intertie. This $700 million undertaking integrated the two major regional power systems and allowed them to take significant economic advantage of their contrasting patterns of energy consumption (U.S. House of Representatives, 1948, pp. 65–69; Merk, 1978, p. 527; Lowitt, 1984, p. 227; Robinson, 1979, p. 82; Kahrl, 1979, pp. 49–50).

Service industries, save for tourism, also made rapid strides during the war years and grew at a phenomenal rate when the hostilities ended. Tourism reemerged after the war as a major sector in the western economy, second only to federal expenditures in importance. Some cities, such as Las Vegas and Reno, existed almost entirely on tourist dollars. Nearly every state west of the plains counted heavily on tourist income, and the West as a region attracted more visitors than any other part of the nation. The boom in automobile sales, the availability of cheap gasoline, the rapid expansion of highway and airport construction with federal funds, affluence and more leisure time, the West's natural beauty, and the expansion of visitor facilities at such national parks and monuments as Yellowstone, Glacier, Grand Canyon, Yosemite, and Mesa Verde – all reflected and contributed to the tourist explosion. By the 1960s, more than 40 million people annually were visiting the national parks alone. But tourism did more than put dollars into local economies. It also attracted corporations in the East and elsewhere, convinced that they had found another western resource for exploitation. They moved in with their chains of gasoline stations, restaurants, motels, and grocery and dry goods stores. "As always," observed Robert Athearn, "the masters of the purse let the colonials hold the money for a brief moment, long enough to give them a feel for it, but then they moved it out for safekeeping" (Jakle, 1985, pp. 185–198, 225–244; Pomeroy, 1977, p. 1186; Athearn, 1986, pp. 150–151).

Banking remained one of the West's most important service industries. Such giants as Bank of America and Crocker Bank, concentrated in California, helped underwrite expansion in aircraft production, aluminum manufacture, shipbuilding, and housing construction. Such large-scale investment, together with the infusion of federal funds, meant that western business was relying increasingly on local or regional rather than eastern financial institutions. The increase in Bank of America's deposits and assets to $5 billion by 1945, making it the world's largest bank, served only as prologue to the massive growth of the postwar years. Also accounting for the rapid growth after 1941, especially on the Pacific coast and in the southwestern Sun Belt, was the emergence of many of the nation's preeminent research and development centers – the Lawrence Radiation Laboratory in Livermore, the Los Alamos Scientific Laboratory, the Jet Propulsion Laboratory in Pasadena, and the RAND Corporation in Santa Monica, among others. These clusters, in turn, attracted numerous aerospace and electronics firms eager to obtain contracts to produce the guidance systems, electronic equipment, rocket engines, and other devices required by the scientific activities (James and James, 1954, pp. 458–477; Hewlett and Anderson, 1962; Koppes, 1982; Baxter, 1946; Gerald D. Nash, 1985, pp. 153–177). The multiplier effect of such developments on urbanization was often staggering.

The growth of western cities, with their expanding circles of satellite communities as well as their investment, commercial, industrial, and cultural penetration of their hinterlands, intensified the kind of imperialism that westerners boldly practiced on one another. As federal funding for defense and water projects allowed many cities to increase dramatically in size, their influence spread, sometimes bringing them into sharp conflict over regional and local markets and over federal contracts. The pacesetters tended to be along the Pacific coast – Los Angeles, San Francisco, San Diego, Seattle – but inland cities such as Denver, Salt Lake City, Phoenix, and Tucson were increasingly asserting themselves over their hinterlands. At the same time, these inland cities competed with less success against the larger coastal cities for federal defense contracts and water projects, and they remained dependent for major investment capital on California and New York banks. Thus, they found themselves in a double bind that still persists. Denver's situation remains typical of much of the West. "Today Denver looks eastward to New York and westward to California for leadership and direction from those companies and money centers that continue to dominate the crossroads of the interior West," note investigative journalists Peter Wiley and Robert Gottlieb (1982, p. 139, *passim*). "Ultimately, Denver remains a secondary center of power whose major role consists of dominating other, more exploited regions of the interior." Less clear but demanding analysis, observes William Robbins, have been the "social costs . . . exacted from tributary regions" and the "human destruction" occurring in the larger urban areas with their "huge underclass[es] of minorities" (1986, pp. 593, 577–597; see also McWilliams, 1949, pp. 342–343, 364; Clayton, 1967, pp. 449–473).

Agriculture: spiraling subsidization

Pearl Harbor almost overnight replaced farmers' concerns about surpluses with worries about how to produce enough to meet the demands of war at a time when agricultural laborers were joining the military or going to work in urban defense plants. Production curbs disappeared in favor of production goals set by Washington for states, counties, and individual farmers, who now experienced a prosperity that had eluded them since the end of World War I. The prosperity was sustained not only by government's high prices and encouragement, but also by technological improvements that allowed farmers to bring in record crops with fewer hands. Artificial fertilizers, improved seeds, more effective control of insects and pests with such agents as DDT, and favorable weather (particularly on the plains, where rainfall remained above average during the war years) increased yields on substantially the same amount of land under cultivation. When growers sought additional workers, a presidential executive order in 1942 and an act of Congress in 1951 (Public Law 78) brought in *braceros* from Mexico, who assured crop harvests, kept labor costs down, and put efforts to unionize U.S. field hands on the defensive (Wilcox, 1947; Schlebecker, 1975, pp. 212–215; Galarza, 1964, 1977).

The prosperity continued into the postwar period, sustained at first by the food needs of nations whose agricultural potential recovered only slowly from the ravages of war and then by the Korean War and eventually the U.S. involvement in Vietnam. Between conflicts, when surpluses built up and foreign competition increased so much that prices threatened to fall, Congress authorized the government to purchase the surpluses or to pay farmers to curb production or to do both. The demand for food in the early postwar years caused farmers to rush additional land into cultivation with little or no attention to soil conservation lessons learned during the Great Depression. When the wet cycle of the 1940s gave way to another inevitable series of dry years in the 1950s, the plains witnessed a return to dust storms and failed farms. The problem, however, was not nearly so severe as during the Great Depression, primarily because the dry spell was shorter but also because many shelterbelts remained, and Washington immediately tried to reinstall principles of soil conservation with such measures as the Watershed Protection and Flood Protection Act of 1954 and the Great Plains Conservation Act of 1956. An added benefit of these and similar steps was a return to grass of large areas poorly suited to cultivation (Matusow, 1967; Rick, 1963; Schapsmeier and Schapsmeier, 1975; Schlebecker, 1975, pp. 277–

315; Worster, 1979, pp. 227–228; Hurt, 1981, pp. 136–156; Merk, 1978, pp. 536–573).

Although nearly all western farmers experienced prosperity during the war and postwar years, some enjoyed better times than others. California continued to be the West's (and the nation's) leader in agricultural production, accounting by the mid-1960s for more than a quarter of the nation's vegetables, fruits, and nuts and more than half of the country's alfalfa seed, pears, lettuce, celery, asparagus, broccoli, and carrots. Agriculture remained the state's primary industry, with crop values in excess of $3.5 billion and a total (including processing, canning, buildings, equipment, and indirect benefits to secondary industries) value of more than $13 billion (*California Magazine*, August 1964). Much of this prosperity rested on the completion of major aqueducts, such as the Colorado River Aqueduct to the southern California coastal plain in 1941 and the All-American Canal to the Imperial Valley in 1942, and on the official start-up in 1951 of the Central Valley Project. An added boost came from intensified mechanization – improved mechanical cotton pickers, lettuce-picking machines capable of determining ripeness, experimentation with other machines to pick grapes, tomatoes, melons, and asparagus – and from development of crop strains that could be harvested mechanically. The continued trend toward large-scale landholding in California and throughout the West and the nation also encouraged mechanization, as did the labor-organizing efforts of César Chávez, who began winning wide public support for his efforts to improve working conditions for field hands in the mid-1960s. The end of the *bracero* program in 1964 and some successful strikes further encouraged Chávez until he was set back by the growing number of legal and undocumented workers from Mexico, increased mechanization, and determined opposition of large landholders (Kahrl, 1979, pp. 38–56; Jelinek, 1982, pp. 78–95, 102–104).

Farm size became a bitter issue in California and national politics after federal project water was introduced into California's Central and Imperial valleys. That issue, together with California's dramatic postwar growth, led to construction of the world's largest water project. Sparking the confrontation over farm size was the provision in the 1902 reclamation law forbidding farmers from using federal water on more than 160 acres (or 320 acres for a married couple) and requiring them to divest themselves of acreage in excess of the maximum within ten years. Many Imperial Valley farmers claimed an exemption on the grounds that they were irrigating their lands before the federal projects were built, and

Central Valley farmers stymied efforts to break up their holdings by using federal water for only 160 acres (or 320 acres) and pumping groundwater for the remainder. Reliance on groundwater promised an uncertain future – a message that came across clearly to Central Valley growers as the rapidly falling water table led to sharp increases in pumping costs, land subsidence, and the specter of exhausted aquifers. To check such developments, to avoid the federal acreage limitation, and to bring more land under cultivation, Central Valley farmers joined with urban and agricultural interests of southern California to win voter approval of a massive state water project in 1960. The $1.75 billion undertaking called for a series of dams and canals to move 1.5 maf from northern California through the Central Valley, where much of the supply would be available to farmers until needed farther south. Because the project was state funded, farmers were not subject to the federal 160-acre limitation. By the late 1960s, project water reached the lower Central Valley, where its availability weakened efforts to maintain the acreage limitation on lands supplied with federal water. Finally, in 1982, Congress bowed to corporate-farm pressure by increasing the acreage limitation to 960 acres for an individual or a legal entity benefiting 25 persons or fewer and 640 acres for a legal entity benefiting more than 25 persons (XCVI Stat. L. 1263, 1982; U.S. Senate, 1976; Koppes, 1978, pp. 607–636; Kahrl, 1979, pp. 49, 50, 51).

Pacific Northwest farmers also prospered during and after the war, although not quite so spectacularly as farmers in California. Agriculture continued to be the main source of income, and contributing significantly to the prosperity was Grand Coulee Dam. Completed on the eve of World War II, it permitted eventual cultivation of nearly a million acres, all of which were distributed in small parcels of 40 acres (or 80 to a married couple) (Sundborg, 1954; Bureau of Reclamation, 1964; Robinson, 1979, pp. 63–64). This pattern is in sharp contrast with the many estate-like farms in California. The difference stems from the lack of irrigation in the Grand Coulee area prior to the project's completion, the absence of groundwater as an alternative source and a way of skirting the federal acreage limitation, and the federal government's determination to prevent big landholders from profiting at public expense.

As in the Pacific Northwest and California, farm profits in the intermountain West proved best wherever water was available for irrigation, primarily in parts of Colorado, Arizona, Utah, and New Mexico. On the plains, major irrigation works remained less in evidence than elsewhere, although significant development followed congressional approval of the Missouri River Basin Project in 1944. A product of cooperation (sometimes hard to discern) between the Bureau of Reclamation and the Army Corps of Engineers, the plan was designed to control floods and stabilize navigation, as well as to generate hydroelectricity and provide water for farms, cities, and industries. The irrigation dimension developed at a slower pace than the other features, but when completed, it is supposed to provide water for 4.3 million acres (Missouri Basin Interagency Committee, 1952; Scheele, 1969; Robinson, 1979, pp. 83–85). A greater boost to irrigation agriculture on the plains came from increased use of groundwater, especially that in the vast Ogallala aquifer underlying parts of six plains states. Following the war, and especially during the drought of 1950–1956, pumping operations expanded dramatically until by the 1980s more than 15 million acres were being farmed with water from the Ogallala and related aquifers (Englebert and Scheuring, 1984, pp. 122, 360; Green, 1973, pp. 145–238).

Stockraising and mineral production: the continuing presence of Washington

Stockgrowers, like farmers, experienced boom times from the 1940s into the 1960s. The decline in unemployment and rising salaries enabled more people to eat beef regularly, a trend that quickened during the postwar boom years of industrial and urban growth. The eight million men and women in the military forces and the subsequent postwar demand for beef at home and in battle-ravaged parts of the world stimulated a sharp rise in beef prices, prosperity for ranchers, and increases in herds, ranch sizes, and the use of scientific methods in breeding and feeding livestock (Schlebecker, 1953, pp. 169–237).

Similarly, mining corporations prospered and increased production in response to the wartime demand for vital metals: copper, lead, zinc, manganese, chrome, vanadium, molybdenum, and (especially during the postwar years) uranium (Gerald D. Nash, 1985, p. 21; 1973, pp. 203–204, 243). The West remained the nation's leading minerals producer into the postwar years, which were characterized by a major shift away from underground operations. By the mid-1950s, when mine shafts in many locales had been sunk more than a mile and were too costly to maintain, corporations began following the pattern established earlier at Bingham in Utah and Morenci in Arizona by turning to open-pit, or strip, mining. This approach significantly lowered labor and other operational costs by permitting a high degree of mechanization that also left deep scars on the land (Merk, 1978, pp. 497–498; Gerald D. Nash, 1973, pp. 242–243).

Petroleum prospered greatly during these years, becoming a billion-dollar-a-year western enterprise that met more than half the nation's demand. Following the war, petroleum emerged as the West's most important mineral in response to the airline, automobile, and highway construction boom, as well as new technologies and the discovery of vast new oil pools, especially along the western coastline and in Alaska. The postwar years, in addition, witnessed the emergence of natural gas, a by-product of petroleum, as a major western energy source. Responding to the urban and industrial boom and to improved technologies, Texas became the nation's major producer, but significant western contributors also included Oklahoma, New Mexico, and California (Bureau of Mines, 1962, vol. II, pp. 317, 326; Gerald D. Nash, 1968, pp. 157–210).

Prosperity came to mining largely because of government intervention. Massive federal purchases during the war were followed by such peacetime measures as the Mineral Stockpiling Program in 1955 and the Natural Gas Act of 1956. Further stimulating production was Washington's continued liberal policy (under the Mineral Leasing Act of 1920) of leasing federal lands (except national parks) containing such resources as oil, natural gas, and coal. By the mid-1960s, some 64 million acres, mostly in the West, had been leased to oil and gas developers. Eastern Congressmen in 1963 finally achieved repeal of the Silver Purchase Act, the most flagrant example of mining special-interest legislation, but mineral producers had good reason to thank Washington for the flush times they enjoyed after 1941. Those times did not last for western metals mining. High production costs, low prices, poor-quality ore, the switch from copper wire to fiber optics, federal and state environmental protection legislation, and, especially, foreign competition and a federal tax policy that encouraged conglomerates to purchase and then shut down mines – all led to sharp declines in the 1970s. By the 1980s, four of every ten jobs in the western mines, mills, and smelters had been lost, and the industry that had launched the creation of the modern West was on the brink of collapse (*Business Week*, 1984, pp. 64–68; *Washington Post*, Apr. 15, 1985; *New York Times*, Mar. 26 and 28, 1985; *Los Angeles Times*, May 4, 1985; Merk, 1978, p. 613; Michael Malone, 1986, pp. 455–464).

The water wars renewed and congressional authority paramount

Urban and industrial expansion meant increased demands for the West's dearest resource: water. Those demands together with agriculture's traditionally heavy consumptive uses – 80–90 percent of available water is used by agriculture – greatly intensified competition among westerners. The most bitter conflicts, as in earliest years, involved the

Colorado River, although westerners everywhere battled over local supplies.

As California completed its aqueducts to the Colorado and experienced spiraling population and economic growth, the six other states relying on the river stepped up demands for projects of their own. Their insistence grew more urgent in 1944, when the United States negotiated a treaty allocating 1.5 maf of the Colorado's flow to Mexico (Hundley, 1966, chaps. 5, 6). Among the basin states other than California, only Colorado had secured congressional authorization for a major undertaking (the Colorado–Big Thompson Project), but the project remained unfinished until 1955 and was just a token of what Colorado and the other Upper Basin states sought. Their ambitions received a severe jolt in 1946, when the Bureau of Reclamation announced that there was insufficient water for all the projects sought by the Upper Basin. The Bureau refused to endorse any projects until the upper states agreed on their individual water rights. The 1922 Colorado River Compact had only half-quieted disputes over the river. It had divided water between the Upper and Lower basins but not among the states of either basin. The pressure now coming from the Bureau, together with California's opposition to any new projects as threats to its own water uses, persuaded the upper states to hammer out their differences in a compact in 1948 and then, despite California's resistance, finally to get much of what they wanted from Congress. The Colorado River Storage Project Act of 1956 authorized construction of Glen Canyon Dam, which would generate revenue through the sale of electricity to build a dozen so-called participating projects serving agricultural and urban needs throughout the Upper Basin (Bureau of Reclamation, 1946, p. 21; Breitenstein, 1949, pp. 214–216, 225; Mann, 1975, pp. 166–167).

Attention now shifted to the lower Colorado River Basin, where California and Arizona had been fighting over their share of the river since the 1920s. In the 1950s, the conflict reached the boiling point when California blocked Arizona's attempts to obtain congressional approval of its long-sought Central Arizona Project. Arizonans reacted bitterly. The Arizona economy and population had mushroomed during the war years with spectacular consequences for the Phoenix and Tucson areas, where urban and agricultural growth was overtaxing local water supplies. Mining of groundwater had led to precipitous declines in water tables and to damaging land subsidence in some areas. Arizona leaders believed that only imported Colorado River water would save the state from certain disaster, and they became increasingly fearful that not enough water remained for them. Recent flow measurements indicated that the negotiators of the 1922 Colorado River Compact had overestimated the river's runoff. In the meantime, the other basin states had been pulling ahead of Arizona in the quest for water. Way out front was California, which had secured projects in the 1928 Boulder Canyon Act and now vigorously opposed Arizona's attempts to get water. Mexico had obtained treaty rights to water in 1944. Upper state projects were about to be authorized. Alarmed by these developments and unable to make headway in Congress, Arizona in 1952 turned to the Supreme Court for help (Hundley, 1975, pp. 282–302).

Following one of the longest, most complicated, and most expensive trials in history, the Supreme Court in *Arizona* v. *California* (373 U.S. 546, 1963) sided with Arizona in 1963. In doing so, however, the court held that Congress has authority over interstate rivers and the power virtually to command the states to obey its orders. Of less significance for the West, if not for Arizona, the decision cleared the way for the billion-dollar Colorado River Basin Project Act of 1968. This measure authorized not only the Central Arizona Project (CAP) – a 241-mile aqueduct with appurtenant dams and pumping stations – but also, as a result of horse trading to obtain needed congressional votes, five projects in the Upper Basin (Hundley, 1975, pp. 305–306, 333–334; LXXXII Stat. L. 885, 1968; Ingram, 1969; for an earlier though quickly forgotten precedent for the 1963 decision, see Littlefield, 1987). Although more than two decades would pass before the CAP or most other projects authorized during these years would be completed, the promise of construction renewed growth, even as some commentators began increasingly challenging the appropriateness of that growth in areas where the demand for water far exceeded the dependable supply.

While the West's traditionally important extractive industries boomed during the war and postwar years, the economy also underwent tremendous diversification. Newer businesses emerged in nearly all parts of the West, with especially notable results in California, the Pacific Northwest, and major areas in the Southwest. In the growth of manufacturing, as in agriculture, stockraising, and mining, the government played a decisive role, pumping into the western economy $29 billion in war contracts and an additional $7 billion for plant construction, military training camps, supply depots, and other troop facilities. In many areas, the federal government became the single largest employer. During the 15 years following the war, Washington poured an additional $150 billion into the West, reaffirming the region's heavy dependence on federal largess (and control) and continuing a practice that persists today. Of a piece with increased federal economic power was the monumental jump in congressional authority over western watercourses resulting from the *Arizona* v. *California* decision. But greater authority, through either budgetary allocations or the courts, was not necessarily to mean more informed or effective leadership. Equally significant, the impact of these expenditures on the West was (and remains) remarkably uneven. California, Washington, Arizona, Colorado, and New Mexico received the lion's share, while the rest of the West, and even some of the states receiving massive federal funding, retained vestiges of the traditional colony: a supplier of raw materials not only for the industries of the East but also for those elsewhere in the West and increasingly in Asia as well. Wyoming "continued to be what it had always been," observed T.A. Larson in a common complaint during the postwar years, "primarily a producer of raw materials to be exported for processing and consumption elsewhere" and a state where "regular infusions of federal money were essential to keep the economy viable" (Larson, 1978, p. 507; Robbins, 1986, pp. 588–597; Gerald D. Nash, 1985, p. 218; 1973, pp. 201–202, 211, 248; Brubaker, 1959, pp. 288–290, 313–316). Yet those receiving federal funding viewed warily the greater outside control accompanying the dollars. Here, it seemed, was a special kind of colonialism. Instead of just raw materials, some believed they were surrendering their skills and natural environments to a military-urban-technological complex in the service of the national government. It was at once a government both generous and all powerful, characteristics perhaps best epitomized in its position as the region's biggest absentee landowner and investor. More than symbolic of Washington's authority was its selection of the West as the place to test the first atomic bomb.

A questioning of values, 1960s–1980s

By the 1960s, the West's rapid postwar growth had caused many observers to challenge the wisdom of much that had occurred. Delight with the general prosperity notwithstanding, less savory consequences of a booming population and economy were emerging: urban congestion; water and air pollution; shrinking wilderness areas; destruction of natural habitats; decimation of many species of animals, plants, and fish; and heightened competition for increasingly scarce and often despoiled land and water resources. Some westerners vigorously sought to correct these abuses and found allies elsewhere in the United States who viewed the West as an area that held treasures belonging to the whole nation.

Preservation versus conservation

Concern about the environment was nothing new, but the attitudes and actions of U.S. citizens during the post-World War II years differed significantly from those of previous generations. In earlier years, beginning particularly in the late nineteenth century and then climaxing first with the Progressive movement of the early 1900s and then again during the New Deal era of the 1930s, the emphasis was on conserving natural resources for future *use* and transforming arid or otherwise marginal lands into productive acreage. This kind of thinking led to demands for careful management of resources and such actions as the Forest Reserve Act (1891), the Reclamation Act (1902), the Governors' Conference on the Conservation of Natural Resources held at the White House (1908), creation of the National Park Service (1916), the Taylor Grazing Act (1934), and the Soil Conservation Act (1935). It also spurred the development of such massive multipurpose public works as Hoover and Grand Coulee dams and California's Central Valley Project. It produced as well such measures as the 1936 Pittman-Robertson Act, aimed at ensuring the continued presence of scarce game animals for future hunters (see, for example, Hays, 1959; Cart, 1972, pp. 113–120; Dodds, 1965, pp. 75–81; Rakestraw, 1972, pp. 271–288).

After World War II, the emphasis on conservation for *use* persisted, but it was challenged and often eclipsed by an increasing concern about the quality of the environment and the interrelationship of all forms of life with their habitats. Such words as "ecology" and "land ethic" became popular. According to this new perspective, wild animals warranted preservation not as game for future hunters but as occupants of a special place in the ecosystem. Forests, rangelands, canyonlands, mountainous areas (regardless of whether valuable minerals or some other resource was present) deserved preservation for similar reasons and because they could serve as living examples of how to restore environmentally damaged places. Threats to health posed by pollution of air, water, and soil had to be eliminated (see, for example, Hays, 1985, pp. 198–241; Roderick Nash, 1982, pp. 200–271). The nation grew increasingly receptive to these ideas in the postwar years and especially in the 1960s, 1970s, and 1980s. The leadership, not surprisingly, came initially from the West, where most of the country's remaining natural resources and wilderness areas are located and where growth rates exceed the national average and threaten to exhaust the resources and wilderness.

The myth of environmental reform

Among the first major attempts at environmental reform were those in the West's largest city, Los Angeles, where smog emerged as a serious problem as early as the 1940s. So foul had become the combination of inversion layer, automobile and airplane exhaust, industrial smoke, and black clouds belched by hundreds of thousands of backyard incinerators that local leaders in 1947 created the Los Angeles County Air Pollution Control District, the first agency of its kind. Although opinion differed (and differs) on the district's effectiveness, the battle proved uphill against the constantly increasing number of automobiles, the principal polluter. When by 1960 the number of cars in the city had reached 4 million and smog had become a serious health hazard – as well as a visual blight in such other California cities as San Francisco, Oakland, Berkeley, Riverside, and San Bernardino – the state legislature passed the nation's first law requiring automobile exhaust control devices (Krier and Ursin, 1977, pp. 52–66, 137–148).

Californians led the West and the nation in polluting their air, but they were not alone. Phoenix, Tucson, Albuquerque, Seattle, Denver, Colorado Springs, and nearly every major western city soon experienced similar problems. No other western state followed California's example with legislation mandating exhaust control equipment for cars, perhaps in part because in 1963 Congress enacted a Clean Air Act for the nation, followed in 1970 by a tougher Clean Air Amendments Act. The creation in 1970 of the Environmental Protection Agency (EPA) reflected a high point in attempts to control air and other forms of pollution. Thereafter, under pressure from automakers, Congress relaxed the provisions of the clean air legislation, and EPA stopped insisting that states test motor vehicles annually or lose funds for highway construction and sewage treatment plants. As a result, smog plagues nearly every urban area, particularly in southern California, where it first emerged. In 1979, Los Angeles had its most serious smog siege in 25 years, and as late as 1987, a congressman publicly accused the agency responsible for improving the air quality of Los Angeles, Orange, Riverside, and San Bernardino counties of having lost control of the situation (Krier and Ursin, 1977, pp. 8–9, 203, 235–236; *Los Angeles Times*, Oct. 29, 1979, Feb. 14, 1987).

Threats to fish, wildlife, and wilderness posed by massive invasion of the native habitat by farmers, subdividers, and tourists also aroused alarm in the immediate postwar years both regionally and nationally. In the 1950s, Congress sought to reverse the decline of hard-hit species with the Dingell-Johnson Act (1950), which provided federal aid to states for the restoration of threatened fish species, and the Fish and Wildlife Coordination Act (1958), which authorized studies of the effects of dams and aqueducts on animals and fish to determine needed remedial action. Such measures were enthusiastically supported by the Sierra Club, Wilderness Society, and other preservationist groups arguing for both protective legislation in general and for recognition of society's need for as much natural wildlife and as many plants as possible. Among their greatest successes were the Wilderness Act of 1964, the Wild and Scenic Rivers Act of 1968, and the National Trails Act of 1968. The Wilderness Act established a National Wilderness Preservation system (similar to the National Park system created in 1916) for keeping some land, initially 9 million acres, permanently wild. A similar desire to preserve at least some of the nation's rivers led to passage of the Wild and Scenic Rivers Act (Roderick Nash, 1982, pp. 200–237; Frome, 1974).

Such victories were usually less than they seemed. Most western rivers had already been filled with so many dams and aqueducts that they scarcely resembled their natural state. The Colorado River, for example, has carried virtually no water to the Gulf of California for more than two decades. The Wilderness Act (LXXVIII Stat. L. 890, 1964) prohibited prospecting and mining in the reserved areas, but only after 1984, and even then prior claims could be developed and the president could authorize construction of roads, dams, and power plants in the name of the national interest. Attempts to protect fish and to improve the environment generally – for example, the Clean Water Act (1972), the Toxic Substances Control Act (1976), and the Resource Conservation and Recovery Act (1976) – seemed too weak to halt the toxic pollution of many western bays, estuaries, and streams. The announcements in 1985 and 1986, for example, of dangerous levels of DDT and PCBs in the fish in Santa Monica Bay and elsewhere along the southern California coast revealed a health hazard to people as well as to fish and led to public warnings against consuming fish caught in the area (Tarr, 1985; see recent issues of *Cry California, Not Man Apart*, the *Los Angeles Times*, and *Santa Monica Outlook*). By this time, toxic-waste poisoning of the land, atmosphere, and water had become as commonplace in the West as in the rest of the nation.

One form of pollution that westerners believed they had escaped was acid rain. Then, in early 1985, the World Resources Institute released a study showing that acid rain threatened to become as serious a problem in the West as it had long been in the East. Caused by the reaction of sulfur and nitrogen oxides from industrial and automobile pollution reacting with water in the atmosphere, acid rain had appeared earlier in eastern areas, where precipitation and industrialization are greater. By the 1980s, automobiles and oil refineries, especially in California, together with copper smelters in the West and in northern Mexico, had contributed to a significant rise in the acidity of lakes and streams in the

Colorado Rockies, southeastern Arizona, southern California, the San Francisco Bay area, California's Central Valley and Sierra Nevada, and the Puget Sound area and Cascade Range of the Pacific Northwest. The changes, observed one scientist, "are the same as people noticed in the East before forests began dying" (*Los Angeles Times*, Mar. 29, 1985).

Such setbacks and compromises reflect the continuing vitality of the traditional western emphasis on intense resource exploitation. They also reveal that advocates of conservation for use – the utilitarians – still command a powerful following. At times, the utilitarians and the preservationists clashed head-on during the postwar years with no clear-cut victory for either side. In the 1950s, they battled furiously over an attempt to build a dam at Echo Park on a major tributary of the Colorado River. The proposed dam, which would have flooded Dinosaur National Monument, was finally dropped from congressional legislation, but the measure that gained approval in 1956 authorized a dam in another remarkable locale, Glen Canyon. Preservationists viewed the outcome as far from a victory (Mann, Weatherford, and Nichols, 1964; Stratton and Sirotkin, 1959; Roderick Nash, 1982, pp. 209–219, 229; Porter, 1963). Equally clouded were the results of another bitter confrontation over dams in the 1960s, this time a proposal to locate one reservoir in Marble Canyon, just east of the main gorge of the Grand Canyon, and another in Bridge Canyon, west of Grand Canyon. The dams were to generate electricity that would provide revenue for building the Central Arizona Project and pump water through the aqueduct. After a lengthy battle that captured national headlines, environmentalists were successful in having the dams removed from the authorizing legislation, and a coal-fired power plant was substituted. By the 1980s the CAP was nearing completion, and the coal-fired plant was spewing pollutants into the atmosphere over Cedar Breaks, Grand Canyon, Zion, and Bryce (*Los Angeles Times*, Feb. 9, 1975; Runte, 1979, p. 185; Roderick Nash, 1982, pp. 227–235). As in the earlier battle over Echo Park, neither side was a clear winner.

Pressure for massive new water projects lessened in the 1970s and 1980s as costs soared, as environmentalists urged wiser use of scarce supplies, and as the population growth rate slowed. Although the rate of population increase slackened, the number of residents continued to climb, driven partially by the Middle Eastern oil crisis which produced new job opportunities in parts of the West; by the arrival of refugees from Southeast Asia; and by the continued arrival of undocumented workers from Mexico. Westerners numbered 53 million in 1970 and 65 million only ten years later. The largest concentrations remained in the traditionally popular urban areas of the Pacific coast and Southwest (U.S. Department of Commerce, 1975, vol. I, pp. 24–37; 1984a, p. 12).

More people meant demands for additional residential housing and jobs that fostered continued urban sprawl despite attempts to hold land developers and subdividers in check. Millions of westerners and residents elsewhere responded warmly to the wilderness crusade and expansion of recreation areas, but their sheer numbers threatened to destroy what they sought to preserve and enjoy. By the early 1970s, guided wilderness trips had become a booming business, and many national parks had been transformed into small urban areas beset by congestion, smog, and crime. Devotees seemed about to love wilderness out of existence, and nervous park administrators responded by requiring limited camping reservations and banning cars from places where they had earlier been welcome. "Unless new technology rescues us," confessed a ranger recently at Yosemite, where the annual number of visitors exceeds 3 million, "we may have to impose a hotel-like reservation system on *all* access" (emphasis added) (*Los Angeles Times*, July 30, 1985, May 25, 1986; Roderick Nash, 1982, pp. 316–341).

In the countryside, the food demands of the Vietnam war prompted some farmers and stockgrowers to return to earlier unwise practices. Many of those practices continued during peacetime as some farmers sought to combat falling prices, especially for grain, as they had done so often – and unsuccessfully – before by increasing production. Such was the case in the mid-1980s when many farmers, especially those who had mortgaged themselves heavily to buy additional land when times were flush and the lands expensive, tried to avoid bankruptcy through overproduction. (See, for example, *Los Angeles Times*, May 13, 1985.) Congress has sought to correct some of the grosser abuses of the land. In 1964, before the Vietnam war, Congress had authorized the Secretary of the Interior to reexamine the acreage covered by the Taylor Grazing Act and to establish criteria for classifying the land for grazing, residences, or industry. In the same year, Congress created the Federal Land Law Review Commission to bring order to the myriad and often conflicting federal land use laws. Six years later, the commission reported more than 100 recommendations aimed at improving environmental quality and land use planning, although not all the recommendations were implemented and the overall results were far from satisfactory (U.S. Public Land Law Review Commission, 1970; Merk, 1978, p. 615). By the mid-1980s, dust storms had returned to parts of the plains, and soil erosion had reached a critical level throughout the nation. The Conservation Foundation reported in 1985 that soil erosion was costing the

United States $6 billion annually in river and harbor dredging, reduced recreational capacity, flood damage, water treatment, and decreased cropland productivity. In the Palouse area of eastern Washington, for example, the great loss of soil has reduced wheat production 20 percent. Such losses, together with the harmful effects on wildlife, made soil erosion a leading, if not the prime, source of water pollution (*Los Angeles Times*, May 7 and Dec. 25, 1985, Jan. 10, 1986).

Water use and abuse

With a current western population in excess of 70 million, with increasing urban and industrial demands for limited (and in many areas already oversubscribed) water supplies, and with little prospect of huge new projects to increase water supplies, the West is entering a new era. The emphasis is now on reallocation and better management of existing supplies. Demands are mounting for new attitudes toward water: for the elimination of legal uncertainties and contradictions, for more flexible water management institutions, for water quality enhancement, and for increased conservation, especially in agriculture, the heaviest user.

The need for new approaches is essential, the more so because water scarcity has combined with pollution to pose a particularly grave threat. Among the most serious concerns is salinity. All water in nature contains some salts, but the concentration increases sharply as the volume of water decreases because of evaporation losses at reservoirs and intense irrigation practices. The higher salt content leaves downstream farmers with few options, none of them attractive. They must switch to more salt-tolerant crops, irrigate fewer acres with the same volume of water, obtain more water to grow the same crops produced earlier, initiate an expensive engineering system to remove the salts, or adopt some combination of these steps. To ignore the problem and allow salt to build up in the soil means reduced productivity and eventually the loss of the land for agriculture altogether – as has happened in Utah, southern Arizona's Gila Valley, California's Central Valley, and elsewhere. Some westerners are seeking solutions based upon cooperation among otherwise competing interests. The Metropolitan Water District of Southern California, for example, has recently offered to cover costs involved in reducing salinity and seepage losses in the Imperial Valley in exchange for the water that is conserved (*Los Angeles Times*, Apr. 10, May 7, Dec. 25 and 29, 1985, Jan. 10 and 17, 1986).

Because agriculture already accounts for about 85 percent of western water use, the trend toward increased salinity worries not only farmers but also urban and industrial users downstream. Saline water means increased costs due to corrosion of pipes and heaters, higher water-softening bills, more soap consumption, and additional treatment of water for industrial and drinking purposes. Its adverse effects on wildlife can be catastrophic. At Nevada's Stillwater Management Area biologists in 1987 blamed increased salinity for the death of seven million fish in two months. Water pollution has also adversely affected international relations. On the Colorado River, the heavily saline flow crossing the border into Mexico created a nasty international situation until the United States agreed in 1973 to improve the water to Mexico's satisfaction. Remedial action included U.S. construction of one of the world's largest desalination plants, now scheduled to become operational by the end of the 1980s (International Boundary and Water Commission, 1982, p. 30; Hundley, 1975, pp. 316–317; Valantine, 1984). Although the plant (and recent increased precipitation in the Colorado River Basin) should significantly lessen the threat to Mexico, salinity remains a major problem for westerners nearly everywhere.

So does agricultural drainage water carrying off such toxic chemicals as selenium, found naturally in many soils, and the residue from chemical fertilizers and pesticides. A recent Interior Department report indicates that nine western wildlife refuges have unusually high concentrations of such wastes, and in California's San Joaquin Valley the wastes have created a massive problem. The valley contains little natural drainage, so contaminated groundwater builds up and must be withdrawn to keep the threatened land, ultimately perhaps 400,000 acres, in production. Although a master drain has been proposed, economic and environmental obstacles – how to finance the water removal and where to dump the waste without adverse effects – have prevented construction. In the meantime, farmers divert their drainage into nearby uncultivated lowlands, including marshlands on well-traveled migratory bird routes, causing widespread deaths and birth deformities among waterfowl. In early 1985 the Bureau of Reclamation, citing the Migratory Bird Treaty Act, announced that it would shut off the flow of irrigation water to 42,000 acres that were sending toxic-waste water to Kesterson National Wildlife Refuge. Farmers loudly protested, and in less than two weeks the order was rescinded, with the understanding that government and agricultural interests would devise a solution (*Los Angeles Times*, Mar. 16, 18, and 29, May 21, Dec. 12, 1985, Feb. 23, Mar. 13 and 23, May 2, 1986; Bureau of Reclamation, 1979).

Despite the reprieve, if the drainage problem remains unresolved, the long-term outlook for farmers is only slightly less pessimistic than the present prospects for the waterfowl. No less concerned are residents along the delta where the San Joaquin River empties. Recent tests there indicate high levels of selenium. In response to such toxicity, as well as to increased threats from other wastes to the environmentally sensitive bay and delta areas, the California Supreme Court in September 1986 upheld a lower court decision ordering the state Water Resources Control Board to pay more attention to water quality. The order could force state authorities to decrease the water available to farmers and other users. "But . . . the court also says we have to balance the needs of one area against . . . [those] of another and of one use against another," observed the chief counsel for the Department of Water Resources. "This is going to be a long and difficult process" (*Los Angeles Times*, June 10, 1985, May 29 and Sept. 19, 1986).

Agriculture's heavy water uses have come under increasing fire from many urban residents who view them as not only often environmentally destructive but also wasteful. They have called for elimination of the subsidies that allow farmers to pay significantly less than city dwellers pay for water and have insisted on stricter definitions of beneficial use. As early as 1890, Wyoming legislated the allowable volume of water to irrigate a specific acreage, but in most states the courts and legislatures have established flexible definitions reflecting soil types, evaporation rates, and other variables. Those definitions, argue the critics, should be reexamined in the light of newer competing needs and redrafted with greater precision (Pring and Tomb, 1979; Dunbar, 1983, pp. 215–216).

Farmers as well as urban residents have created massive problems for themselves in many areas by relying heavily on groundwater. Accelerating the intense use of such water has been the appropriation of nearly all the West's surface flows. In this heightened competition, groundwater has been extracted in many areas far faster than nature can replenish it, thus guaranteeing an eventual shortage. In California alone, such overdrafts on underground aquifers have contributed measurably to the state's water deficit, which had reached 2.4 maf by the mid-1970s (California Department of Water Resources, 1974, p. 2). In other words, Californians, as well as westerners in numerous other locales, annually use much more water than is available on a permanent basis. Another dramatic example is what has been happening to the Ogallala aquifer, the large reservoir underlying parts of six states in the heart of the high plains. There, farmers have been mining water at the rate of nearly 8 maf annually, but the natural recharge is less than 200,000 acre-feet a year. Such overdraft, projections indicate, will result in taking more than 5 million acres out of irrigation in the next 35 years, with an accelerated decline in irrigation thereafter (Engelbert and Scheuring, 1984, pp. 78,

122–123, 293–294, 360–362). Much of this land will simply be abandoned; other acreage will become grazing land or will be put into dry-land farming, in which only careful attention to earlier lessons learned can keep it in production.

A long-recognized by-product of heavy groundwater mining is severe environmental damage through land subsidence and erosion. Some states, most notably Arizona and New Mexico, have recently made significant advances in improving the chaotic laws governing groundwater, but many areas are still struggling with the issue. Another unintended result of intensive water mining in both urban and rural areas is increased chemical contamination of underground supplies because of inadequate waste disposal. Some wells have been cleaned up, but many have been shut down, and the water of still others has been blended with supplies brought in from elsewhere. The net result has been increased water use, much of it in the form of bottled water demanded by people who do not trust anything coming from a tap. In southern California alone, one of three homes uses bottled water as the principal source of drinking water, compared with one in seventeen for the United States as a whole. Reacting against contaminated aquifers, Arizona enacted legislation in the spring of 1986 to protect its well water from toxic waste and inspired other westerners to work for similar protection (Dunbar, 1983, pp. 153–191; *Los Angeles Times*, Jan. 16 and May 15, 1986).

The groundwater problem is only one of many instances in which flawed legal and administrative institutions have frustrated attempts to manage western water resources effectively and to plan intelligently for the future. An issue of fundamental importance is the "use it or lose it" principle of the appropriation doctrine, which threatens an appropriator with the loss of a right to water unused because of conservation efforts. Equally telling are the legal uncertainties surrounding such questions as the status of water as a commodity transferable across state lines to the highest bidder, the extent of the unexercised riparian right in those states that recognize the principle, the status of water retained within streams for fishery or aesthetic purposes, and the extent of federal reserved rights, especially Indian water rights. The last is a particularly vexing issue because the West possesses more Indian reservations (55 percent) and more reservation Indians (nearly 75 percent) than any other section of the country. The Indians are asking the courts to award them title to vast quantities of water now going to non-Indian farms and cities. The legal precedents are formidable. The Supreme Court held in 1908 that Indian reservations possess a special water right that exists regardless of whether the water is being used. Then, in 1963, the Supreme Court defined the extent of that right – enough water for all the "practicably irrigable" acreage on a reservation (*Winters* v. *United States*, 207 U.S. 564, 1908; *Arizona* v. *California*, 373 U.S. 546, 1963; Hundley, 1978, pp. 455–482; Hundley, 1982, pp. 17–42; see also Bowden, Edmunds, and Hundley, 1982, pp. 163–182; Dunning, 1982). Because irrigated agriculture is among the heaviest users of water, these decisions are potentially earthshaking, although the continuing legal battles over their precise meaning leave their ultimate significance unclear.

Besides the legal uncertainties, there is jurisdictional confusion and conflict among the managers of western water resources. Federal, state, regional, and local agencies battle incessantly over prerogatives or feigned prerogatives. For example, California officials managing the State Water Project have clashed repeatedly with federal operators of the Central Valley Project (CVP) over water quality standards, the prices charged to customers, and many other issues. Federal subsidization of water has been a special sore point because subsidized farmers are charged only about a tenth of what state project farmers must pay. Also vexing have been conflicts over water quality standards, especially those for San Francisco Bay. A major reason for a California attempt in 1982 to build a $5.6 billion peripheral canal in the Sacramento–San Joaquin delta was to overcome the Bureau of Reclamation's refusal to acknowledge that CVP water could be used to prevent saltwater intrusion into the bay. When voters rejected the canal at the polls, California water and environmental leaders negotiated a compromise, offering the bureau excess capacity in the state aqueduct in exchange for a water quality pact. Although this so-called coordinated operating agreement received congressional and presidential approval in 1986, others are trying to persuade Congress to allow California to take over operation of the CVP and merge it with the State Water Project (*Los Angeles Times*, May 13 and 19, Aug. 4, Sept. 10 and 16, 1985, Sept. 17, Oct. 9 and 10, Nov. 3, 1986; Kahrl, 1979, p. 56; Bowden, Edmunds, and Hundley, 1982, pp. 170–173, 176–181; Storper and Walker, 1984; Willey, 1985). The outcome of these complex and controversial maneuvers is far from clear, as is true of innumerable similar clashes on all governmental levels throughout the West.

Such confrontations should not be allowed to obscure significant advances in water management and environmental protection during the last 20 years. On balance, however, those advances seem more piecemeal and inconclusive than systematic and decisive. On the other hand, the dramatic increase in the number of people working on behalf of the fragile western habitat testifies to significant change. So does the outpouring of writing on behalf of nature by Wallace Stegner, Rachel Carson, Donald Worster, Raymond Dasmann, Roderick Nash, David Brower, and other latter-day Aldo Leopolds and John Muirs. There is also growing recognition that technology offers only temporary respite from water shortage. Even the most impressive of dams will eventually silt up – if they do not first collapse, as has already happened at Teton in Idaho and elsewhere – and the best sites have already been taken (Goldsmith and Hildyard, 1986; Worster, 1985, pp. 308–311, *passim*; Reisner, 1986). Especially telling has been Congress's refusal to authorize a major new water project since 1972, the rejection by California voters of the peripheral canal in 1982, and Congress's approval (over a presidential veto) of a major new Clean Water Act in 1987. These signs may in the long run prove ephemeral, but they remind us that the fate of the West (and the nation) has ultimately depended upon the consent of the governed.

Overshadowing even the most impressive environmental achievements and implicitly mocking them has been the relentless growth in western population since the mid-nineteenth century. From a few thousands of newcomers, the West grew to 11.2 million by 1900, doubled its population during the next 20 years, and then more than doubled it again to 44 million by 1960. The next two and a half decades witnessed an additional 21 million people in the West, and experts predict a population approaching 100 million by the year 2000. Such numbers alone underscore the fragility of the modern West's artificial civilization of abundance and pointedly remind westerners that they share with the world a fundamental question about values: When does the quantity of lives significantly affect the quality of life?

To understand the modern American West is to understand its past. The West has grown extraordinarily fast, much more rapidly than regional and national leaders have been able to adjust legal and social institutions governing human behavior. At the outset of settlement, the emphasis was on survival, and the local and then national laws governing critical resources reflected that reality by placing a premium on exploitative prowess. The example of water, the most important resource, tells it all: first in time meant first in right. Water was a commodity to be used for survival and profit, not hoarded and certainly not preserved for aesthetic or recreational purposes. By the mid-twentieth century, these ideas, together with the West's natural wealth, technological and scientific advances, and the extraordinary infusion of federal dollars and influence, had catapulted the region to the forefront of the world's economically powerful areas. Institutional change had accompanied this growth, but it had been slow and disorderly. Concern for recreation, wildlife, and the environment could be found in federal, state, and local legislation, but so too could the traditional emphasis on using the land, water, and other resources even when such use meant environmental degradation. A recent reminder has been the so-called Sagebrush Rebellion in which many westerners demand that the region's biggest landlord, the federal government, transfer its holdings to state governments whose restrictions on development are likely to be more lax (Wiley and Gottlieb, 1982, pp. 300–301). Not surprisingly, this call for rebellion has been loudest in areas where federal expenditures have not kept pace with the outlays made in such places as Los Angeles–San Diego, Phoenix–Tucson, San Francisco–Oakland, Seattle, and Albuquerque. Although the relative loss of federal dollars may benefit the environment by slowing growth, it also weakens the government's leverage in effectively enforcing environmental laws.

The tenacity of the exploitative tradition becomes clearer upon realization that it antedates settlement of the modern West. It extends back nearly four centuries to that first frontier along the Atlantic coast. The New World was viewed as a place of abundance, an area to be subdued and exploited. There seemed no need to worry about conserving soil or timber or minerals or water or any other natural resource when the supply appeared limitless and richer frontiers were just beyond the horizon. Such belief propelled settlement across the Appalachians, into the Old Northwest and Old Southwest, and onto the eastern prairies. When the waves of pioneers reached the Mississippi River Valley, their vision blurred momentarily as they contemplated the vast arid and semiarid expanse before them, but their confidence rapidly returned with the discovery of precious metals in California and elsewhere. Despite the sparse water supply, they and national leaders devised institutions that sought to compensate for the scarcity and to promote rapid growth.

By nearly every economic indicator, the West by the mid-twentieth century had achieved success. Ironically, that success had transformed much of the region into a reflection of the East that earlier generations had sought to escape. The powerful engine in this transformation was attitudes and institutions more in harmony with an earlier frontier era than the present with its vast population, growing evidence of environmental damage, and increasing competition over the region's most critical natural resource – water. If the modern West is to retain or reassert those qualities of life that proved most attractive to earlier generations, the U.S. public must abandon past attitudes that are out of step with present realities. The public must devise customs, laws, administrative structures, and other institutions in harmony with the necessities of today and tomorrow.

References

Arrington, Leonard J. 1958. *Great Basin Kingdom: An Economic History of the Latter-Day Saints, 1830–1900*. Lincoln: University of Nebraska Press.

Arrington, Leonard J. 1969. "The New Deal in the West: A Preliminary Statistical Inquiry." *Pacific Historical Review, 38*, 311–316.

Arrington, Leonard J., and Hansen, Gary B. 1963. *"The Richest Hole on Earth": A History of the Bingham Copper Mine*. Logan: Utah State University Press.

Athearn, Robert C. 1986. *The Mystic West in Twentieth-Century America*. Lawrence: University of Kansas Press.

Balderrama, Francisco E. 1982. *In Defense of La Raza: The Los Angeles Mexican Consulate and the Mexican Community, 1929–1936*. Tucson: University of Arizona Press.

Ball, Carleton. 1930. "The History of American Wheat Improvement," *Agricultural History, 4*, 48–71.

Barth, Gunther. 1975. *Instant Cities: Urbanization and the Rise of San Francisco and Denver*. New York: Oxford University Press.

Baxter, James P., III. 1946. *Scientists against Time*. Boston: Little, Brown.

Benedict, Murray R. 1953. *Farm Policies of the United States, 1790–1950*. New York: Octagon.

Bowden, Gerald, Edmunds, Stahrl W., and Hundley, Norris. 1982. "Institutions, Laws and Organizations." In Ernest Engelbert and Ann Scheuring, eds., *Competition for California Water: Alternative Resolutions*. Berkeley: University of California Press.

Breitenstein, Jean S. 1949. "The Upper Colorado River Basin Compact," *State Government, 22*, 214–216, 225.

Brennan, John A. 1969. *Silver and the First New Deal*. Reno: University of Nevada Press.

Brubaker, Sterling L. 1959. "The Impact of Federal Government Activities on California's Economic Growth, 1930–1956." Ph.D. dissertation. University of California, Berkeley.

Burroughs, John R. 1971. *Guardian of the Grasslands: The First Hundred Years of the Wyoming Stock Growers Association*. Cheyenne: University of Wyoming Press.

Business Week. 1984. Dec. 17, pp. 64–68.

Butt, Paul D. 1960. *Branch Banking and Economic Growth in Arizona and New Mexico*. Albuquerque: University of New Mexico.

California Department of Water Resources. 1974. "The California Water Plan: Outlook in 1974." Bull. 160-74. Sacramento.

California Magazine of Commerce, Agriculture & Industry. 1964. Vol. 54, June, p. 48, and Aug., pp. 23–25.

Cart, Theodore W. 1972. "'New Deal' for Wildlife: A Perspective on Federal Conservation Policy, 1933–1940," *Pacific Northwest Quarterly, 63*, 113–120.

Clayton, James L. 1967. "Impact of the Cold War on the Economies of California and Utah," *Pacific Historical Review, 36*, 449–473.

Cleland, Robert G., and Hardy, Osgood. 1929. *The March of Industry*. Los Angeles: Powell Publishing.

Clements, Kendrick A. 1979. "Politics and the Park: San Francisco's Fight for Hetch Hetchy, 1908–1913," *Pacific Historical Review, 47*, 185–216.

Cole, Donald B. 1948. "Transmountain Water Diversion in Colorado," *Colorado Magazine, 25*, 49–65, 118–133.

Congressional Record. 1913. 63d Congress, 2d sess., 1189.

Cooper, Martin R., Barton, Glen T., and Brodell, Albert P. 1947. *Progress of Farm Mechanization*. U.S. Department of Agriculture Misc. Pub. 630. Washington, D.C.

Dale, E. Everett. 1930. *The Range Cattle Industry*. Norman: University of Oklahoma Press.

Dodds, Gordon B. 1965. "The Historiography of American Conservation: Past and Prospects," *Pacific Northwest Quarterly, 56*, 75–81.

Drache, Hiram M. 1964. *The Day of the Bonanza: A History of Bonanza Farming in the Red River Valley of the North*. Fargo, North Dakota: Interstate.

"Drought and Insects Bring Havoc to Crops," *Newsweek*. 1934. May 19, pp. 5–6.

Droze, Wilmon H. 1977. *Trees, Prairies and People: A History of Tree Planting in the Plains States*. Denton: Texas Women's University.

Dunbar, Robert G. 1983. *Forging New Rights in Western Waters*. Lincoln: University of Nebraska Press.

Dunning, Harrison C. 1982. "Water Allocation in California: Legal Rights and Reform Needs." Research Paper 1982. University of California, Berkeley, Institute of Governmental Studies.

Dykstra, Robert R. 1968. *The Cattle Towns*. Lincoln: University of Nebraska Press.

Elliott, Arlene. 1970. "The Rise of Aeronautics in California, 1849–1940," *Southern California Quarterly, 52*, 1–32.

Engelbert, Ernest A., and Scheuring, Ann F., eds. 1984. *Water Scarcity: Impacts on Western Agriculture*. Berkeley: University of California Press.

Fahey, John. 1986. *The Inland Empire: Unfolding Years, 1879–1929*. Seattle: University of Washington Press.

Fite, Gilbert. 1968. "The Farmer's Dilemma, 1919–1929." In John Braeman, Robert H. Bremner, and David Brody, eds., *Change and Continuity in Twentieth-Century America*. Columbus: Ohio State University Press.

Fite, Gilbert. 1966. *The Farmer's Frontier, 1865–1900*. New York: Holt, Rinehart and Winston.

FitzGerald, D.A. 1935. *Livestock under the AAA*. Washington, D.C.: Brookings Institution.

Foss, Phillip O. 1960. *Politics and Grass: The Administration of Grazing on the Public Domain*. Seattle, Washington: Greenwood Press.

Frink, Maurice, Jackson, W.T., and Spring, A.W. 1956. *When Grass Was King*. Boulder, Colorado.

Frome, Michael. 1974. *Battle for the Wilderness*. Boulder, Colorado: Westview Press.

Fugate, Francis L. 1961. "Origins of the Range Cattle Era in South Texas," *Agricultural History, 11*, 142–157.

Galarza, Ernesto. 1964. *Merchants of Labor*. Charlotte, North Carolina.
Galarza, Ernesto. 1977. *Farm Workers and Agri-Business in California, 1947–1960*. Notre Dame, Indiana: University of Notre Dame Press.
Ganol, John T. 1937. "The Desert Land Act in Operation, 1877–1891." *Agricultural History, 11*, 142–157.
Gard, Wayne. 1954. *The Chisholm Trail*. Norman: University of Oklahoma Press.
Gates, Paul W. 1968. *History of Public Land Law Development*. Washington, D.C.: U.S. Govt. Printing Office.
Giannini, A.H. 1926. "Financing the Production and Distribution of Motion Pictures," *Annals of the American Academy of Political and Social Science, 128* (November), 48.
Giovinco, Joseph. 1968. "Democracy in Banking: The Bank of Italy and California's Italians," *California Historical Society Quarterly, 47*, 195–218.
Glass, Mary Ellen. 1968. "The Newlands Reclamation Project: Years of Innocence, 1903–1907," *Journal of the West, 7*, 55–63.
Goldsmith, Edward, and Hildyard, Nichols. 1986. *The Social and Environmental Effects of Large Dams*. San Francisco: Sierra Club.
Golzé, Alfred R. 1961. *Reclamation in the United States*. Caldwell, Idaho: Caxton Printers.
Green, Donald E. 1973. *Land of the Underground Rain: Irrigation on the Texas High Plains, 1910–1970*. Austin: University of Texas Press.
Greever, William. 1963. *Bonanza West: The Story of Western Mining Rushes, 1848–1900*. Norman: University of Oklahoma Press.
Gressley, Gene M. 1966. *Bankers and Cattlemen*. New York: Alfred A. Knopf.
Hampton, Benjamin B. 1970. *History of the American Film Industry from its Beginnings to 1931*. New York: Dover.
Hargreaves, Mary W.M. 1957. *Dry Farming in the Northern Great Plains, 1900–1925*. Cambridge, Massachusetts: Harvard University Press.
Hargreaves, Mary W.M. 1976. "Land Use Planning in Response to Drought: The Experience of the Thirties," *Agricultural History, 50*, 561–582.
Hayley, J. Evetts. 1953. *The XIT Ranch of Texas*. Rev. ed. Norman: University of Oklahoma Press.
Hays, Samuel P. 1959. *Conservation and the Gospel of Efficiency: The Progressive Conservation Movement, 1890–1920*. Cambridge, Massachusetts: Harvard University Press.
Hays, Samuel P. 1985. "From Conservation to Environmental Politics in the United States since World War II." In Kendall E. Bailes, ed., *Environmental History: Critical Issues in Comparative Perspective*. Lanham, Maryland: University Press of America.
Held, R. Burnell, and Clawson, Marion. 1965. *Soil Conservation in Perspective*. Baltimore: Johns Hopkins University Press.
Hewett, Edgar L. 1930. *Ancient Life in the American Southwest*. Indianapolis, Indiana: Biblo.
Hewlett, Richard G., and Anderson, Oscar, Jr. 1962. *The New World*. University Park: Pennsylvania State University Press.
Hibbard, Benjamin. 1924. *History of the Public Land Policies*. Reissued, 1965, by University of Wisconsin Press, Madison.

Hicks, John. 1931, 1961. *The Populist Revolt.* Lincoln: University of Nebraska Press.

Hoffman, Abraham. 1974. *Unwanted Mexican Americans in the Great Depression.* Tucson: University of Arizona Press.

Hoffman, Abraham. 1981. *Vision or Villany: Origins of the Owens Valley–Los Angeles Water Controversy.* College Station: Texas A&M Press.

Holley, William C., and Arnold, Lloyd C. 1938. "Changes in Technology and Labor Requirements in Crop Production: Cotton." In Works Progress Administration, National Research Project Report A-7. Philadelphia.

Hollon, W. Eugene. 1966. *The Great American Desert.* New York: Oxford University Press.

Hoover, Calvin. 1934. "The New Deal in the United States," *Economic Journal, 44*, 581–583.

Horwitz, Morton. 1977. *The Transformation of American Law, 1780 to 1860.* Cambridge, Massachusetts: Harvard University Press.

Hundley, Norris, Jr. 1966. *Dividing the Waters: A Century of Controversy between the United States and Mexico.* Berkeley: University of California Press.

Hundley, Norris, Jr. 1975. *Water and the West: The Colorado River Compact and the Politics of Water in the American West.* Berkeley: University of California Press.

Hundley, Norris, Jr. 1978. "The Dark and Bloody Ground of Indian Water Rights: Confusion Elevated to Principle." *Western Historical Quarterly, 9*, 455–482.

Hundley, Norris, Jr. 1982. "The 'Winters' Decision and Indian Water Rights: A Mystery Reexamined," *Western Historical Quarterly, 13*, 17–42.

Hurt, R. Douglas. 1981. *Dust Bowl: An Agricultural and Social History.* Chicago: Nelson-Hall.

Hutchins, Wells A. 1971–1977. *Water Rights Laws in the Nineteen Western States.* 3 vols. Washington, D.C.: USDA Economic Research Service.

Ingram, Helen M. 1969. *Patterns of Politics in Water Resource Development: A Case Study of New Mexico's Role in the Colorado River Basin Bill.* Albuquerque: University of New Mexico.

International Boundary and Water Commission, U.S. Section. [1982.] *Joint Projects of the United States and Mexico through the International Boundary and Water Commission.* El Paso, Texas.

Ise, John. 1916. *Our National Park Policy: A Critical History.* Baltimore: Ayer.

Jakle, John A. 1985. *The Tourist: Travel in Twentieth-Century North America.* Lincoln: University of Nebraska Press.

James, Marquis, and James, Bessie R. 1954. *Biography of a Bank: The Story of Bank of America N.T.&S.A.* New York: Greenwood Press.

Jelinek, Lawrence J. 1982. *Harvest Empire: A History of California Agriculture.* 2d ed. San Francisco: Boyd & Fraser.

Kahrl, William L., ed. 1979. *The California Water Atlas.* Sacramento: Governor's Office of Planning and Research.

Kahrl, William L. 1982. *Water and Power: The Conflict over Los Angeles' Water Supply in the Owens Valley.* Berkeley: University of California Press.

Kinney, Clesson S. 1912. *A Treatise on the Law of Irrigation and Water Rights.* 2d ed., 4 vols. San Francisco: Bender-Moss.

Klose, Nelson. 1950. *American's Crop Heritage: The History of Foreign Plant Introduction by the Federal Government*. Ames: Iowa State College Press.

Knight, Oliver. 1956. "Correcting Nature's Error: The Colorado–Big Thompson Project," *Agricultural History, 30*, 157–169.

Koppes, Clayton R. 1978. "Public Water, Private Land: Origins of the Acreage Limitation Controversy, 1933–1953," *Pacific Historical Review, 47*, 607–636.

Koppes, Clayton. 1982. *JPL*. New Haven, Connecticut: Yale University Press.

Krier, James E., and Ursin, Edmund. 1977. *Pollution and Policy: A Case Essay on California and Federal Experience with Motor Vehicle Air Pollution, 1940–1975*. Berkeley: University of California Press.

Larson, T.A. 1978. *History of Wyoming*. 2d ed. Lincoln: University of Nebraska Press.

Lee, Lawrence B. 1977. "Reclamation and Irrigation." In Howard R. Lamar, ed., *The Reader's Encyclopedia of the American West*. New York: Crowell.

Littlefield, Douglas R. 1987. "Interstate Water Conflicts, Compromises, and Compacts: The Rio Grande, 1880–1938." Ph.D. dissertation. University of California, Los Angeles.

Limerick, Patricia N. 1987. *The Legacy of Conquest: The Unbroken Past of the American West*. New York: Norton.

Lowitt, Richard. 1984. *The New Deal and the West*. Bloomington: Indiana University Press.

Lyons, Eugene. 1964. *Herbert Hoover: A Biography*. New York: Doubleday.

Malin, James C. 1944. *Winter Wheat in the Golden Belt of Kansas: A Study in Adaptation to Subhumid Geographical Environment*. Lawrence: University of Kansas Press.

Malone, Michael, ed. 1983. *Historians and the American West*. Lincoln: University of Nebraska Press.

Malone, Michael. 1986. "The Challenge of Western Metal Mining: An Historical Epitaph," *Pacific Historical Review, 55*, 455–464.

Malone, Thomas E. 1965. "The California Irrigation Crisis of 1886: Origins of the Wright Act." Ph.D dissertation. Stanford University.

Mann, Dean E. 1975. "Conflict and Coalition: Political Variables Underlying Water Resource Development in the Upper Colorado River Basin," *Natural Resources Journal, 15*, 166–167.

Mann, Dean E., Weatherford, Gary D., and Nichols, Phillip. 1964. "Legal Political History of Water Resource Development in the Upper Colorado River Basin." Lake Powell Research Project Bull. 4. Los Angeles.

Mattison, Ray H. 1951. "The Hard Winter and the Range Cattle Business," *Montana: The Magazine of Western History, 1* (October), 5–21.

Matusow, Allen J. 1967. *Farm Policies and Politics in the Truman Years*. Cambridge, Massachusetts: Harvard University Press.

McKibben, Eugene G., and Griffin, R. Austin. 1938. "Changes in Farm Power and Equipment: Tractors, Trucks, and Automobiles." In Works Progress Administration, National Research Project. Report A-9. Philadelphia.

McKinley, Charles. 1952. *Uncle Sam in the Pacific Northwest*. Berkeley: University of California Press.

McWilliams, Carey. 1949. *California: The Great Exception*. New York: Greenwood Press.

Meinig, Donald W. 1955. "The Growth of Agricultural Regions in the Far West, 1850–1910," *Journal of Geography, 54*, 221–232.

Merk, Frederick. 1978. *History of the Westward Movement.* New York: Alfred A. Knopf.

Metropolitan Water District of Southern California. 1947. *Metropolitan Water District Act.* Los Angeles.

Meyer, Michael C. 1984. *Water in the Hispanic Southwest: A Social and Legal History, 1550–1850.* Tucson: University of Arizona Press.

Miller, William H. 1951. "Agriculture in the High Plains: The History of a Struggle against Environment." Ph.D. dissertation. University of California, Berkeley.

Miner, Craig. 1986. *West of Wichita: Settling the High Plains of Kansas, 1865–1890.* Lawrence: University of Kansas.

Missouri Basin Interagency Committee. 1952. *Missouri River Basin Development Program.* Washington, D.C.

Moeller, Beverly. 1971. *Phil Swing and Boulder Dam.* Berkeley: University of California Press.

Nash, Gerald D. 1968. *United States Oil Policy, 1890–1964.* New York: Greenwood Press.

Nash, Gerald D. 1973. *The American West in the Twentieth Century: A Short History of an Urban Oasis.* Englewood Cliffs, New Jersey: Prentice-Hall.

Nash, Gerald D. 1985. *The American West Transformed: The Impact of the Second World War.* Bloomington: Indiana University Press.

Nash, Roderick. 1982. *Wilderness and the American Mind.* 3d ed. New Haven, Connecticut: Yale University Press.

Nelson, Bruce. 1946. *Land of the Dacotahs.* Reissued, 1964, by University of Nebraska Press, Lincoln.

New York Times. 1985. Mar. 26 and 28.

Nichols, Roger L., ed. 1986. *American Frontier and Western Issues.* New York: Greenwood Press.

Olsen, Michael L. 1970. "The Beginnings of Agriculture in Western Oregon and Western Washington." Ph.D. dissertation. University of Washington.

Osgood, Ernest S. 1929, 1957. *Day of the Cattleman.* Chicago: University of Chicago Press.

Ostrom, Vincent. 1953. *Water & Politics: A Study of Water Policies and Administration in the Development of Los Angeles.* Los Angeles: Johnson.

Paul, Rodman W. 1963. *Mining Frontiers of the Far West, 1848–1880.* Reissued, 1974, by University of New Mexico Press, Albuquerque.

Paul, Rodman W. 1977. "Mining, Metal." In Howard R. Lamar, ed., *The Reader's Encyclopedia of the American West.* New York: Crowell.

Paul, Rodman W., and Etulain, Richard W., compilers. 1977. *The Frontier and the American West.* Arlington Heights, Illinois: Harlan Davidson.

Paxson, Frederick L. 1946. "The Highway Movement, 1916–1935." *American Historical Review, 51*, 236–253.

Pelzer, Louis. 1936. *The Cattleman's Frontier.* Glendale, California: Arthur H. Clark.

Pisani, Donald. 1984. *From the Family Farm to Agribusiness: The Irrigation Crusade in California and the West.* Berkeley: University of California Press.

Pomeroy, Earl. 1957. *In Search of the Golden West: The Tourist in Western America*. Reissued, 1985, by Howe Brothers, Salt Lake City, Utah.
Pomeroy, Earl. 1965. *The Pacific Slope*. New York: Alfred A. Knopf.
Pomeroy, Earl. 1977. "Tourist Travel." In Howard R. Lamar, ed., *The Reader's Encyclopedia of the American West*. New York: Crowell.
Porter, Eliot. 1963. *The Place No One Knew: Glen Canyon on the Colorado*. San Francisco: Sierra Club.
Pring, George W. and Tomb, Karen A. 1979. "License to Waste: Legal Barriers to Conservation and Efficient Use of Water in the West," *Rocky Mountain Mineral Law Institute, 25*, art. 25.
Quiett, Glenn C. 1934. *They Built the West: An Epic of Rails and Cities*. New York: Cooper Square.
Rae, John B. 1965. *The American Automobile*. Chicago: University of Chicago Press.
Rae, John B. 1968. *Climb to Greatness: Aircraft Industry, 1920–1960*. Cambridge, Massachusetts: MIT Press.
Rakestraw, Lawrence. 1972. "Conservation Historiography: An Assessment," *Pacific Historical Review, 61*, 271–288.
Rasmussen, Wayne D., and Baker, Gladys L. 1972. *The Department of Agriculture*. New York: Praeger.
Raup, Hallock F. 1932. "The German Colonization of Anaheim, California." *University of California Publications in Geography, 6*, 123–146.
Reisner, Marc. 1986. *Cadillac Desert: The American West and Its Disappearing Water*. New York: Viking.
Richardson, Elmo R. 1959. "The Struggle for the Valley: California's Hetch Hetchy Controversy," *California Historical Society Quarterly, 38*, 249–258.
Rick, Spencer A. 1963. *United States Agricultural Policy in the Post-War Years, 1945–1963*. Washington, D.C.
Rickard, Thomas A. 1921. *Concentration by Flotation*. New York: John Wiley.
Rickard, Thomas A. 1932. *A History of American Mining*. New York: McGraw-Hill.
Robbins, William G. 1986. "The 'Plundered Province' Thesis and the Recent Historiography of the American West," *Pacific Historical Review, 55*, 577–597.
Robinson, Michael C. 1979. *Water for the West: The Bureau of Reclamation, 1902–1977*. Chicago: Public Works Historical Society.
Runte, Alfred. 1979. *National Parks: The American Experience*. Lincoln: University of Nebraska Press.
Saloutos, Theodore. 1969. "The New Deal and Farm Policy in the Great Plains," *Agricultural History, 43*, 345–355.
Schapsmeier, Edward L., and Schapsmeier, Frederick L. 1975. *Ezra Taft Benson and the Politics of Agriculture: The Eisenhower Years*. Danville, Illinois: Interstate.
Scheele, Paul E. 1969. "Resource Development Politics in the Missouri Basin: Federal Power, Navigation, and Reservoir Operation Policies, 1944–1968." Ph.D. dissertation. University of Nebraska.
Schlebecker, John T. 1953. "Grasshoppers in American Agricultural History," *Agricultural History, 27*, 91.

Schlebecker, John T. 1963. *Cattle Raising on the Plains, 1900–1961*. Lincoln: University of Nebraska Press.

Schlebecker, John T. 1975. *Whereby We Thrive: A History of American Farming, 1607–1972*. Ames: Iowa State University Press.

Shannon, Fred. 1945. *The Farmers' Last Frontier: Agriculture, 1860–1897*. New York: Farrar & Rinehart.

Shideler, James. 1957. *Farm Crisis, 1919–1923*. Berkeley: University of California Press.

Shinn, Charles H. 1884, 1965. *Mining Camps: A Study in American Frontier Government*. New York: Harper and Row.

Simms, D. Harper. 1970. *The Soil Conservation Service*. New York: Praeger.

Smith, Duane A. 1967. *Rocky Mountain Mining Camps: The Urban Frontier*. Reissued, 1974, by University of Nebraska Press, Lincoln.

Smith, Karen L. 1986. *The Magnificent Experiment: Building the Salt River Reclamation Project, 1890–1917*. Tucson: University of Arizona Press.

Storper, Michael, and Walker, Richard A. 1984. *The Price of Water: Surplus and Subsidy in the California State Water Project*. Berkeley: Institute of Governmental Studies, University of California.

Stratton, Owen and Sirotkin, Phillip. 1959. "The Echo Park Controversy." Inter-University Case Program 46. University, Alabama.

Sundborg, George. 1954. *Hail Columbia: The Thirty Year Struggle for Grand Coulee Dam*. New York: Macmillan.

Tarr, Joel A. 1985. "The Search for the Ultimate Sink: Urban Air, Land and Water Pollution in Historical Perspective." In Kendall E. Bailes, ed., *Environmental History: Critical Issues in Comparative Perspective*. Lanham, Maryland: University Press of America.

Treadwell, Edward F. 1931. *The Cattle King*. Reissued, 1981, by Western Tanager, Santa Cruz, California.

Trimble, William J. 1914. *Mining Advance into the Inland Empire*. Reissued by Johnson Books, Boulder, Colorado.

U.S. Department of Agriculture. 1935. *Yearbook of Agriculture, 1935*. Washington, D.C.

U.S. Department of Agriculture. Bureau of Agricultural Economics. 1938. *Livestock on Farms, January 1, 1867–1919*. Washington, D.C.

U.S. Department of Commerce. 1919. *Statistical Abstract of the United States, 1919*. Washington, D.C.

U.S. Department of Commerce. 1922. *Statistical Abstract of the United States, 1921*. Washington, D.C.

U.S. Department of Commerce. 1935. *Statistical Abstract of the United States, 1935*. Washington, D.C.

U.S. Department of Commerce. 1965. *Statistical Abstract of the United States, 1965*. Washington, D.C.

U.S. Department of Commerce. 1975. *Historical Statistics of the United States: Colonial Times to 1970*. 2 vols. Washington, D.C.

U.S. Department of Commerce, 1984a. *Current Population Reports: Local Population Estimates*. Washington, D.C.

U.S. Department of Commerce. 1984b. *Statistical Abstract of the United States, 1984*. Washington, D.C.

U.S. Department of the Interior. 1926. *Annual Report, 1926*. Washington, D.C.

U.S. Department of the Interior, Bureau of Mines. 1962. *Minerals Yearbook, 1961*. Vol. II: *Fuels*. Washington, D.C.

U.S. Department of the Interior, Bureau of Mines. 1982. *Minerals Yearbook: Centennial Edition, 1981*. Vol. I: *Metals and Minerals*. Washington, D.C.

U.S. Department of the Interior, Bureau of Reclamation. 1946. *The Colorado River: A Comprehensive Department Report on the Water Resources of the Colorado River Basin for Review Prior to Submission to the Congress*. Washington, D.C.

U.S. Department of the Interior, Bureau of Reclamation. 1964. *The Story of the Columbia Basin Project*. Washington, D.C.

U.S. Department of the Interior, Bureau of Reclamation, California Department of Water Resources, California State Water Resources Control Board, and San Joaquin Valley Interagency Drainage Program. 1979. *Agricultural Drainage and Salt Management in the San Joaquin Valley*. Fresno, California.

U.S. Department of the Interior, Geological Survey. 1909. *Mineral Resources of the United States: Calendar Year 1908*. Washington, D.C.

U.S. Department of Transportation. 1965. *Highway Statistics: Summary to 1965*. Washington, D.C.

U.S. House of Representatives. 1923. "Colorado River Compact." H. Doc. 605. 67th Congress, 4th sess.

U.S. House of Representatives. 1948. "Hoover Dam Documents." H. Doc. 717. 80th Congress, 2d sess.

U.S. Public Land Law Review Commission. 1970. *One Third of the Nation's Land*. Washington, D.C.

U.S. Senate, Select Committee on Small Business and Committee on Interior and Insular Affairs. 1975–1976. *Hearings on Will the Family Farm Survive in America*? 94th Congress, 1st and 2d sess.

Valantine, Vernon. 1984. Colorado River Board of California. Interview. May 15.

Washington Post. 1985. Apr. 15.

Webb, Walter Prescott. 1931. *The Great Plains*. Reissued, 1981, by University of Nebraska Press, Lincoln.

Wells, Merle. 1974. *Camps and Silver Cities*. Rev. ed. Boise, Idaho: Bureau of Mines and Geology.

White, Gerald T. 1977. "Oil Industry." In Howard R. Lamar, ed., *The Reader's Encyclopedia of the American West*. New York: Crowell.

White, Gerald T. 1980. *Billions for Defense: Government Financing by the Defense Plant Corporation during World War II*. University, Alabama: University of Alabama Press.

Wiel, Samuel C. 1911. *Water Rights in the Western States*. 3d ed. 2 vols. San Francisco, California. Reissued by Ayer, Baltimore.

Wilcox, Walter W. 1947. *The Farmer in the Second World War*. Ames, Iowa. Reissued by Da Capo Press, New York.

Wiley, Peter, and Gottlieb, Robert. 1982. *Empires in the Sun: The Rise of the New American West*. Tucson: University of Arizona Press.

Willard, James F., ed. 1918. *The Union Colony at Greeley, Colorado, 1869–1871*. Boulder: University of Colorado Historical Collections.

Willey, Zach. 1985. *Economic Development and Environmental Quality in California's Water System*. Berkeley: University of California Press.

Williamson, Harold, et al., 1963. *The American Petroleum Industry*. Vol. II: *The Age of Energy, 1899–1959*. Reissued, 1981, by Greenwood Press, Westport, Connecticut.

Worster, Donald E. 1979. *Dust Bowl: The Southern Plains in the 1930s*. New York: Oxford University Press.

Worster, Donald E. 1985. *Rivers of Empire: Water, Aridity, and the Growth of the American West*. New York: Pantheon.

3 The Central Valley of California

CHARLES V. MOORE AND
RICHARD E. HOWITT

The conflicts and control of water in California have been the source of political careers, personal fortunes, and Hollywood movies. Water has characteristics of both public and private property. Over the past century, water management and development have shifted from small local institutions to predominately large public agencies. The impetus for changing institutions came from physical, economic, and technical shifts; the implementation was manifest political power.

This chapter presents a brief overview of the physical and economic characteristics of California's Central Valley, followed by a historical overview of economic development in the valley, including water law and the major actors in the development and management of water. Emerging problems in the Central Valley water industry are then presented, and the policy solutions to these problems are developed. These conclusions lead to the prediction that California's water institutions are about to change substantially again.

Physical and economic setting

Hydrologically, the Central Valley is divided into three basins, the Sacramento, San Joaquin, and Tulare Lake, based on the drainage areas of rivers and streams. In their unimpaired state, all Central Valley rivers drained to the ocean through the Sacramento–San Joaquin Delta and San Francisco Bay. Owing to reservoir and aqueduct construction, the Tulare Lake Basin is closed, with outflows occurring only in extremely wet years.

Interbasin transfers have concentrated on moving water from the northern half of the state, which receives 75 percent of the precipitation, to the southern half, which receives only 25 percent but contains two-thirds of the population.

The first major transbasin diversion occurred in 1934, when the city of San Francisco constructed the Hetch Hetchy aqueduct, taking water from the Sierra Nevada across the Central Valley to San Francisco. The initial transbasin diversion for agricultural purposes was the federally

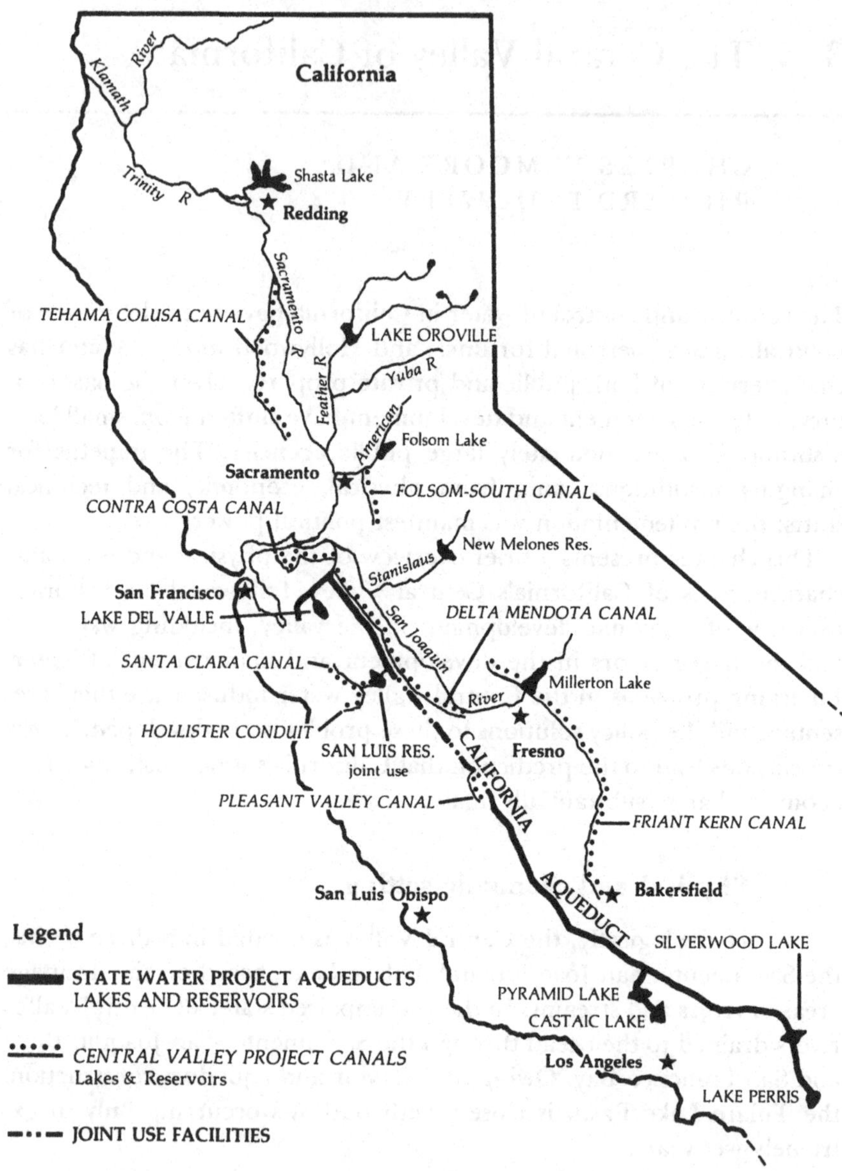

Figure 3.1. Major features of the State Water Project and the Central Valley Project (adapted from California Department of Water Resources, 1983).

constructed Friant-Kern Canal, finished in 1949, which took water from the San Joaquin River. The canal ships 1.2 million acre-feet (maf) south along the east side of the San Joaquin Valley into the Tulare Lake Basin. (See Figure 3.1.) The state-sponsored State Water Project (SWP) commenced deliveries in 1968. Conceived as a combined agricultural and municipal industrial (M and I) project, SWP water is transferred south from the Sacramento Basin along the west side of the San Joaquin Valley. Agricultural deliveries are made to the San Joaquin Basin and the Tulare Lake Basin. Major M and I deliveries are pumped over the Tahacapi Mountains and into the Los Angeles Basin. The planned capacity of the SWP is 4.2 maf, of which approximately 2.2 maf are currently developed.

On the valley floor, mean annual rainfall ranges from 22 inches (56.01 cm) at Red Bluff in the northern Sacramento Valley to 13.37 inches (33.96 cm) at Stockton in the center to 6.36 inches (16.15 cm) at Bakersfield at the southern end of the valley. More important is the annual precipitation in the mountains containing the headwaters of the major rivers feeding into the valley. For example, at the Pitt River Powerhouse near Mount Shasta at the north end of the valley, mean annual precipitation is 74.9 inches (190.3 cm); at Soda Springs in the central area, at an elevation of 6,750 feet (2,057 m), where the snowpack remains until midsummer, the precipitation is 52.76 inches (134.01 cm). Farther south in the Sierras above Fresno, precipitation at Huntington Lake, elevation 7,020 feet (2,140 m), is 32.45 inches (82.42 cm).

Approximately 60 percent of California's water supply comes from surface sources. In the Central Valley, 23 major reservoirs store winter runoff for irrigation use. These storage reservoirs range in capacity from 10,000 acre-feet up to 4.5 maf at Lake Shasta on the Sacramento River. In addition, riparian and appropriative water rights holders divert the natural flow of water courses.

The remaining 45 percent of the water supply comes from groundwater sources recharged from precipitation and from percolation from streams, overirrigation, return flows, and artificial recharge basins. Groundwater utilization includes an estimated 1.5 maf of overdraft in the San Joaquin Valley. The California Department of Water Resources has defined eight subbasins within the San Joaquin and Tulare Lake Basins as critically overdrafted. The current average rate of overdrafting is 5 percent of total water supplies. However, preliminary studies (discussed below) suggest that some of this overdrafting still benefits the valley economically.

Salinity of streamflow, measured in parts per million (ppm), is lowest near the headwaters. Maximum concentrations of salts for the water year are less than 90 ppm for inflows to Lake Shasta at the northern end of the valley, reaching only 128 ppm before entering the Sacramento–San Joaquin Delta.

Inflows to the San Joaquin Basin from the Sierra Nevada range show equally low salt concentrations; however, owing to the high proportion of irrigation return flows, salt concentrations are higher for streams on the valley floor. For example, the maximum salt concentration at Millerton Lake on the San Joaquin River is 51 ppm. Fifty miles downstream, the maximum concentration reaches 300 ppm. Salts continue to accumulate, reaching 1,320 ppm just south of the confluence with the Merced River. As the San Joaquin River flows farther north, return flows are diluted by other major streams, and the salt concentration drops to below 320 ppm near Stockton in the central portion of the Central Valley.

Salinity affects crop yields directly. Irrigation water of less than 480 ppm does not harm even salt-sensitive crops. From 480 ppm to 960 ppm, such sensitive crops as stone fruit and vegetables are damaged. Detrimental effects are observed in most commercial crops when the concentration ranges between 960 and 1,920 ppm. When salinity exceeds 640 ppm, irrigation water must be carefully managed and subsurface drainage provided.

Little information is available on the quality of groundwater in the Central Valley, but deep percolation of saline surface water exacerbates the salt concentration problems, especially in a closed basin such as the Tulare Lake Basin.

Climate, soils, and water supply are the physical determinants of cropping patterns in the Central Valley. (Table 3.1 presents 1980 data on the crop acreage by hydrologic basin.)

Heavy soils, cool nights, and ample water supplies are conducive to rice production in the Sacramento Basin. Low water costs in the Sacramento Basin have encouraged expansion of crops with high water requirements such as irrigated pasture. The high proportion of small grains in the Sacramento Basin is due in part to higher average rainfall in the winter growing season; irrigation is supplemental in dry years.

In the San Joaquin and Tulare Lake basins, the longer growing seasons and warm nights are ideal for cotton and alfalfa production. Higher water costs, especially in the Tulare Lake Basin, make irrigated pasture uneconomical. The long, rainless summer and fall in the San Joaquin and Tulare Lake basins are conducive to grape production, especially for raisins, which are sun dried in the field.

In the last half of the nineteenth century, Central Valley producers grew primarily for the local market because transcontinental transportation was limited. Once a reliable rail system and later an interstate highway system were completed, national markets opened. The last half of

Table 3.1. *Irrigated crop acreage by hydrologic basin (thousand acres)*

Crop	Sacramento Basin	San Joaquin Basin	Tulare Lake Basin
Small grains	399	275	500
Rice	491	41	13
Cotton	0	197	1,239
Sugarbeets	59	66	39
Corn	140	207	95
Other field crops[a]	190	211	151
Alfalfa	105	181	319
Pasture	359	301	67
Tomatoes	108	60	38
Other truck crops[b]	32	86	115
Almond-pistachio	94	187	126
Other deciduous[c]	178	146	153
Citrus-olives	14	8	166
Grapes	5	230	480
Total crop acres	2,176	2,142	3,384
Double crop	92	80	72
Total land area	2,084	2,062	3,312

[a]Dry beans, safflower, milo, sunflower, etc.
[b]Potatoes, melons, lettuce, onions, etc.
[c]Walnuts, peaches, prunes, plums, etc.
Source: California Department of Water Resources, 1983, p. 144.

the twentieth century has brought expansion into international markets. In 1983, 22 percent of the rice crop was exported, while 15 percent of the raisins and 18 percent of fresh orange crop went through California ports. The international share is expected to increase, even though California is the most populated state in the union.

The range of returns to irrigated water varies as widely as the crops; although $40–$60 per acre-foot is a common median 1980 value for the valley, the return ranges from $180 per acre-foot for high-value fruit crops to $0 per acre-foot for some irrigated pasture uses. More recently, as returns from rice and cotton have fallen, less irrigated land has been used for these crops. Market forces limit the expansion of irrigated acreage planted in high-value crops. On balance, the market outlook for California's crops suggests a reduction in irrigated acreage over the next ten years.

All major population centers in California are hydrologically connected to the Central Valley. The State Water Project's California Aqueduct runs the length of the Central Valley to deliver water to the Los Angeles Basin. The San Francisco Bay area's water supplies originate in the Sierra Nevada, with pipelines and canals traversing the Central Valley. Thus, statewide population growth competes directly for the Central Valley's riparian water supplies. Table 3.2 presents demographic changes by hydrographic area in the state.

Overall state population growth between 1972 and 1980 was 15 percent. In absolute terms, the Santa Ana, San Diego, and Los Angeles basins – all in the south coastal area – grew the most. Within the Central Valley, population increased by more than 750,000 over the 12-year period. Owing to the high summer temperatures and the lack of water meters, gross daily per capita water consumption in the Central Valley is high compared to that in coastal cities. Bakersfield, Fresno, and Sacramento report a per capita consumption of about 300 gallons per day, while Corning, in the Sacramento Basin, reports a per capita consumption of almost 400 gallons per day.

The population growth in the Greater Los Angeles area and the evolution of the Colorado Water Compact will mean that the Metropolitan Water District may use its entitlement to an additional 0.33 maf of the State Water Project firm yield. Currently, this water is being purchased as low-cost "surplus" water by agencies in the southern San Joaquin Valley (California Department of Water Resources, 1983). Compared with the total agricultural water use in the Central Valley of 21.4 maf per year, this 1.5 percent reduction in firm supplies over the next ten years is not a major problem. Even locally, the increasing urban entitlements on the San Joaquin Valley would reduce supplies only 2.4 per-

Table 3.2. *Population growth in California, by hydrographic basin, 1972–1980*

	Population		
Basin	1972	1980	% growth
North Coast	363,000	459,000	26
San Francisco	4,475,000	4,790,000	7
Central Coast	833,000	1,005,000	21
Los Angeles	7,398,000	7,927,000	7
Santa Ana	2,364,000	2,974,000	26
San Diego	1,529,000	2,068,000	35
Sacramento	1,311,000	1,674,000	28
San Joaquin	805,000	1,014,000	26
Tulare	989,000	1,178,000	19
North Lahontan	44,000	61,000	39
South Lahontan	245,000	303,000	24
Colorado River	237,000	320,000	35
Total	20,593,000	23,773,000	15

Source: California Department of Water Resources, 1983, p. 41.

Table 3.3. *Characteristics of farms with irrigated land, by basin, 1982*

	Sacramento Basin	San Joaquin Basin	Tulare Lake Basin
Total number of farms	8,931	11,746	13,981
Total irrigated acres	1,617,359	1,539,178	3,241,881
Average farm size (acres)	69.1	131.0	231.9
Number of farms over 1,000 acres	795	499	763
Irrigated acres in farms over 1,000 acres	880,958	707,399	2,104,326
Number of farms over 2,000 acres	380	228	377
Irrigated acres in farms over 2,000 acres	593,462	430,519	1,658,729
Percentage of farms over 1,000 acres	8.9	4.2	5.5
Percentage of irrigated land in farms over 1,000 acres	54.5	45.9	64.9
Percentage of farms over 2,000 acres	4.2	1.9	2.7
Percentage of irrigated land in farms over 2,000 acres	36.7	28.0	51.2

Source: U.S. Department of Commerce, 1982.

cent. (Methods of accommodating these supply shifts are discussed later in the chapter.)

Aggregate statistics do not give a clear picture of the structure of agriculture in a region as large as the Central Valley. Cotton, grain, and rice farms tend to be larger than specialty crop farms raising deciduous fruits, nuts, and vegetables. One indication of the size and distribution of irrigated farms in the region is the concentration of control of land. Table 3.3 presents data on farm size characteristics by basin in the valley floor counties.

As Table 3.3 indicates, the Tulare Lake Basin contains the largest farms and is marked by the most concentrated control of land by operators. Dominated by more than 1.2 million acres of cotton production, 2.7 percent of the farms encompass 51.2 percent of the irrigated land in the Tulare Lake Basin. Although the average farm in the Sacramento Basin is smaller (only 69.1 acres), the largest 4.2 percent of the farms contain 36.7 percent of all the irrigated land. This situation probably reflects the importance of rice production in the basin.

Historical development

The virgin landscape of the Central Valley consisted of a high proportion of swamps, saline and alkaline lands, and riparian forest along the tributary streams. Early settlers in the first half of the nineteenth century used the native grass to feed livestock. Hides and tallow for export were the main source of cash for the Spanish and Mexican land grantees. The gold rush increased the value of livestock when gold miners demanded beef.

The Homestead Act of 1862 spurred cultivated agriculture through sale of the public domain to people willing to settle and improve the land. An additional impetus came when the California legislature repealed an earlier law denying compensation for crops damaged by livestock.

Early irrigation development in the San Joaquin Valley was modeled after southern California's canal colonies. Land obtained under the Green Act of 1868 was divided into 20- and 40-acre parcels and sold to new settlers (Kahrl, 1978). But developers retained the right to deliver water at whatever price the market would bear. Canal colonies became more popular after the first transcontinental railroad was completed in 1869. Railroads were granted 11,580,000 acres of California land to help pay construction costs and thus entered the business of water development and colonization. In this unregulated era, railroads obtained a double benefit, selling land and water and setting the freight rates. Many of these early colonies failed because of conflicting water rights, high water charges, and the inability to transport production to market economically. However, by 1878 there were more than 1,000 miles of canals in Fresno County alone.

In the 1870s, a series of drought years signaled the demise of the cattle industry on the valley floor. Many years of heavy rainfall in the 1880s induced rapid expansion of acreage on which wheat was grown; by 1889, California produced 40 million bushels of wheat annually (Kahrl, 1978), carried by train and river boat to Port Costa near San Francisco Bay for export. California ranked second in total wheat production, with rain-fed wheat coming from the Sacramento Valley and irrigated wheat from the San Joaquin Valley.

Shallow water tables and the availability of artesian wells aided in the expansion of irrigated land, but as increasing numbers of farmers used the groundwater aquifiers, water tables began to fall. With the invention of the internal combustion engine and the shallow well pump, irrigation moved beyond riparian and gravity flow service areas. By 1906, 597 irrigation pumps were in operation. Four years later, the number had increased to 5,000 pumps; by 1920, there were 11,000 pumps; and by 1930, 23,000. In 1940, 1.5 million acres were using groundwater as the primary irrigation source. In many areas, pump technology, combined with the declining real cost of energy (fossil or electrical), meant that groundwater was a more cost-effective water source than surface delivery.

The Wright Act (1887), which provided for formation of irrigation districts, strengthened the salability of bonds for irrigation districts and mutual water companies. The new legislation districts were democratically organized, and each landowner had one vote. Large landowners who carried the burden of debt service based on their acreage pressed for, and obtained, additional enabling legislation that allowed creation of water districts where voting rights were based on assessed valuation of lands in the district. By 1970, there were 892 public water service organizations in California.

An overview of water costs and use in the Central Valley discloses two phases of development and three groups of irrigation districts. Development of water in the Central Valley can be broadly divided into the local development era (1887–1935) and the governmental development era (1935–1985). The lowest-cost water districts are old districts holding appropriative water rights on local streams. They include many districts along the Sacramento River, districts in the Sacramento–San Joaquin Delta, and those in the east side of the San Joaquin Valley. In the median-cost group are federal water project districts along the west side of the Sacramento Basin, districts receiving water from the the Delta-Mendota, Madera, and Friant-Kern canals in the San Joaquin Valley. The highest costs are faced by State Water Project districts in Kern County. Within this group, individual districts may have abnormally high water costs at the farm head gate owing to costs within the district.

The irrigation district is the central institution in rural California water development. During the first 50 years (1887–1935) of local water development, the water district was the developer and seller of water. In the last 50 years, the development of state and federal interbasin projects has changed the role of the irrigation district into a local retailer of water developed elsewhere. But these districts continue to hold their own water rights.

An irrigation district is a public corporation organized under general state laws primarily to provide irrigation water for lands within its boundaries. It is empowered to issue bonds and to tax, and it derives its revenue chiefly from assessments on the land. The lands of an objecting minority may be included and assessed if they will benefit from district improvement.

Initially, irrigation districts' revenue source was assessment on the land to benefit. This approach provided a reliable stream of income for servicing debt and paying operating expenses. By California law, district assessments are based on the land's fair market value. In part, districts relied on *ad valorem* assessments because in the early years the technology for measuring and metering water was crude and expensive.

Water tolls, charges based on the volume of water delivered, are provided for in the Wright Act of 1887. The use of water tolls is at the discretion of an individual district's board of directors. By 1955, 60 percent and 48 percent of total receipts to irrigation districts in the Sacramento Valley and the San Joaquin Valley, respectively, were from tolls (Brewer, 1969).

Water prices and use in Central Valley irrigation districts depend largely on the water rights or entitlement held by a district. By 1985, districts formed prior to 1940 had paid off the loans and bonds used to finance the construction of main canals and distribution systems. Irrigation districts are nonprofit public corporations, and some, such as Modesto Irrigation District in the San Joaquin Basin, charge nominal water costs because they also have hydroelectric power revenues. (Modesto charged $1 per irrigated acre for an entitlement of 4 acre-feet per acre in 1984.)

Irrigation development in the Sacramento Valley trailed the San Joaquin by several decades, partly because much of the area has low-lying heavy soils and because the large flood flows that periodically inundate the valley are unmanageable. Although the steam-powered tractor was available, these large, cumbersome vehicles were used as stationary power sources. Horse-drawn implements could not work the adobe clay soils. Not until 1912 was rice found to be adaptable to the soils and climate of the Sacramento Valley and farm-mechanization technology available to exploit this discovery.

With the advent of such heavy water-using crops as rice and alfalfa in the Sacramento Valley, the demand for irrigation water increased rapidly. Canals such as the Glenn-Colusa and the Butte-Sutter were built for intrabasin transfers by gravity flow. These canals were constructed by irrigation districts that held appropriative water rights in trust for district landowners, a shift away from the mutual ditch companies (with transferable shares) that flourished when the San Joaquin Valley was first developed.

Until this time, all irrigation water development had been at the private or public district level. The state was pushed into water development planning when a privately drawn statewide water plan was published in 1920. Heavily supported by agricultural interests in the Central Valley, the proposed plan included a large dam at the head of the

Sacramento River and the first transbasin transfer between the Sacramento and San Joaquin basins. Such an ambitious scheme required large pumps to lift water out of the Sacramento–San Joaquin River Delta and push it south to the more arid and overdrafted San Joaquin Valley.

In 1931, the state engineer published a modified plan that obtained the same result; it was adopted by the state legislature in the depression year of 1933. The state was unable to obtain funding, and in 1935 the federal Bureau of Reclamation took over the project, now called the Central Valley Project (CVP). Water was transferred not only from the Sacramento Basin south through the Delta Mendota Canal, but also from the San Joaquin Basin south to the Tulare Lake Basin through the Friant-Kern Canal. The transfer technology had been revolutionized with the capability to cast large concrete canal sections and lift large volumes of water at a low unit cost using hydroelectricity.

With construction of the CVP and later the SWP, long distance water transfers became the norm. To obtain water service, a number of water districts had to band together, urban and agricultural negotiating as a single public entity. The era of individual action had passed. The era of interdependency with respect to politics, economics, and engineering had arrived. In Kern County, for instance, the Department of Water Resources (DWR) insisted on a single general contractor for the entire county, so the Kern County Water Agency was formed to negotiate and sign a master contract. This umbrella district in turn contracted with 19 member units. (The voters of the county had to ratify the master contract underwriting the financial obligation to the state.)

Owing to the relatively small size of Kern member units, only one of which completely overlays a groundwater basin, groundwater recharge and management are conducted by the Kern County Water Agency. Thus, the member units are interdependent with respect to surface water deliveries and deliveries through the aquifer as well.

Entry of the federal government into water development changed not only the scale and control of projects but also the cost of the water to landowners. In June 1902, Congress involved the federal government directly in the construction and management of land reclamation projects through passage of the Reclamation Act. The act created an agency within the Department of the Interior, currently known as the Bureau of Reclamation, to administer this program.

The full costs of early Bureau of Reclamation projects had to be paid by landowners within five years, later lengthened to ten years. Congress, recognizing the difficulty new settlers faced in repaying project costs while also repaying loans for private development, extended the

Table 3.4. *Increase in land values owing to project water and estimated subsidy per acre, 1978*

California project	Excess land value per acre[a]	Current market land price per acre	Increases per acre in land value[b]	Estimated subsidy per acre[c]
Glenn-Colusa	$1,200	$1,700	$500	$ 101
Westlands	550	1,500	950	1,422[d]

[a]Includes values of land and irrigation improvements, except irrigation pumps.
[b]Measured as the difference between current market land price and excess land values.
[c]Retroactive to year of initial construction.
[d]Average for San Luis Unit.
Source: Moore, Wilson, and Hatch, 1982.

project repayment period to 40 years in the 1926 Omnibus Adjustment Act.

Federal subsidies to irrigated agriculture were available at the outset. The 1902 act required landowners to pay all project costs except interest, which was charged only for delayed or late payments. The subsidy for an interest-free loan for these five-year repayment periods is small relative to that after the 1926 Omnibus Adjustment Act, which spread repayment over 40 years.

In the Interior Appropriations Act of 1926, Congress came to grips with land speculation in federal projects. Owing to the large subsidy granted to irrigators in these projects, significant economic rents were captured by initial landowners. Congress ordered the Secretary of the Interior to require recordable contracts to sell the excess land of landowners before they were eligible for water services.

Water pricing policy by the Bureau of Reclamation can create additional subsidies. One 1939 amendment to the Reclamation Act allowed the Secretary of the Interior to defer payment of construction charges and adjust installments downward to the water user's ability to pay. The Bureau of Reclamation has used this provision as authority to set water rates equal to 75 percent of the estimated ability to pay (Bureau of Reclamation, 1984), with no provision to recapture the deferred payments.

Federal subsidies to irrigation are not distributed uniformly across irrigation districts. Table 3.4 displays results of two methods of estimating the magnitude of irrigation subsidies in two Central Valley case study districts. The right-hand column shows the calculated subsidy based on the difference between project costs (including interest) and contractual payments by the district to the Bureau of Reclamation. Column 3 presents the increase in land value based on appraised values of irrigated land without regard to the project (excess land value) and the 1978 current market price for the same land. The difference (column 4) reflects the portion of the subsidy captured by the landowner. In Westlands Water District, the largest district, the calculated lump-sum present-value subsidy is $1,422 per acre, whereas land value enhancement was only $950 per acre, indicating that the federal government spent $1.50 for each $1 of benefits captured by landowners. On the other hand, in Glenn-Colusa, a rice-growing district in the Sacramento Valley, benefits of $5 were captured for each $1 of federal subsidy.

The State Water Project in 1960 continued moving water from the Sacramento Valley south to the more arid San Joaquin and Tulare Lake basins. Landowners disturbed over the 160-acre limitation associated with Bureau of Reclamation water projects politically backed a state-sponsored project, even though their water costs would increase.

The major thrust of the California Department of Water Resources is planning for and implementing water deliveries within the state. Water rights and permits and water quality monitoring and enforcement are the function of the Water Resource Control Board, a separate entity.

The department's key facility is the 600-mile long, 4.23 maf State Water Project. In the 1930s and 1940s, it was widely accepted that the Central Valley Water Project was the ultimate solution to California's water problems. This optimism failed to account for the rapid influx of people into California, especially southern California during and after World War II, and for the dissatisfaction of landowners in the Tulare Basin with the acreage limitations attached to CVP water.

The plan for the SWP was completed and presented to the California legislature in 1957. A $1.75 billion bond issue was passed by the voters in 1960, and deliveries began in 1962. Water crossed the mountains into southern California in 1971.

Water pricing to SWP contractors is significantly different from that used by the Bureau of Reclamation in the CVP. Because almost all SWP water is "wheeled" through the Sacramento–San Joaquin Delta, the delta has become the basing point for all water charges throughout the system. The delta water charge includes the amortized cost of all facilities above the delta, including the main conservation storage facility, Oroville Dam on the Feather River.

The SWP used a close approximation of incremental cost pricing, including interest, to determine contractor charges. Contractors along the California Aqueduct, which runs from the delta to southern California, are charged an annual sum that is adjusted to reflect changing operating costs along the aqueduct.

Average water costs are significantly higher to SWP contractors than to CVP contractors. In 1985, the average farm head-gate cost of water in Westlands Water District, a CVP contractor, was about $16 per acre-foot. The average farm head-gate cost in Lost Hills Water District, a SWP contractor just a few miles south of Westlands, was $35 per acre-foot in 1985. Farther along the aqueduct, farm head-gate costs in Wheeler Ridge–Maricopa Water Storage District were about $65 per acre-foot that year. (Surplus or unused and entitlement water is delivered to existing contractors at the pumping cost.)

When the Bureau of Reclamation assumed responsibility for constructing and operating the CVP and the SWP, the era of individuals or irrigation districts filing for water rights and constructing transport facilities to their lands ended. Water from both federal and state projects is allocated among users not on a basis of first in time, first in right but on a long-term contractual basis. The length of the water delivery contract is generally based on the length of the repayment period for the project, usually 40 to 50 years. Water users in the CVP and SWP are not water rights holders but water contractors. Water delivery contracts have reduced the probability of litigation over vested water rights, but some commentators believe that centralizing allocation decisions may have increased the power of special interest groups.

Since 1902, groundwater law in California has been based on correlative rights. These rights are analogous to surface riparian rights when an overlying landowner holds his rights in common with all other landowners in a basin. Groundwater extracted from a basin must be put to a reasonable and beneficial use, and only surplus water may be exported from the basin.

In the case of long-term groundwater overdraft where withdrawals exceed natural recharge, an injured overlying landowner may petition the court to apportion groundwater pumpage to the safe yield of the aquifer. Southern California experience has shown that adjudicating a groundwater basin takes decades and can cost tens of millions of dollars. No groundwater basins in the Central Valley have sought relief by this route. A less costly approach to the overdraft problem is for all parties to agree to a management plan. The state has empowered water service agencies to enter into such agreements, appoint a water master, and tax all groundwater extractions to purchase imported water for recharge or in lieu of pumping.

As the real cost of energy declined and technology improvements in pumping plant design were developed, the real cost of groundwater exploitation decreased and thus overdrafting increased. Real energy costs for irrigation pumping declined steadily throughout the 1930s, 1940s, 1950s, and 1960s but rose sharply after the OPEC oil embargo of 1973. Between 1973 and 1980, the index of prices paid for agricultural production inputs increased about 11 percent annually while energy costs for pumping increased 25 percent annually.

Owing to the groundwater laws noted above and the high cost of resolving groundwater conflicts through the courts, groundwater management in the Central Valley has been local. The largest groundwater management unit is the Kern County Water Agency in the Tulare Lake Basin. During 1981, the district percolated 107,000 acre-feet into the underground aquifer. Individual irrigation districts that are the major overlying water management units in a groundwater basin also manage their own recharge programs. For example, the Arvin-Edison Water Storage District, also in Kern County, has percolated 900,000 acre-feet since 1966 through its 650 acres of spreading ponds. In drought years, no water is recharged and withdrawals are increased under Arvin-Edison's program of conjunctive use of groundwater and surface water.

Changes in the institutions and scale of water development over the past hundred years have left a checkerboard of water districts, water rights, control, and pricing in the Central Valley. With shifts in demand for water, economic pressure will increase for the flexible development of the current institutions to reconcile differing historical rights with the efficient allocation of water.

Emerging water problems in the Central Valley

The major emerging problems for irrigated agriculture in California's Central Valley can be classified as urban demand for increased surface supplies, groundwater overdraft, and the effect of residuals from agriculture on water quality. In recent years, the highest priority has shifted from supply augmentation to water quality management. However, within the Central Valley these emerging problems are independent.

Urban supply pressures on irrigated agriculture will come from two sources. Growth of urban areas in the valley is predicted to continue at its present rapid rate. In terms of net supply, however, the effect is largely offset by the consequent reduction in irrigated acreage. Standard suburban development accommodates approximately three families per acre; at the current high levels of domestic water use, suburban development and irrigated agriculture use about the same amount of water. Urban expansion in the valley has occurred mostly on the best agricultural land. (The California Department of Water Resources estimates that 700,000 acres of potentially irrigable land are as yet undeveloped in the Central Valley.) Despite the displacement of prime agricultural land by urban growth, expansion of the irrigated area will be constrained by poor land quality and low profit margins in agriculture. Thus, the effect of urban growth in the valley is not anticipated to increase the scarcity of irrigation water supplies significantly.

Growth in water demand in the Greater Los Angeles area will affect agricultural supplies in the valley. In Chapter 6, Vaux cites a projected increase under the existing price structure of 0.55 maf by 2000. In addition, the supplies from the Colorado River are going to be phased down 0.810 maf. Vaux shows that this maximum supply requirement of 1.36 maf can be substantially reduced by changes in the pricing structure. Nonetheless, agriculturalists in the San Joaquin Valley must anticipate a reduction of the state project water currently purchased at "surplus" rates; 326,000 acre-feet is surplus only because the Metropolitan Water District is not currently exercising its full entitlement. Reducing the amount of available surplus water will effectively increase the average price of water to several districts on the west side of Kern County. It is not clear that these districts could afford to purchase additional SWP water at full rates if it were available. Unless farmers' real profits increase substantially, additional full-price supplies at $60 per acre-foot and above will remain unsold.

The growth in urban demand over and above the urban entitlements currently used in the valley will raise the opportunity cost of water used in agriculture. Agricultural water that could be sold to augment urban supplies will substantially increase in value, given the difference between its current supply cost and the cost of new urban supplies. Given the current outlook for farming profits, this alternative sale of water should be viewed more as an opportunity for increased revenues than as a problem for valley agriculture.

The Central Valley of California is fortunate to have both usable groundwater aquifers under most agricultural regions and substantial average recharge from deep percolation and runoff from the surrounding mountains. Groundwater extraction in the Central Valley averages approximately 4.7 maf, about 22 percent of the total Central Valley supply. The overdraft (withdrawals minus total recharge) is estimated at 1.33 maf per year, 6.2 percent of Central Valley annual use. The severity of the annual overdraft varies widely from 1 percent of total supplies in the northern Sacramento Basin to an unsupportable 11.5 percent in the southern Tulare Lake Basin.

From the viewpoint of a physical mass balance, an average overdraft can be resolved only by augmenting surface supplies or removing land from irrigated production. The State Water Plan of 1963 takes such a viewpoint and adds 1.5 maf of surface supplies as a "need" to resolve the overdraft problem. Two assumptions are implicit in this policy: first, the existence of physical overdraft, *per se*, is socially bad, and, second, replacing the overdraft with additional new surface supplies is the most effective control method. However, the available evidence does not support these assumptions, and both the timing and quantity of new surface supplies required to control the overdraft for the long-term benefit of Central Valley agriculture and the environment are considerably exaggerated.

In the Central Valley, the only significant environmental impact of increasing groundwater depth from overdrafting is land subsidence – an effect rarely studied, much less quantified or costed. Substantial land subsidence can cause loss of aquifer holding capacity and replacement or repair of well casings and canal linings. In extreme cases, damage to buildings and roads has resulted. Nonetheless, because there are no cumulative effects, the lack of evidence on the costs of aquifer subsidence indicates that it is not currently a major factor. Groundwater quality in the Central Valley is not affected by overdraft rates, so this effect can also be removed from the calculations.

The remaining cost of groundwater overdraft in the valley is the cost of additional pumping required to extract the groundwater from ever-increasing depths. This notion is consistent with an accepted definition of critical groundwater overdraft as that which "threatens to cause significant adverse environmental, social or economic impact" (California Department of Water Resources, 1983).

The tradeoffs in optimal groundwater policy cannot be fully appreciated without a brief discussion of the economics of externalities and stocks and flows. Groundwater stocks are analogous to stocks of capital in a bank. Instead of a stream of interest payments from the capital, increased groundwater stocks yield an annual flow of cost savings in pumping. In this context, the first economic management issue can be stated as, "At what depth of groundwater does one stop overdrafting and stay in steady state?" This depth is calculated by balancing the value of net returns from pumping now and using an acre-foot of water to grow crops against the present value of the stream of additional future pumping costs that a fall in the average groundwater depth owing to overdrafting would cause. Of course, the net returns must be based on the marginal present and future uses of groundwater. Thus, unless the real marginal net return from irrigation is expected to grow faster than the real discount rate, conserved groundwater would not be used for irrigation in the future. In this calculation, the value of leaving the stock of groundwater in place and reaping the cost-savings benefits equals the financial return from liquidating the stock.

A second problem is that although the benefits from irrigation accrue to the individual farmer, the savings from pumping costs are a common property. Called an externality problem by economists, the asymmetry of individual and regional benefits is the main reason why groundwater can be excessively overdrafted.

When the value of groundwater used for irrigation exceeds the present value of pumping costs saved, overdrafting an aquifer is optimal for both the individual and society. However, once this point is exceeded, there is a range of groundwater depths at which it is optimal for the region to be in steady state. Yet, individual incentives still point to overdrafting.

The third and newest problem area is water quality. Currently, the debate, accusations, and rumors concerning irrigation water quality in California's Central Valley dwarf the contentious topics of surface and groundwater supplies. Speculation is rampant partly because so few hard data are available. In addition, political action has been as subtle as a loose cannon on a rolling deck, with juggling of responsibility between state and federal agencies and with commands and countermands of water supplies in quick succession.

The trace element selenium and the Kesterson drainage and wildlife area have become familiar in the western and national press. This dramatic example, itself serious, is a symptom of a rapidly growing water quality problem in the southwestern part of the Central Valley.

The residual products of irrigated agriculture in the Central Valley can be categorized broadly as salts (usually chlorides) and toxics (trace elements, heavy metals, and pesticide residuals). The first should be further divided into deep aquifer salinization and saline perched groundwater; toxic contamination can be divided into contaminated surface waters (such as those at Kesterson) and contaminated groundwater that affects drinking water supplies.

The rapidly increasing importance and cost of disposing of residuals have already begun to alter irrigation management practices and thus indirectly the demand for water in parts of the western San Joaquin Valley that are subject to salinity and trace element problems. In addition, the supply of new irrigable land is affected because most of the undeveloped irrigable land is on the west side of the valley. Given its attendant salinity and selenium problems, this area is unlikely to be developed because removing the additional residuals that soils of marine origin would produce is so expensive.

Salinity buildup in the Central Valley has been the subject of few economic research studies. Clearly, the magnitude of crop salt damage and the relationship of changes in salinity to crop yields is central to assessing the problem. To date, research on this topic is sparse, but it does show the magnitude of the problem. In 1974, Moore, Snyder, and Sun estimated the total annual cost of a one-part-per-million increase in the salt content of the Colorado River at the Imperial Dam at $24,000. This figure was a short-run estimate from a static analysis; the long-run damage function remains unknown. (This short-run damage estimate did not include the cost of subsurface tile drainage and disposal.)

The best way to remove salts from the plant root zone is through underground tile drain pipes. Tile lines may be spaced 1,000 feet apart in a homogenous sandy soil with good hydrolic conductivity and sunk down to 100 feet for heavy clay soils with poor lateral water movement to the drain lines. Thus, the capital cost of installing tile drains can range from a low of $100 per acre up to $700 per acre within a fairly small geographic area.

The cost of disposal of saline subsurface flows may match or exceed that of on-farm drainage. In California, two engineering approaches have been taken. For the federal San Luis Project, a concrete-lined drain approximately 250 miles long to transport flows to the Sacramento–San Joaquin River Delta has been proposed. Ninety miles have already been constructed, but the discharge permit has been deferred because high concentrations of selenium were detected in the waste water. The estimated capital cost per acre drained was $600 in 1967 dollars. The second engineering approach uses locally constructed evaporation ponds. Both systems require a collective local decision to finance the district collector drains and to transport tile effluent to either the ponds or the master drain. A few irrigation districts in the San Joaquin Valley are idling up to 16–18 percent of their poorest irrigable land for use as evaporation ponds. The estimated capital cost for one such system was $800 per acre (1985 dollars) drained in addition to the costs of private on-farm tile drainage.

Where irrigation is upslope from the valley trough, natural subsurface water movement can be sufficient to drain the higher elevation lands. However, these flows compound the problem of draining lower lands, where soils tend to be a heavier texture and are more expensive to drain.

The Clean Water Act of 1977 assigns the liability for damages from water degradation to the emitter. In other words, water users are assigned a property right to clean water. For nonpoint pollution surfaces, such as natural subsurface return flows, the extremely high transaction costs of monitoring and enforcement explain why the major thrust of such activities has been pressing for adoption of the best management practices.

The transaction costs of identifying subsurface flows change with location. In the typical situation where the upslope farm operations do not need to invest in underground tiling but the operators at lower elevations do install tile, the nonpoint pollutant is converted to a point source pollutant at the tile drain's discharge point. The transaction cost of enforcing quality standards or imposing effluent charges drops significantly if liability is placed on the almost innocent second party, even though much of the externality is created by upslope operators.

A less publicized problem caused by salts in deep-percolating agricultural water is the slow salinity buildup in groundwater aquifers. This problem has the same common property characteristics as groundwater quantity, but the effects take longer to appear because of the time water takes to percolate down to aquifers and the cumulative nature of salinity damage.

Until recently, California's irrigated agriculture was spared high costs and severe restrictions on the disposal ability of residual by-products and drainage water. However, public awareness of the toxic potential of agricultural by-products has sharply increased, and additional constraints and costs on toxic by-products of irrigated agriculture are likely to appear soon.

Monitoring of groundwater for toxic substances has been ongoing, especially where wells are used for drinking water. Pollution by pesticides and nitrogen has been recorded in sporadic episodes and some regions. So far, however, frequency and magnitude of these toxic pollution occurrences have not been sufficient to justify modifying irrigated agricultural practices.

A more insidious and severe type of toxic pollution has recently become apparent in the Central Valley. In 1985, the discovery of deformed and dead bird embryos at the Kesterson Wildlife Refuge precipitated the disclosure that selenium was being carried into Kesterson in the agricultural drainage water and concentrated by the closed system and the food chain. The source of the selenium is marine-based soils 20 or more feet beneath the topsoil. As the leaching fraction of the applied water percolates through this layer, the selenium and possibly other trace elements are taken up. The concentration initially taken up does not seem excessive, but when concentrated in drainage systems, selenium reaches toxic levels.

Initially, it was thought that the high selenium-bearing subsoils were restricted to an area called the Panoche Fan. However, tentative evidence is emerging that toxic levels could be building up in areas all along the west side of the valley north and south of Kesterson. Only a small area of approximately 8,000 acres has tile drains that lead directly to the San Joaquin drain and Kesterson. Because of subsurface lateral flows, the area contributing effluent has been estimated as the 49,000 acres adjacent to the tiled area.

In the face of political and legal pressure over this suddenly emerging problem, the initial response of public agencies did not contribute to a rational analysis. Agricultural interests want to keep the farmland in production and develop local evaporation ponds to receive the drainage. The federal executive branch, through the Department of the Interior, announced in March 1986 that it would immediately stop supplying water to farmers in the area. Given the urgency of cotton planting, the problem quickly became a political football tossed from the Congress, the executive branch, and the state-level Water Quality Control Board. After considerable political maneuvering, both irrigation water and restricted drainage capacity have been restored to the area.

Emerging problems for the irrigated agricultural industry in the Central Valley can be seen as urban competition for existing surface supplies, overdrafting of groundwater aquifers, and degradation of water and environmental quality by residual toxics and salinity. These problems are of such recent political and economic significance that they threaten agricultural production at current levels. Despite the importance of many traditional structural solutions to these problems, several nonstructural solutions show promise for resolving these conflicts.

Alternative policy solutions to the emerging water problems

In the hundred years of water development in the Central Valley, the inevitable response to an imbalance between water supply and demand has been to build additional facilities to increase supplies. The scale and institutions have changed, but the solution to problems has invariably been structural. Under certain conditions, nonstructural policy solutions that reallocate existing supplies or change technology, and thus modify the supply-demand imbalance, may be more effective.

Structural solutions to shifts in water demand are rational when the marginal cost of developing and distributing additional water is equal to, or lower than, existing developed supplies. Because these costs *are* lower, the structural solutions have an appealing technical simplicity. As urban population and industry grow, increased supply capacity meets the demand at historical consumption levels. Groundwater overdraft is solved by supplying cheap surface water in those areas to replace it. Residual problems are solved by diluting the present volume of contaminated return flow with new surface supplies where a drainage outlet exists (for example, in northern San Joaquin Valley). For areas such as the Tulare Lake Basin, salt and trace element extraction methods such as reverse osmosis or evaporation are envisaged for removing already-accumulated residual loads. However, cost estimates for additional water development show that the phase of decreasing average costs for water has ended and that the marginal cost of new supplies is rapidly increasing. In addition, the growing awareness of the environmental costs of new water supplies has substantially increased the political cost of expanding supplies.

During the initial phase of large interregional transfer projects, the

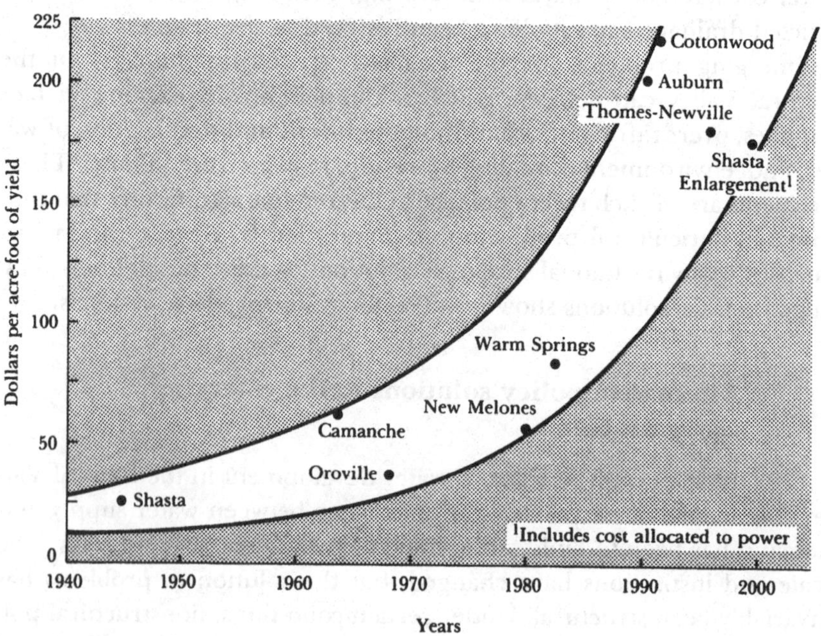

Figure 3.2. Historical and projected costs of water supply facilities in 1980 dollars (California Department of Water Resources, 1983).

Table 3.5. *Average cost of SWP entitlement water*

Year	Santa Clara Valley Water District (SCVWD)[a]	Kern County Water Agency (KCWA)[a]	Antelope Valley–East Kern Water Agency (AVEK)	Metropolitan Water District of Southern California (MWDSC)
1980	43.49	29.09	56.25	123.31
1985	90.54	54.27	217.68	240.51
1990	121.98	80.23	279.94	275.87
1995	122.51	81.27	259.93	272.14
2000	126.13	80.42	234.81	274.34
Percentage change	+190	+176	+317	+122

[a]With SWP surplus water included, the 1980 cost figures are $40.13 for SCVWD and $18.84 for KCWA. The 1985 figures would be $85.88 and $47.87, respectively. The percentage change would be +214 and +327, respectively.

Sources: California Department of Water Resources, 1983, tables B-5B and B-22; staff calculations.

inherent advantages of developing the first dam sites and the economies of scale in water transportation ensured that increased capacity came at a constant or decreasing average cost. For the Central Valley from 1940 to 1975, surface water supplies had a constant average capital cost. In 1980, the construction cost of water projects began to increase sharply. A comparison of Oroville Reservoir to the proposed Cottonwood Creek project shows an increase in annual cost per acre-foot of yield of approximately $40–$218 in constant 1981 dollars. (Figure 3.2 illustrates the shift from a relatively constant marginal cost to rapidly increasing marginal costs.) In addition to the increased construction cost, the energy cost embodied in transported water will inevitably increase when current power contracts expire between 1985 and 1990. As Table 3.5 shows, projected cost increases in constant 1980 dollars run between 122 percent and 317 percent over the next 20 years for selected SWP contractors.

Projections of the demand for irrigation water reflect the assumption that the value of the agricultural product from additional irrigated acreage will always exceed the costs substantially. In short, if potentially irrigable land is available, the agricultural community will develop it. The continuation of past expansionary trends has led the California Department of Water Resources to project substantial increases in irrigated crop acreage between 1980 and 2010. Over these 30 years, pasture acreage shows a 10 percent decline, and smaller declines are shown for grain and sugar beets. All other crops increase, some of them substantially: rice, 15.6 percent; cotton, 18.4 percent; corn, 24.4 percent; tomatoes, 49.3 percent; truck crops, 23 percent; and grapes, 31 percent. Because the projected market share expansion roughly equals historical yield increases over time in such key crops as cotton, these additional California acreage forecasts seem optimistic (California Department of Water Resources, 1983). However, the requests for entitlement water by contractors in 1983 were even more optimistic.

The 1983 requests for firm entitlement water increase 85 percent – from 1.567 maf for 1984 to 2.906 maf for 1989. Examination of past requests shows that requests are substantially scaled down as the delivery date approaches (California Department of Water Resources, 1983). More recently, agricultural spokesmen have been expressing reservations about agriculture's ability to pay for additional supplies from conventional reservoir storage projects. This viewpoint has been bolstered by the current cost-price squeeze facing California farmers and the realization of the ill effects of acreage expansion on depressed and inelastic markets.

In 1983, agricultural contractors to the SWP faced a quandary regarding the costs and restrictions of the peripheral canal and the alternative Through Delta Facility. Like the Central Valley project, the State Water Project is not complete. The major bottleneck in the system is the Sacramento–San Joaquin Delta. To repel salt water moving upstream from San Francisco Bay into delta agricultural areas and into the intakes of the California Aqueduct and Delta Mendota Canal, approximately 0.75 maf of water (carriage water) must be provided. The Department of Water Resources has urged a restructuring of channels within the delta to reduce the quantity of carriage water, thus allowing an additional 500,000 acre-feet to be exported to meet contractual entitlements. The Through Delta Facility had an incremental cost of about $68 per acre-foot; if averaged in with the cost of existing facilities, the average cost of contract entitlement water would have increased about $9 per acre-foot.

The problem faced by agricultural SWP contractors in Kern County is balancing the higher probability of receiving their full entitlement against a significantly higher cost over the 40-year payout period. SWP contracts provide for southern California gradually to build up deliveries as population grows. Unused southern California entitlement may be delivered to Kern County agricultural contractors at the marginal cost of pumping along the aqueduct. In dry years, agricultural users will absorb all shortfalls up to 50 percent. Thereafter, shortfalls will be apportioned equally between urban and agricultural users.

The enlarged Through Delta Facility, including additional pumping plant capacity, can be viewed as an insurance policy against drought. In 1990, there is a 75 percent probability of Kern County's receiving the full agricultural entitlement using existing facilities. However, by investing in the enlarged facility, the probability of receiving a full entitlement increases to about 94 percent. By the year 2000, assuming that southern California expands use of its entitlement, there is only a 15 percent probability of Kern County's receiving its agricultural entitlement with existing facilities. With the enlarged facilities, the comparable probability is 45 percent.

Given the projected costs of the Through Delta Facility, growers who have no groundwater and depend on entitlement water would have to be willing to pay 7 percent of their present benefits as an insurance premium if they want the enlarged facility built by 1990 (see Moore and Gossard, 1984, for a complete description of the model used and the detailed calculations). Moore and Gossard show, however, that even the most conservative operator in an area without groundwater would want to have the enlarged facility built by the year 2000 because the expected net benefits with the enlarged facility are by then slightly higher than with the existing facility.

Operators with groundwater would in all probability want the project constructed by 1990, assuming that all the system's capacity to deliver surplus water in wet years can be used for recharge and in lieu of pumping. The case for having the project operational by the year 2000 is not as clear because the expected benefits are similar.

The Kern County Water Agency, historically a strong supporter of SWP expansion, faces an internal conflict because benefits are unequally distributed over time and space. The incremental cost of augmenting supply by structural policies – $70–$120 per acre-foot – exceeds the marginal value of water in agriculture threefold or more. Thus, the ability to use these new supplies to dilute drain water or substitute for groundwater overdraft is infeasible if the beneficiaries pay the full incremental cost of supplies. For the foreseeable future, living with the problems clearly seems preferable to paying for the structural solutions available.

Not only are the monetary costs of adding to the water supply high, but environmentalists have been actively pointing out the potential environmental damages associated with additional development. Further, the political cost of moving additional water from areas of origin to other parts of the state has increased rapidly in the past decade.

In June 1981, California voters rejected a peripheral canal proposal to redirect water around the delta from the Sacramento River to the pump stations at the head of the aqueduct. The vote brought together environmentalists who opposed any additional water exports to the south and a few large farmers in the San Joaquin Valley who were opposed to the water quality standards in the delta. It also froze additional water development on north coast rivers. Over 90 percent of the voters in northern California voted against the measure. Two years later, the state administration proposed an alternative Through Delta Facility, which has failed thus far to pass the California legislature. Direct agricultural beneficiaries were split partly because of project costs.

Given the poor prospects facing structural policy solutions, politicians are giving nonstructural solutions increasing attention. By definition, such policies yield incentives for changing the place, type, or technology of water used to increase effective supply or reconcile supply and demand. The four nonstructural policies reviewed here are reallocation of existing supplies through water trading, local groundwater control, conservation and quality management of irrigation, and the institutional modifications needed to implement these policies.

Quasi-markets and transfers of water have been advocated recently as nonstructural means of balancing supply and demand within the Central Valley and between the Central Valley and urban areas. If water transfers are to be effective, two conditions must be met. First, the costs and value in use of the water must differ by appreciably more than the transport costs between the two regions. Second, the seller must benefit from transfers by lower water costs or cash payments; meeting this condition requires an effective marginal cost pricing institution. The first condition is substantiated in the Central Valley by both the steep projected price increases for expanded supplies shown in Figure 3.2 and the range of prices lower than the existing SWP costs in all other regions of the Central Valley.

The effectiveness of price as an allocator of agricultural supplies is also influenced by farmers' water use in response to water price changes – what economists call the price elasticity of demand. The price elasticity of demand for irrigation water cannot be estimated directly from Central Valley data because of the lack of price movement. However, models of aggregate farmer response do allow estimation of the elasticity of the derived demand for water over a price range. In 1976 dollars, the elasticity is 1.5 in the price range of \$25–\$35 and falls to −0.46 in the \$35–\$45 per acre-foot range (Howitt, Watson, and Adams, 1980). These results show that agricultural water users would be responsive to price changes in the current range for most users. Thus, pricing by market systems would be an effective allocation method.

Table 3.6. *Annual benefits from trade*

	Transfer direction	Quantity (million acre-feet)	Benefit	
1980	IMP–LA	1.09	IMP	$ 13,150,850
			LA	25,462,400
	SA–LA	0.31	SA	108,500
			LA	7,241,600
	NA–SFB	0.69	NA	686,550
			SFB	20,344,650
			Total	66,994,550
1995	IMP–LA	1.25	IMP	15,043,750
			LA	59,893,750
	SA–LA	0.44	SA	253,000
			LA	21,082,600
	NA–SFB	0.92	NA	984,400
			SFB	59,110,000
			Total	156,367,500
2020	IMP–LA	1.28	IMP	12,198,400
			LA	73,344,000
	SA–LA	1.00	SA	1,110,000
			LA	57,300,000
	NA–SFB	1.07	NA	1,177,000
			SFB	73,878,150
			Total	$219,007,550

IMP = Imperial Valley; LA = Los Angeles region; NA = northern valley agriculture; SA = southern valley agriculture; SFB = San Francisco Bay area.
Source: Vaux and Moore, 1983.

Two studies of potential water markets in the Central Valley have been published. In the first, Vaux and Howitt (1984) consider both intravalley and intervalley trade between agricultural and urban regions. The second study, by Gossard et al. (1982), considers water trades between agricultural districts in the southern part of the Central Valley.

The interregional trading model shows that the combination of trade and marginal cost pricing can equate the aggregate urban and agricultural supply and demand through 2020, with only a slight (100,000 acre-foot) increase in structural supply statewide. Statewide benefits from trade in 1980 dollars increase from $67 million per year in 1980 to $219 million per year in 2020. (See Table 3.6.)

Trade does not take place between the Sacramento Valley and the southern San Joaquin Valley in this model because of high transport costs. However, the Sacramento Valley does sell water to the San Francisco Bay area in all scenarios, and the southern San Joaquin Valley sells water to the Los Angeles Basin. Owing to the increased value, but not cost, of water in the Central Valley from trade and marginal cost pricing, the growth in demand for water projected under average cost pricing and without trade does not occur. The large efficiency gains from trade produce the unusual policy situation in which all parties are made better off – a political rarity.

The second trading study of the Central Valley addresses the gains realized from trade between two agricultural regions. Specifically, the Modesto Irrigation District and the Turlock Irrigation District are considered exporters, and 12 water districts in the Kern County Water agency are importers. A trade of 150,000 acre-feet per year between the districts proves optimal, yielding net benefits of $2.79 million a year (in consumers' and producers' surplus) to the exporters and $2.38 million a year to the importers. A reduction in irrigated acres of −9.5 percent in the exporting region, predominantly in irrigated pasture and corn, ensured that the secondary economic impacts on the exporting region are low. (Table 3.7 shows the benefits and quantities traded.) The lost value of crops to the exporting regions is a weighted average of the returns per acre-foot from the crop acreage taken out of production. The benefits to the importing regions come largely from substituting cheaper imported water for groundwater rather than from expanding acreage in the Kern area. Because the production model used to predict these responses was not constrained but was in market equilibrium, the substitution of water supplies rather than expansion of acreage supports the hypothesis that the Central Valley is not currently expanding irrigation.

Several writers have noted that owing to the bimodal distribution of rainfall on California's watersheds, what is needed is not additional average year supplies but additional dry-year supplies. A system of contingent demand-transfer agreements could meet these dry-year needs at a lower cost than could expanding additional fixed supply capacity. Contingent demand contracts would be valuable for individuals or regions with firm water rights used on low-value interruptible fodder crops. During normal or wet years, the sellers would grow their crops as usual, but they would also collect an annual retainer payment. The contract would specify that the sellers provide a certain amount of water at a delivered price when a contingent index, such as the DWR rule curve, indicated that a dry year was occurring.

Table 3.7. *Economic effects of trading 153,486 acre-feet at $38.79 per acre-foot*

Seller	Price per acre-foot	Acre-feet sold	Export water revenue	Lost value of crops	Net gain
Modesto Irrigation District	$35.27	83,982	$3,257,182	−$1,776,159	$1,481,023
Turlock Irrigation District	37.74	69,504	2,695,726	− 1,384,189	1,311,540
Total		153,486	$5,952,908	−$3,169,345	$2,792,563

Buyer	Acre-feet bought	Net savings per acre-foot	Net gain
Kern Delta WD, Rosedale–Rio Bravo WSD	14,671	$ 3.54	$ 51,935
Semitropic WSD, Buena Vista WSD	33,673	6.74	226,956
North Kern WSD, Shafter-Wasco ID, South San Joaquin MUD	30,019	15.25	457,789
Arvin-Edison WSD, Wheeler Ridge–Maricopa WSD (part)	15,632	21.88	342,028
Belridge WSD, Berrenda Mesa WD, Lost Hills WD	15,579	6.62	103,133
Wheeler Ridge–Maricopa WSD (part)	43,913	27.50	1,207,607
Total	153,486		$2,389,453

WD = water district; WSD = water storage district; ID = irrigation district; MUD = municipal utility district.

Source: California Assembly Office, 1985.

The key concept of contingent demands is that it is cheaper to remove low-value interruptible crops from production for dry years – on the average, say, of one in five – than to pay the capital costs of a system that is essentially idle for four out of five years. A parallel can be drawn between contingent demand sale and the current system of "surplus" water pricing used by DWR, in which surplus water users pay only the variable costs of delivery. However, with contingent demand, the firm water right remains with the original land area, and conditions for exercising the contingent demand are clearly specified by outside physical criteria.

Some advantages of contingent demands are that the secondary local impacts of crop removal would be only sporadic and thus would be minimal. In addition, the capacity to convey these trades during dry years would be ensured in the Central Valley. Representative prices and yields indicate that dry-year supplies could be negotiated for 15–25 percent of the cost of construction by contingent trades, and some of the objections to sustained annual water transfers could be avoided.

The evolution of a market for traded water would boost agricultural conservation by increasing the opportunity value of water. Although traded water is controversial, a consensus is emerging over two aspects of conservation in the Central Valley. First, substantial subsurface flows on the west side of the valley are not being beneficially reused, and in many places they contribute to the salinity and toxic problems. Second, agricultural conservation is influenced as much by farm management as by capital equipment, a point that implies that farm-level incentives for improved management are required for conservation. Water markets for conserved water will partially provide such incentives. The cost of water to the farmer will not change, but the value of water saved will be a significant incentive for conservation as long as it can be sold.

Discussion of the introduction of water trading invariably raises the objection that existing institutions would prevent it. A central argument of this chapter is that legal and economic institutions have and will adapt to new opportunities. The history of Central Valley water development and the adoption of trading systems in other western states support this view (Rand Corporation, 1978). In addition, legal barriers to trade in the Central Valley appear to be few, and they are mainly questions of legal interpretation (California Assembly Office, 1985). The effect of recent amendments to California water law and a range of legal questions posed by intervalley trades is addressed in Gossard et al. (1982). The state water code and local district rules and regulations are examined for constraints to trade. According to this report, if it can be demonstrated that the trade can be made without damaging other legal water users, the local economy, or the environment "unreasonably," then there appear few legal barriers to trade in the Central Valley. The legal section concludes:

> If it is in the economic interests of [Turlock Irrigation District, Modesto Irrigation District, and Kern County Water District] (or some of their individual members) to arrange for a trade of water from the districts to the agency, there is nothing in state law, the rules and regulations of the districts, or the agency's contracts with DWR and its member units to prevent or even discourage such a trade.

Evidently, the barriers to trading water are on the more elemental level: the uncertainty and cost of court challenges and the difficulty of quantifying consumptive use in many locations. Despite these uncertainties, the physical and economic rationale for water trading seems to be rapidly increasing in California; if continued, this thinking will gradually modify the attitudes and practices that currently impede water trading.

The degree to which groundwater basins should be overdrafted is a complex economic and hydrologic decision. The naive view that all physical overdrafting is socially wasteful and should be controlled in the near future was conclusively rejected by California voters in the 1982 Proposition 13 initiative. The intricacies of the problem restrict solutions to local basins, and the relevant policy question is not *whether* overdrafting ceases but *when*. Given the uncertainty that dominates many of the estimates of the hydrologic and economic parameters needed for policy decisions, a policy horizon of 50 years would seem to be the maximum.

If the cost of overdrafting exceeds the benefits, regional groundwater management can, optimally, modify individual pumping decisions through incentives. Methods advocated range from regulation to pump taxes; however, all require that the steady state rate of pumping be quantified and allocated among overlying aquifer users. The allocation can involve high legal costs if the safe yield of the aquifer is strongly disputed.

Three studies have been made of the economic cost of overdraft in parts of the Central Valley. A study of four groundwater basins – Tule, Kaweah, Kings, and Madera – on the east side of the San Joaquin Valley (Howitt, 1979) showed the differences between physical and economic measures of critical overdraft. If physical criteria are the measure, the basins are being critically overdrafted.

Using the regional returns to irrigation water obtained from a production model, the study calculates the effects of declining groundwater levels on regional profitability. (See Table 3.8.) The results showed that the four basins classified as critical using the physical criteria had not yet reached the critical economic level of overdraft – that is, the groundwater stock at which pumping should be restricted to the steady state level to maximize long-run economic benefits. Interestingly, the time needed to reach steady state varied substantially. Two basins were relatively close at 12 and 18 years. The other two basins were far from the steady state point at 36 years and 62 years at the current rates of overdraft. The benefits of remaining at steady state once it is reached are appreciable, ranging from a present value of \$153–\$453 per acre.

Policy alternatives for six basins in Yolo County (Sacramento Basin) for both agricultural and urban uses and the conjunctive use of surface water and groundwater are examined in a second study (Noel, Gardner, and Moore, 1980). The results demonstrate a \$53 million economic benefit gained by allocating water optimally in the region. This surplus is discounted over a 30-year horizon. Strikingly, the costs of overdrafting

Table 3.8. *Optimal economic steady state for the region*[a]

	Tule[b]	Kaweah	Kings	Madera[b]
VMP per acre-foot[c]	$40.07	$39.44	$37.15	$36.50
User cost	$19.88	$22.81	$19.88	$16.45
Depth of optimal steady state (feet)	177	155	178	176
Years to steady state[d]	11.5	17.7	62	36

[a]The regions here are the provisional groundwater of the California water management basins outlined in the DWR Bulletin 118-80 in response to Section 12924 of the California Water Code. Background data used are: a discount rate of 8 percent, the agricultural price index for 1979, and a future average energy price of $.06 per kilowatt hour in 1979 dollars.

[b]For these two regions, the detailed analysis unit boundaries did not coincide exactly with the draft report basin boundaries (Howitt, 1979).

[c]The value of marginal product of an acre-foot of water applied to the least profitable crops, obtained from an economic model of the region (Howitt, Watson, and Nuckton, 1979).

[d]Assuming that current average rates of overdraft are maintained.

Source: Howitt, 1979.

vary widely within the same county. The optimal tax rates per acre-foot varied substantially over time and among basins. In 1990, the tax rates varied between $16.55 and $1.22 per acre-foot.

The regional variability of the results from these studies underscores the point that the value of groundwater control can be accurately quantified only on a local basis. At the current levels of overdraft in Yolo County and the east side of San Joaquin, a blanket restriction on overdrafting could impose substantial unnecessary social costs on the Central Valley.

A larger comprehensive study of the whole San Joaquin Valley was commissioned by the Department of Water Resources (1982). This study used interdependent hydrologic and economic models to quantify the valleywide options for conjunctive use. The results substantiate the conclusions of the previous studies. A wide variation among the costs of pumping groundwater in different locations was reported. For example, in the tenth year, Turlock Irrigation District had a regional cost of $8,200 per foot of extra lift, whereas Merced Irrigation District costs exceeded $20,000 per foot of extra lift.

Pumping depths along the west side of the valley are predicted to increase the fastest. In contrast, some northern valley districts showed decreased pumping depths at current efficiencies of irrigation. The Department of Water Resources model runs do not include a private solution that allows direct calculation of the social costs of overdrafting by district. However, the results support the conclusion that a statewide ban on overdrafting would be against society's best interests. In many parts of the valley, continued overdraft in the 40-year planning future is still optimal. Tentative evidence shows that perhaps only half the physically critical basins should be encouraged to plan for a steady state solution in the next 20 years.

If the current 1.5 maf of annual overdraft need only to be reduced by half in the next 20 years, the additional surface water needs could be reduced 0.75 maf. When the costs of additional surface supply replacements are faced, even after averaging them over the district, farmers are unlikely to want to replace the overdraft quantities fully. Clearly, for the foreseeable future the current 1.5 maf figure for overdraft replacement is a significant overestimate. A subjective estimate – based on the assumption that half the current overdraft stabilizes in the next 30 years, that additional supplies average $70 per acre-foot, and that the elasticity of demand for water by farmers is 0.5 – suggests that a more reasonable replacement need would be 0.5 maf. Even this figure could be substantially reduced by pricing policies.

Overdrafting in the Central Valley does have a potential benefit to the water management system: the potential subterranean storage created in the aquifers. The Department of Water Resources groundwater study concluded that aquifer storage could smooth out seasonal fluctuations by short-run overdrafting. Already, many districts and agencies in the San Joaquin Valley recharge groundwater artificially. A substantial reduction in the decline of groundwater in Kern County is attributed to the recharge program. The costs of recharging are allocated by a sophisticated system of zones of benefit. However, the problem of uncontrolled rates of extraction persists.

The Sacramento Valley in the north has high groundwater levels and high-quality water. But many farmers adjoining the Sacramento River have problems with seepage or excessive groundwater. Paradoxically, farmers at the northern part of the distribution system suffer crop losses from high groundwater tables; those in the south face the cost of depleted groundwater. Farmers in the Sacramento River region are seeking relief from the pumping costs of reducing high groundwater levels. This region has both minimal lifts into the distribution system and the potential for an annual winter-recharge gradient. Curiously, the two complementary problems in the north and south have not been linked. The most obvious reason is the resistance of northern farmers to state actions threatening their "area of origin rights." Recently, the concept of private water rights and the transfer and sale of groundwater have received a more sympathetic hearing by the Sacramento Valley farmers. If the surplus in the Sacramento Valley and the shortage in the southern San Joaquin could be linked by secure property rights and incentives, steady state may be achievable at a low cost to the valley. At the moment, the economic incentives for developing such an institution have not been quantified.

Groundwater hydrology studies and the resulting conjunctive management methods will increasingly enable groundwater storage programs to substitute for surface reservoirs in meeting dry-year peak demands. An accepted computer groundwater model can reliably define aquifer-related property rights, much as barbed wire did on the western ranges in the 1880s. The ability of groundwater models to link aquifer recharge actions and agricultural pumping in different areas will make it possible to use groundwater as an off-peak storage and distribution facility. The capacity of groundwater to act as a dry-period water reserve was dramatically demonstrated in the 1976–1977 drought, when farmers increased their pumping 40 percent, greatly offsetting the reduction in surface allocations.

The policy response to the recent concern and litigation over agricultural residuals has been to investigate the potential of structural solutions to the problems of removing salt and toxins from drain water. Although physical removal is unquestionably part of the solution, substantial reductions in the total quantity of residuals can be achieved by nonstructural incentives to make on-farm water management more efficient. The central concept is on-farm conservation of irrigation water.

Structural salt and toxic removal systems include evaporation ponds, reverse osmosis plants, and deep well injection plants. Clearly, some minimum salt load has to be treated by one or all these methods to reach saline steady state, but the current uncertainty on the costs and technology of these structural methods has tempered a rush to construct them. One characteristic is clear: all are expensive to build and operate. Estimates of the annual costs of drainage water disposal vary from $300 to $600 per acre-foot.

Given the high cost, the current policy emphasis is on minimizing the salt or other toxic load. Methods include melding and managing saline irrigation water and using water conservation to reduce the salt and drainage volume.

In the short run, the effects of groundwater salinization can be largely offset by blending groundwater with high-quality surface water in many regions of California (Howitt and Mean, 1981). Mitigative actions can be implemented by crop rotation changes, blending waters from sources of

Table 3.9. *Crop acreage response to change in groundwater salinity, Henry Miller Water District (imported water: 37,500 acre-feet)*

	Average salinity (parts per million)[a]		
	1,280 (960)	2,560 (1,088)	3,840 (1,216)
Acreage for:			
Cotton	9,984	9,942	9,946
Alfalfa	1,123	1,337	1,305
Wheat	7,332	7,158	7,187

[a]Averages for unconfined aquifer and (in parentheses) confined aquifer.

Source: Proceedings of the Thirteenth Biennial Conference on Ground Water, University of California, Davis, California Water Resources Center Rep. 53, November 1981.

different quality, and changing irrigation methods. A model of two water districts in Kern County (Shafter-Wasco and Henry Miller) allowed cropping pattern adjustment, water blending of two or more supply sources, and technology changes.

The two districts represent widely different aspects of Kern County agriculture in terms of perched groundwater, soil type, groundwater salinity, and cropping patterns. Henry Miller District has a significant perched groundwater problem and high average salinity. In 1975 the dominant crops grown there were relatively salt tolerant.

Shafter-Wasco District, on the east side of the valley, has better soil and significantly lower salinity levels. Surface water entitlements comprise a similar percentage of the total 1975 demand in both districts, though Shafter-Wasco has Central Valley Project contracts, and Henry Miller relies on Kern River and State Water Project entitlements.

Increasing unconfined aquifer salinity reduced net farm income for Shafter-Wasco. For the first 700 ppm increase in salinity, the net revenue reduction was negligible. Groundwater salinity projections suggest that the 700 ppm figure will not be exceeded for several decades; thus, the results imply that farmers in districts similar to Shafter-Wasco can offset what are likely to be the average costs of salinity in the near future. Henry Miller District shows no significant crop rotational changes in response to marked salinity increases in the unconfined aquifer. (See Table 3.9.) Consequently, changes in net farm income are insignificant because the high proportion of imported water allows extensive blending of water.

The two examples show that districts under different soil, water, and crop regimes can economically adjust in the short term to above-average

Table 3.10. *Intertemporal model for salinity management*

	1980	1990	2000	2010	2020	2025
Value per acre-foot groundwater	$7.21	$7.38	$7.46	$7.46	$7.30	$7.12
Melded groundwater salinity (parts per million)	612	648	689	726	763	782
Cost of increased salinity per acre-foot pumped (1975 base salinity = 590 parts per million)	$0.62	$1.71	$2.94	$4.11	$4.93	$5.13

Source: Howitt and Llop, forthcoming.

salinity levels. Using blending and rotation to adjust would be much more costly if a district had a high proportion of salt-susceptible crops with poor soil types and low proportions of surface water entitlements.

The long-term economic effects of salinity increases in groundwater aquifers on agriculture in Kern County, under several levels of pumping costs and leaching fractions, were examined by Howitt and Llop (forthcoming). The researchers made some difficult simplifying assumptions. Farm crop rotations in wet districts are assumed to remain in the same proportion as 1975, but the acreage of high-value crops is expanded to accommodate increased future demands. The confined and unconfined aquifers that have significantly different salinity levels and an uncertain degree of independence are assumed melded into a single aquifer. The lag time or memory of the percolation of salts for applied water to the aquifer is ignored in the model equations.

Table 3.10 shows the model results for a 45-year period in which the model minimized the cost of salinity effects, pumping, and surface water cost subject to constraints on land, surface water, and the existing pattern of crop rotation. The optimal management scheme in this model did not include the ability to remove salts from the system. Accordingly, the salinity level of the melded groundwater rose steadily from 590 ppm in 1975 to 792 ppm in 2025, a 0.72 percent annual increase. At this level of melded salinity, it is economically optimal to accept this annual increase rather than to reduce production significantly. Over the period, the optimal rate of groundwater pumping increased slightly despite the increase in pumping lift that this level implies. The imputed net value of groundwater stocks in Table 3.10 changes little over time, and the costs of salinity increases in the melded groundwater steadily rise from 9 percent of the value of water in 1980 to 72 percent in 2025. Thus, if farm level adjustments are ignored and salt continues to leach into the aquifer, the value of groundwater stocks will begin to erode substantially after 40 years.

These preliminary salinization studies should be used with caution. Salinization of a large aquifer is essentially an irreversible action. When combined with long time lags, the conventional economic analysis can overlook blocked intervals of suboptimal salinization.

Many short-run remedies to the problem of toxic agricultural residuals have been suggested, but only two seem economically feasible. The first is to use local and regional collection-evaporation ponds to dispose of the drainage water. Uncertainty about the cost and type of pond rests on the interpretation of whether they will be classified as toxic waste sites, in which case they must be double lined. If this construction is required, the cost of drainage disposal in double-lined ponds will make agriculture uneconomic for all but the highest-value crops. The currently proposed alternative is to locate the evaporation ponds over tight clays with low permeability. One estimate of the annual cost of this approach is $300–$600 per acre-foot of drainage water per year. Given current leaching fractions averaging 20 percent or more, the cost per acre per year under existing technology for the ponds alone is $200–$340 – enough to affect irrigation practice on the west side of the valley greatly. Some engineering studies indicate that with excellent uniformity, the leaching requirement for salt removal to ensure zero loss of yield can be reduced to 5 percent, cutting the drainage disposal cost 75–90 percent.

Technical advances in the methods and equipment for more precise and effective irrigation have been rapid. Substantial improvements in the field application efficiency of irrigation have been achieved by changes in timing and uniformity under experimental conditions. The challenge is to extend these results economically to practical field operations, and a large research effort is currently directed at many aspects of the problem.

At this early stage, the impact of the toxic buildup will cause two shifts in water use in the west side of the valley. First, the rate of introduction of new acreage of irrigated land on the west side will be greatly slowed or stopped altogether. Second, the leaching fraction applied over 1.6 million acres of salt-susceptible area will gradually be reduced from the current 20 percent to as low as 8 or even 5 percent, decreasing the demand for applied agricultural water about 600,000 acre-feet. Both these actions will notably affect the supplies and management of irrigation water in the Central Valley.

As with all technical policy changes, the institutional structure must be changed before the policy can be implemented. Over 40 years, a well-entrenched institutional structure of private firms and public agencies has evolved in the Central Valley to support water supply development and allocation by building additional structures. As shown here, the costs of structural supply augmentations or residuals disposal make them impractical short-run policies. But political and institutional shifts could allow the Central Valley water industry to stay in physical and economic equilibrium over the next four decades.

The coalition of agricultural interests in the San Joaquin Valley and the urban developers in southern California who united to exert pressure for continued surface water development is showing signs of disintegration. Among the southern California urban interests, opinions differ over using traditional structural or nonstructural (market and transfer) methods to obtain future supplies. The unity of Central Valley agricultural interests is also faltering. The last two political debates on the peripheral canal and the Through Delta Facility elicited diverse views from the agricultural industry for and against the propositions (Gwynn, Thompson, and L'Ecluse, 1983).

Complicating this fragmentation of interests is the emergence of environmental interest groups as a major influence on California water policy. The environmental groups, however, are not unified in their policy prescriptions, which range from legal activism to encouraging private market action. Paradoxically, a study by the Environmental Defense Fund helped trigger negotiations for the transfer of Imperial Irrigation District water to the Los Angeles Metropolitan Water District.

In the water development era of the 1960s and 1970s, constant water development costs and high returns to irrigation in Central Valley agriculture justified the widely held view that the expanding development of water strengthened the Central Valley industry and the state. However, in the 1980s and 1990s, the Central Valley faces sharply increasing structural costs, severe market competition, and accumulated salt and toxic residuals. Under these conditions, the Central Valley and California as a whole will increasingly turn to nonstructural alternatives to balance changing water demands with existing supplies.

Water rights laws and institutions have tended to increase the security of tenure of those who have been given the right to capture, divert, and use the waters of the state. In the past, security of tenure implied tying water to specific geographic locations, which prevented water from being allocated to its use.

Before nonstructural policy options can be implemented, property rights must be improved and clarified. For surface water, pre-1914 appropriative rights did not have to be recorded, and post-1914 rights were, until 1969, recorded only as seasonal flow rates. Riparian rights have yet to be quantified, and because they attach to specific land parcels, they are nontransferable.

Property rights should serve as an information system, allow the rights holder to exclude other users, and be transferable and enforceable. If the property rights on irrigation and drainage water are weak or imprecise, the cost of additional information, bargaining, and enforcement will increase. These transaction costs have become a major reason why markets fail to be formed.

A four-step approach would minimize water transfer transaction costs. First, the irrigation district should become the local dealer, taking possession of the water property rights, if not already thus held. Second, the district should create a membership pool like those formed by wheat- and rice-marketing cooperatives. A cap or limit should be placed on the pool, and if it is oversubscribed, individual landowners should have a *pro rata* share in the pool. Third, the irrigation district should underwrite the contract for delivery of water from the pool to the buyer, collecting a fee to cover actual administrative expenses. Fourth, following a suggestion from the California Governor's Commission to review the water law, a three- to five-year trial should be initiated to determine the extent, if any, of downstream third-party injury. If unreasonable injury results, the contract could be abrograted or renegotiated.

Use of the irrigation district as a water-marketing cooperative greatly reduces the cost of determining how much water can be legally transferred. The costs of operating a water court to determine for each parcel of land the quantity of evapotranspiration and return flow from other basins, added to the administrative costs of a multitude of individual transactions, could be excessive. This point has been demonstrated in the water courts created in Colorado.

The state should continue to protect the property rights of third parties who do not participate in market transactions. The California Water Resource Control Board (CWRCB) would retain its authority to protect downstream return flow users who have rights to these flows. In current practice, CWRCB approves changes in location of diversion points, timing, and use. This system protects third parties from uncompensated damages if their rights are impaired by water transfers out of the basin.

Improving property rights in surface water to encourage transfers through a market system may not capture the full benefits of the gains from trade if property rights in groundwater remain uncertain. If surface water is transferred only to result in severe groundwater overdrafts and the attendant costs imposed on surrounding pumpers, then property rights in groundwater must also improve. The state would need to take the lead in minimizing the transaction costs of groundwater basin adjudication by collecting and processing information on groundwater hydrology, extraction, and recharge.

Market brokerage functions could be handled by state and federal water agencies or the private sector. The public sector agencies would have first knowledge of the location and propensities of potential buyers and sellers. Because these agencies also control the plumbing – the canals and aqueducts necessary to transfer water from one location to another – they could also serve as a clearinghouse for bringing potential buyers and sellers together to help form a market.

Externalities from agricultural residuals such as salts, nitrates, and pesticides would be managed by enforcement of property rights to clean air and water. Calculation of the optimum levels of agriculture residues in the Central Valley is currently restricted by high information and contracting costs. Owing to economies of size in information collection and processing, the public sector can provide this service most efficiently and economically. In the case of toxic materials, the public sector must continue data collection as a basis for setting environmental standards for these residuals.

In summary, state and federal water agencies now face relatively high costs for additional developed supplies and an agricultural demand curve that, at least for now, has shifted downward. Water of low value in use is currently tied closely to specific geographic areas or irrigation districts. Significant benefits from trade to both buyers and sellers have been demonstrated if markets can be created to effectuate transfers. The nonstructural policy option would transform water agencies from predominantly builders and allocators of new supplies to providers of information and impartial brokers for existing supplies. Providing information to minimize the cost of market creation would replace construction of new facilities as the primary agency thrust. Understandably, the strongest resistance to this nonstructural policy may come from the technological bureaucratic establishment, not from the water users themselves.

References

Brewer, Michael F. 1969. "Water Pricing and Allocation with Particular Reference to California Irrigation Districts." California Agriculture Experiment Station, Giannini Foundation, Mimeo Rep. 235.

California Assembly Office of Research. 1985. "Water Trading: Free Market Benefits for Exporters and Importers." Rep. 058-A.

California Department of Water Resources. 1980. "Ground Water Basins in California." Bull. 118-80. Sacramento.

California Department of Water Resources. 1982. "The Hydrologic-Economic Model of the San Joaquin Valley." Bull. 214. Sacramento.

California Department of Water Resources. 1983. "The California Water Plan: Projected Use and Available Water Supplies to 2010." Bull. 160–83. Sacramento.

Gossard, T., et al. 1982. "Water Trading in the San Joaquin Valley." Working paper. University of California, Davis, Department of Agricultural Economics.

Gwynn, D., Thompson, O., and L'Ecluse, K. 1983. "The California Peripheral Canal: Who Backed It, Who Fought It." *California Agriculture*, Jan.–Feb., 22–24.

Howitt, R.E. 1979. "Is Overdraft Always Bad?" *Proceedings of the Twelfth Biennial Conference on Ground Water.* University of California, Davis, California Water Resources Center Rep. 45.

Howitt, R.E., and Llop, A.A. Forthcoming. "Integrated Quantity and Quality Management for Supplies for Irrigation Water." Working paper. University of California, Davis, Department of Agricultural Economics.

Howitt, R., and Mean, P. 1981. "An Economic Approach to Ground Water Quality Management." *Proceedings of the Thirteenth Biennial Conference on Ground Water.* University of California, Davis, California Water Resources Center Rep. 53.

Howitt, R., Watson, W., and Adams, R. 1980. "Crop Production and Water Supply Characteristics of Kern County." University of California Giannini Foundation Information Series 80-1, Berkeley.

Howitt, R. E., Watson, W.D., and Nuckton, C.F. 1979. "Efficiency and Equity of Managment of Agricultural Water Supplies." University of California, Giannini Foundation Information Series 79-1, Berkeley.

Kahrl, William L. 1978. *The California Water Atlas.* State of California Office of Planning and Research, Sacramento.

Moore, Charles V., and Gossard, Thomas. 1984. "Benefits and Costs of a Through Delta Facility to Kern County." Working Paper 84–86. University of California, Davis, Department of Agricutural Economics.

Moore, C.V., and Howitt, R.E. 1984. "Water Issues in California." Paper presented to the J.H., Western Agricultural Economics Association, San Diego.

Moore, C.V., Snyder, J.H., and Sun, P. 1974. "Effects of Colorado River Water Quality and Supplies on Irrigated Agriculture." *Water Resources Research*, *10*(7).

Moore, Charles V., Wilson, David L., and Hatch, Thomas C. 1982. "Structure and Performance of Western Irrigated Agriculture." University of California, Giannini Foundation Information Series 82-2, Bull. 1905.

Noel, J.E., Gardner, B.D., and Moore, C.V. 1980. "Optimal Regional Conjunctive Water Management." *American Journal of Agricultural Economics*, *62*(3), 491–498.

Rand Corporation. 1978. *Efficient Water Use in California.* Santa Monica.

U.S. Department of Commerce. 1982. *Census of Agriculture.*

U.S. Department of the Interior, Bureau of Reclamation. 1979. *Agricultural Drainage and Salt Management in the San Joaquin Valley.*

U.S. Department of the Interior, Bureau of Reclamation. 1984. *CVP Rate Setting Policy Proposal.* Sacramento, California.

U.S. General Accounting Office. 1980. *Effects of Pricing Water from Federal Projects at Full Cost.* App. I, *Auburn-Folsum South Unit Case Study.* San Francisco Regional Office.

Vaux, H., Jr., and Howitt, R. 1984. "Managing Water Scarcity: An Evaluation of Interregional Transfers." *Water Resources Research, 20*(7), 785–792.

Vaux, H., Jr., and Moore, C. 1983. "An Evaluation of Inter-regional Water Trade." Working paper. University of California, Davis, Department of Agricultural Economics.

Wichelns, D. 1984. Unpublished paper. University of California, Davis, Department of Agricultural Economics.

4 Land and water management issues: Texas High Plains

RONALD D. LACEWELL
AND JOHN G. LEE

Irrigation developed on the Texas High Plains using water from an exhaustible groundwater resource (the Ogallala aquifer). Pumping from the Ogallala in this region has diminished the resource to the point that some areas have already made a transition back to rain-fed or dryland agricultural production. This transition from an intensive agriculture to an extensive one (irrigated to dryland) offers a unique opportunity for study in the United States, one that can provide lessons for other regions that eventually will face such a transition.

The focus of this case study is the implications of aquifer mining. They include impacts on agriculture, soil erosion, and present policy options most likely to effect a smooth transition to dryland farming considering minimizing long-term environmental degradation as well as political and economic feasibility.

Description of the region

The Texas High Plains is a nearly level to undulating semiarid region that includes approximately 35,000 square miles in 42 counties. Average annual rainfall ranges from 14 to 21 inches, and the growing season varies from 180 to 220 days. Elevation ranges from 3,000 to 4,000 feet above sea level.

The major soil resource areas include the hardlands (54 percent of the total), composed of fine-textured clays and clay loams, such as the Pullman and Mansker series; the mixedlands (23 percent of the total), composed of the medium-textured loams and loamy sands, such as the Portales, Olton, and Amarillo series; and the sandylands (23 percent), composed of the coarse-textured sands, such as the Brownfield and Tivoli series (Godfrey, Carter, and McKee, 1967). The sandyland soils are particularly sensitive to wind erosion in the absence of crop residue or canopies provided by growing crops.

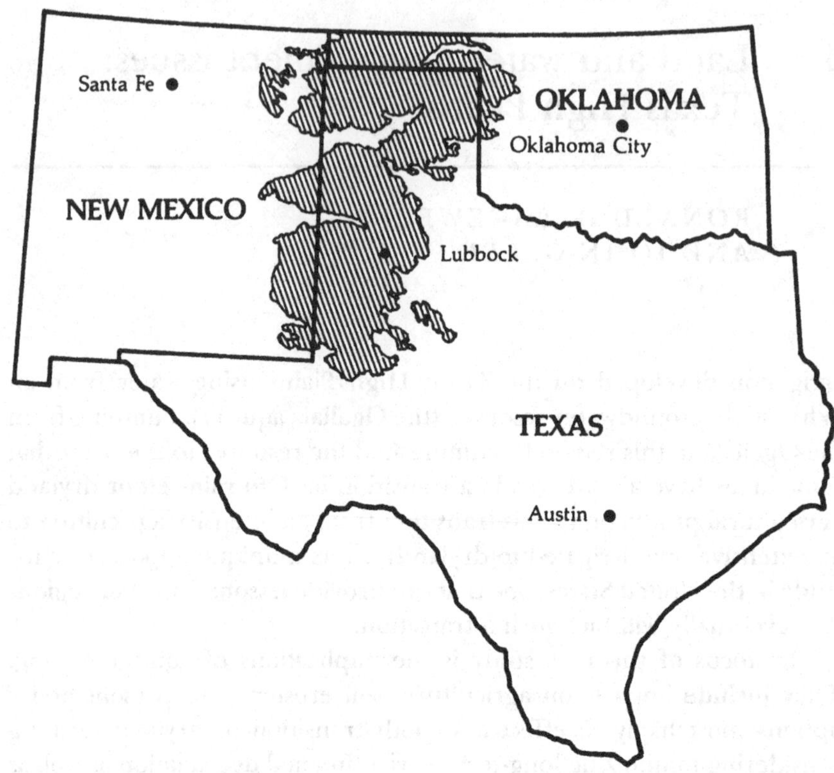

Figure 4.1. The Ogallala aquifer in the High Plains of Texas (High Plains Associates, 1982, pp. 2–3).

The major water resource in the region is the Ogallala aquifer. (See Figure 4.1.) It underlies some 35,450 square miles in Texas (Luckey, Gutentag, and Weeks, 1981). Recharge to the Ogallala in Texas is limited, amounting to only about 0.2 inches per year, or 372,000 acre-feet. Currently, an estimated 400 million acre-feet (maf) of water are in the Texas Ogallala (Knowles, 1984).

Surface water sources include the Colorado, Canadian, Red, and Brazos rivers and minor streams. River flows are relatively low except during high incidence of moisture in the area or upstream. Surface water availability is basically insignificant compared to the Ogallala. Only a small amount of irrigation is from surface sources. However, through reservoirs, some communities use surface water for municipal purposes.

Farming and ranching encompass about 97 percent of the land area, including 10.7 million acres in cropland and 10.4 million acres in pasture and rangeland. About 5.2 million of the cropland acres are irrigated with groundwater from the Ogallala aquifer (Texas Department of Water Resources [DWR], 1981). This is a semiarid region, and irrigation increases most crop yields several-fold. Crops account for 52 percent of the cash receipts from agriculture in the region, and livestock accounts for 42 percent. The remaining 6 percent is accounted for by government payments.

The major crops produced on the Texas High Plains are cotton, wheat, grain sorghum, and corn. From 1971 to 1982, cotton accounted for approximately 34 percent of the harvested acreage, wheat for 25 percent, grain sorghum for 23 percent, and corn for 8 percent; the remaining 10 percent of the acreage was divided among some 20 other crops. Of this remaining 10 percent, more than half is for catch crops (for example, soybeans and sunflowers). A catch crop is planted late in the season after the primary or planned crop is lost to such weather factors as hail and wind. The rest is divided between relatively high-value cash crops (sugar beets, potatoes, etc.) and grass crops (hay, ensilage, pasture, etc.) (Texas Crop and Livestock Reporting Service, 1971–1983). Table 4.1 presents average dryland and irrigated planted acreage for major crops in the period 1980–1983. In addition to crop production, the Texas High Plains is a major beef cattle–producing region, with more than three million head of cattle marketed annually from feedlots between 1971 and 1982.

The regional economy is primarily agriculture and petroleum based (High Plains Associates, 1982). As in the Ogallala, oil and gas are exhaustible resources with diminished production. This fact has serious implications for the small rural communities that have developed to service oil and gas and irrigated agriculture. Employment across the region is 25 percent in irrigated agriculture industries, 20 percent in petroleum, 10 percent in dryland agriculture, and 45 percent in other

Table 4.1. *Average dryland and irrigated crop acreage on the Texas High Plains, 1980–1983*

	Acres planted[a] (thousands)	
Crop	Dryland	Irrigated
Corn	176.25	418.45
Cotton	1,988.20	1,796.80
Grain sorghum	824.37	696.65
Wheat	2,258.60	1,123.85

[a]Does not include more than 20 crops planted on about 10 percent of the total cropland.
Source: Texas Crop and Livestock Reporting Service, 1980–1983.

industries. These other industries are primarily agriculture and petroleum based, including feedlots, meat-packing plants, oil seed mills, textile mills, oil and gas pipe, and related service sectors.

The small communities are mostly dedicated to agriculture; they provide cotton gins, grain elevators, implements, fertilizer and other inputs, and finance. Selected exceptions include a junior college at Levelland and a sucker rod manufacturer and implement manufacturing plant at Lockney. In the Hereford area in the western part of the region, a nuclear waste disposal site is being evaluated. If this site is selected, it would be a major employer and would have a significant economic impact on the area. However, there are concerns related to possible contamination of the groundwater in the Ogallala over the next several decades. Currently, the manufacturing plants in small communities constitute a small part of the overall regional economy. Thus, the small communities on the Texas High Plains will reflect the economic state of the agricultural and petroleum industries at the regional and national levels.

The largest urban areas are Lubbock, with a population of 173,974, and Amarillo, with a population of 149,230 (*Texas Almanac*, 1983). These cities have supplemented their groundwater supplies with surface water from Lake Meredith on the Canadian River and other reservoirs. They have a broad economic base beyond agriculture, oil, and gas. This economic base includes regional retail trade, finance, light manufacturing, and specialized health services, along with several government installations. Although these cities will be affected by declining agricultural output as the region returns to dryland production, the effects will be much less severe than those experienced by small rural communities.

History of the area

Farming on the Texas High Plains began under dryland conditions. Dryland farming developed to a few thousand acres in the 1880s. However, in the late 1880s a drought began that severely retarded further development of farming. In 1889 the Canadian River ceased to flow, and sandstorms accompanied the winds. Many farmers began leaving the area, and as a result of the drought, mortgage problems on the remaining farms grew serious as farmers encountered several seasons of crop failures (Green, 1973). Thus, the sensitivity of the land to the semiarid environment was demonstrated almost a century ago.

Dryland farming again flourished after the drought and continued until the 1930s. Once again drought and the dust bowl affected the High

Table 4.2. *Irrigated acreage on the Texas High Plains, 1936–1981*

Year	Acres irrigated	Year	Acres irrigated
1936	80,000	1954	2,909,000
1937	160,000	1959	3,887,000
1938	200,000	1964	4,094,000
1939	230,000	1969	4,507,000
1940	250,000	1974	4,485,000
1945	550,000	1978	4,542,000
1949	1,548,000	1981	5,200,000

Sources: For 1936–1945, Hughes and Magee, 1960; for 1949–1978, U.S. Department of Commerce; and for 1981, Texas DWR, 1981.

Plains farmer (Heimes and Luckey, 1982). Again in the 1950s, a lengthy drought was experienced. In this semiarid region, years of low rainfall are frequent, and periods of major drought over several years have occurred three times since the late 1880s. The Texas High Plains is subject to relatively high average wind speeds. The average monthly wind speed is about 13 miles per hour, reaching 15.6 in March. In the early afternoon, the average wind speed is greater than 18 miles per hour. Winds can exceed 50 miles per hour over several hours (Elliott, 1977). Drought in the region in conjunction with high temperatures and wind speeds leaves the soil vulnerable to erosion under dryland crop production. Some signs suggest that history could repeat itself. Wind damage in late 1984 and early 1985 was reported as the most extensive since the 1950s (Cordonier, 1985). Annual wind erosion, not considering deposition, for the sandyland soils in the southern Texas High Plains is estimated at 23–108 tons per acre under dryland conditions (Lee et al., 1985b). Over 50 years, an average annual erosion rate of 73 tons per acre equals 20 inches of top soil lost. This condition has implications for sustained productivity and eventually may even affect success in revegetation of grass cover. A recent study indicated a yield loss of 6 percent for each inch of top soil lost (Brown et al., 1984).

Currently, irrigation provides sufficient quantities of biomass (residue) and crop canopies to reduce wind erosion substantially. Irrigated acreage has been accelerated by periods of drought. Irrigated acreage for the Texas High Plains is shown in Table 4.2 for 1936–1981.

Because of the development of more than 5 million irrigated acres, pumping large volumes of water from the exhaustible Ogallala has lowered the static water level. Recent estimates of annual pumping rates for the Texas High Plains range from 5 to 8 maf, depending upon prices

Table 4.3. *Water level changes to 1980, Texas High Plains*

Incremental change (feet)	% of region
+25 to +10	1
+10 to −10	23
−10 to −25	34
−25 to −50	17
−50 to −100	18
−100 to −150	7

Source: Luckey, Gutentag, and Weeks, 1981.

and rainfall patterns. Table 4.3 indicates the magnitude of decline in the water level for Texas. The southern region of the High Plains was the first to develop irrigation rapidly because the aquifer is shallow. In turn, Texas has gone much further in dewatering the Ogallala. More than a 25 percent decrease in saturated thickness has occurred in 29 percent of the aquifer, and more than a 50 percent decrease has occurred in 8 percent of the aquifer in Texas (Luckey, Gutentag, and Weeks, 1981). Overall, approximately 23 percent of the Ogallala in Texas was dewatered by 1980, for a total of 110 maf pumped from an initial nearly 500 maf.

The increased lift and reduced saturated thickness associated with mining the Ogallala increase pumping costs and decrease well yields. The significance of these changes is exaggerated by economic conditions in production agriculture. Input prices, relative to product prices, were lower in the 1950s and 1960s than they are today (Lacewell and McGrann, 1982). Of particular significance is the rise in energy costs for pumping. Since the early 1970s costs have increased approximately 400 percent (Ellis, Lacewell, and Reneau, 1985).

Because of higher pumping costs, increased water depth (lift), and lower well yields, producers adjusted cropping patterns and the quantity of water applied per acre. High water use crops such as corn and soybeans were deemphasized in favor of wheat and cotton (Condra, 1984). Further, inexpensive adjustments to increase irrigation efficiency were implemented, for example, shortening row lengths for gravity flow systems, converting to low-pressure sprinklers, and replacing worn sprinkler nozzles (Ellis, Lacewell, and Reneau, 1985). Rising pumping costs resulting from large price increases for energy, narrow profit margins, and increased price risk are reflected in the marginal value product (MVP) of irrigation water. The MVP of irrigation water for the Texas

High Plains was estimated at $5.98 per acre-foot in 1969 (Beattie, Frank, and Lacewell, 1978), but by 1977 it had increased to $19.67 per acre-foot (Beattie, 1981).

Farmers also significantly increased the number of sprinkler systems as the importance of water application control and distribution efficiency increased. Development focused on efficiency of sprinkler systems and their energy requirements. New farming systems began to emerge: minimum tillage, rotating a row crop such as cotton or sorghum with wheat or other small grains, and using chemicals for weed control to reduce the number of implement trips across the field. A major objective in all cases was to cut costs and maximize the use of pumped and natural water. Thus, parts of the region have been involved in the transition from an irrigated to dryland agriculture for many years.

Adjustment issues

The Texas High Plains increased cropland acreage and developed a highly productive agricultural region based largely on irrigating from the Ogallala. By 1980 the region had exhausted more than one-fifth the water available in the aquifer. With recharge essentially negligible, continued mining of the aquifer will continue to reduce water availability and increase pumping lifts. Storage in the Ogallala in Texas is projected at 342 maf in 2000 (a 19 percent reduction from 1980) and 260 maf by 2030 (a 34 percent reduction) (Knowles, 1984). The declining water table is estimated to support only about 55 percent of current irrigated acres in 2000 and 35–40 percent in 2030 (Grubb, 1981). This loss means that dryland crop production acreage will vastly increase over the next several years.

Continued pumping will result in a decline of the water level in the Ogallala. Increasing lift and relatively expensive energy can be expected to maintain an upward pressure on the cost to pump. Because the Ogallala aquifer is basically stock in nature, that is, it is nonrenewable and depletable, the aquifer could be physically exhausted. However, because well yields become so small at low levels of saturated thickness and costs rise, pumping stops with some recoverable water left in the aquifer.

Economic exhaustion of the aquifer then becomes the focal point. Economic exhaustion occurs when pumping water for irrigation costs more than its worth to crop production. When it is not profitable to continue irrigating, the land reverts to dryland or is abandoned for crop production. Although cropland on the Texas High Plains has not been abandoned in the past, it has been in other regions. Land abandonment can have serious economic and environmental consequences. A primary environmental concern is the greater potential for wind erosion on unprotected soils. Their annual erosion rate can more than double those of soils protected by crops and surface residues.

Critical to the reversion from irrigated to dryland crop production are the use of supplemental water to maintain irrigated acreage, adoption of new technology and management practices, and changes in cropping patterns.

Supplemental water alternatives

With an exhaustible and declining water resource, much interest has focused on providing supplemental water to the Texas High Plains. In addition to interbasin transfer of water, technologies to release capillary water remaining in the desaturated zone of the aquifer are being investigated. This alternative could delay or offset exhaustion of the Ogallala.

Water transfer plans have been formulated for decades. They range from an ambitious plan to use excess water in Alaska, the Northwest Territory, and the Rocky Mountain region of Canada and distribute it to water-deficit areas of Canada, the United States, and northern Mexico. Termed the North America Water and Power Alliance, the overall plan would generate 60–180 million kilowatts of electricity and 75 maf of water annually at an estimated cost of several hundred billion 1982 dollars (Banks, Williams, and Harris, 1984). Environmental, economic, and political issues make such an ambitious plan unlikely in the foreseeable future.

The Mississippi River was also viewed as a potential source of water for the Texas High Plains. Although the study indicated technical feasibility for the diversion, a high cost per unit of delivered water was also projected (for example, $330 per acre-foot). The capital cost was estimated at $19.5 billion in 1982 dollars (Banks, Williams, and Harris, 1984). Again, economic feasibility could not be satisfied.

Recently, an analysis was conducted on the feasibility of taking water from the Arkansas and White rivers in Arkansas and transporting it southwestward across Arkansas and northeast Texas and then to the southern High Plains of Texas. Two sizes of transfer facilities were considered (1.55 maf and 8.68 maf per year). The total cost was estimated at $5.3 billion for the small facility and $20.6 billion for the large facility. The costs per acre-foot delivered to the southern High Plains were $490 and $441 in 1977 dollars for the small and large facilities, respectively (High Plains Associates, 1982). These costs per acre-foot do not include distribution to the farm gate or on-farm distribution costs. The value of water for irrigation of crops from the Ogallala is estimated at \$40–\$80 per acre-foot (Office of Technology Assessment, 1983). A \$60–\$80 value per acre-foot in 1982 dollars was estimated in the study that evaluated interbasin transfers (High Plains Associates, 1982).

The large discrepancy between the cost to deliver water and its value in production suggests little opportunity for this or any new large-scale water development in any region in the foreseeable future. Further, with high energy requirements of water transfer, costs would increase significantly as energy costs rise. A serious question of surplus water in the basin of origin was indicated, particularly considering future needs. Finally, such major diversions of water could result in irreversible environmental impacts (Office of Technology Assessment, 1983). Thus, large interbasin transfers of water to the Texas High Plains are unlikely.

Another alternative for supplemental water relates to capillary water in the Ogallala. The Ogallala formation in the Texas High Plains contains about 40 percent water when saturated (Wyatt, 1984). Some 15 percent will drain by gravity and, thus, can be pumped by wells, leaving 25 percent held by capillary forces.

The High Plains Underground Water Conservation District No. 1, in conjunction with Texas Tech University, initiated a study in 1980 of potential methods for recovering this capillary water (Wyatt, 1984). Of several procedures considered, an air drive method was selected for field testing. In 1982, field tests were performed whereby air was injected into the Ogallala formation. Measurements of the saturated thickness showed an increase, and 160 days after the tests an additional 406 acre-feet of water was indicated to be in storage.

The results of subsequent tests are also positive. A cursory economic analysis of water released indicates a cost of about \$50 per acre-foot. However, the water remains in the aquifer and must therefore be pumped. Under current prices and average groundwater availability, pumping 1 acre-foot would cost approximately \$30. This cost, coupled with the injection cost, totals about \$80 per acre-foot. It is significantly less than interbasin water transfer plans. This research represents an exciting potential for new technology that may extend irrigation in the region, further extending the period of transition from irrigation to dryland. In addition, for an area that has ceased to irrigate and wishes to establish pasture for livestock grazing, wildlife, and soil conservation, it is possible that sufficient water could be released to allow ample irrigation for establishing grass in this semiarid region. Only that level of irrigation necessary to establish the grass would be expected. Furthermore, because returns from livestock grazing are limited, a subsidy for irrigation may be necessary to provide incentives for the farmer to establish an adequate grass or natural cover to provide protection against wind erosion.

Technology and management

Even with a successful program of releasing capillary water in the previously mined aquifer, the Ogallala is still exhaustible and other alternatives are necessary. An important factor in maximizing the value of the water that remains and in using natural precipitation to its fullest extent relates to new technology and soil and water management strategies. Most on-farm management strategies that improve water use efficiency also tend to reduce soil erosion. This section discusses alternative soil conservation and water management practices involving irrigation distribution systems and the estimated impacts of management and technology on the Texas High Plains. Because water is the major limiting natural resource for agricultural production on the Texas High Plains, much of the technology developed is designed to improve use of existing water supplies. Water-conserving techniques intended to retain and use the water in the field more effectively or improved application of irrigation water may be the major objective. Each may contribute to maintaining farmers at a higher level of production and net revenue during and after the transition to dryland production while providing for control of water and wind erosion.

Management options

The purpose of conservation techniques aimed primarily at water is to maximize crop production by minimizing the amount of water that is lost. As much of the water as possible that arrives on the field as precipitation or irrigation must be kept in the root zone as soil moisture. Major avenues of loss are runoff, evaporation, and percolation. Several technologies are discussed in detail in Ellis, Lacewell, and Reneau (1985).

Runoff can be reduced by changing the topography of the land either by leveling and terrace construction or by tillage practices. Conservation and contour bench terracing for dryland grain production not only increased the average yield but also reduced yield variability and, on steeper slopes (more than 2 percent), reduced soil erosion.

Another landforming technique that is returning to use after being developed in the 1930s and nearly abandoned by 1950 is the use of microbasins, also known as furrow dams. Microbasins are small earth dams that are mechanically placed 4–10 feet apart in each crop row. There are thousands of these small dams across an acre of cropland. They catch and retain rainfall, thereby reducing surface runoff. The original intent was to control wind erosion and save water during fallow periods, but a number of problems arose: poor weed control; difficulty with dam emplacement, seedbed preparation, and tillage; and the inability to demonstrate yield increases. For these reasons, furrow dams were abandoned in favor of stubble-mulch tillage, terracing, and other conservation practices (Clark, 1979). With advances in machinery, chemical weed control, and changes in use from fallow periods to dryland and irrigated summer crops, furrow dams have been shown to be effective in reducing runoff and increasing both soil moisture and crop yields (Lyle et al., 1981; Lyle and Bordovsky, 1980, 1986). In dryland grain sorghum trials at Bushland, Texas (1975–1977), 0.7–3.3 inches of additional rainfall was retained on the field, increasing yields 25–40 percent (Clark, 1979). Dryland cotton studies have shown increases of 11–25 percent (Runkles, 1980).

The value of surface residue in controlling wind and water erosion has long been recognized. The need to control soil erosion by wind, particularly in the drier portions of the Great Plains during the drought of the 1930s, led to development of stubble-mulch farming. Stubble-mulch tillage refers to plowing and cutting only the roots of plants while leaving the biomass above ground. This is now the basic tillage method in many dryland areas and represents an early form of limited tillage (Unger and McCalla, 1980). Johnson and Davis (1972) reported long-term (25 years of data) wheat yield gains of 17 percent using stubble-mulch tillage compared to conventionally tilled (turning biomass beneath the soil) continuous wheat. Managing wheat residue to keep it on the surface has been shown to increase precipitation storage 40–80 percent (Greb, Smika, and Black, 1967; Unger and Weise, 1979). Grain yields for subsequent crops increased 16–38 percent.

Minimum or conservation tillage has been shown to be effective in the control of wind and water erosion (Zingg and Whitfield, 1957; Unger and McCalla, 1980) and in the reduction of air and water pollution (Unger and Box, 1972). Compared with other conservation methods, it is purported to conserve 20 percent more soil moisture and increase wheat yields 12 percent (Jones, Unger, and Fryrear, 1985). Apparently, small amounts of residue, such as 1,800 pounds per acre, are sufficient to control wind erosion in most areas if the residues are maintained on the field surface. Dryland cotton, however, does not produce enough residue to control wind erosion. For dryland cotton acreage, tillage becomes the effective method of wind erosion control if there is sufficient moisture to produce a cloddy surface (Jones, Unger, and Fryrear, 1985). Thus, to retain moisture so that tillage practices are effective, strategies must be adopted to preserve rainfall in the soil. Soil conservation practices on cropland are generally designed to maximize use of rainfall. Hence, most soil conservation practices are by necessity important water conservation practices.

An effective long-term wind erosion control practice is a return of the land to grass, particularly the more sandy soils. Much of this land is not suitable for intensive dryland farming (Stewart and Harman, 1984). The need is for revegetation, which will be more difficult unless it is done while a particular field is still being irrigated.

This concept of revegetation of dryland cropland to grass presents a problem. Clearly, grassland is less sensitive to wind erosion than cropland, especially cotton. But to irrigate one additional year just to establish grass for cattle in a cow-calf operation is not a wise economic decision. Huszar (1985) considered wheat and cow-calf production shifts in Colorado from 1977 to 1983. Basically, wheat yields on dryland acres have increased 71 percent, whereas it still takes 30–40 acres of grass per cow per year. Although costs have risen for both, the bottom line is that net returns to land and other fixed inputs are about $70 per acre for wheat compared to $3 per acre for cow-calf production. These figures suggest a strong economic incentive for wheat production over cow-calf production. Although a year-round cow-calf operation is replaced by the wheat, farmers often graze young cattle, termed stockers, on the wheat for three or four months. These stockers come from ranches throughout the country. Typically, the stockers then go to a feedlot for finishing. After the stockers are taken off wheat, the wheat matures and the grain is harvested.

A situation similar to that in Colorado exists between cotton production and cow-calf production in the southern High Plains. Even though the region is subject to periodic drought and production of dryland crops suffers significant wind erosion, the incentive for the farmer is dryland crop production. Thus, to accomplish revegetation of grass to control wind erosion and use in a cow-calf operation, either government regulation or a substantial annual subsidy to farmers would be necessary.

Distribution systems

Crop production is possible in the High Plains without irrigation, but irrigation allows much higher production levels. Nonetheless, because irrigation is a major user of energy, it constitutes a large portion of the variable cost for irrigated crops, and it is depleting a nonrenewable resource (groundwater). Research on improving the efficiency of pumping and application of water continues. Statewide, on-farm irrigation water use efficiency (the amount of irrigation water stored in soil for plant use versus the amount that is applied) is estimated at about 60–70 percent (Wyatt, 1984). With advanced techniques, efficiencies up to 98 percent have been demonstrated (Lyle and Bordovsky, 1986).

Two main types of irrigation systems used on the High Plains are flood, or furrow, irrigation on hardland soils and sprinkler systems predominant on mixed or sandy soils. Techniques designed to increase the efficiency of furrow irrigation include alternate furrow irrigation, furrow diking, surge flow, and automated furrow systems. Alternate furrow application permits reduction in irrigation size and coverage of a larger area in a timely manner. It results in greater lateral water movement in the soil and reduced deep percolation losses.

Alternate furrow irrigation can also be used with furrow diking or row dams in the nonirrigated furrows to reduce rainfall runoff. Significant yield increases for both cotton and grain sorghum have been obtained by adding basin tillage to alternate furrow irrigation (Clark, 1979; Lyle and Dixon, 1977).

Surge flow application is designed to deliver large surges of water to the furrow on an intermittent cycle to reduce percolation losses at the upper end of the field and hence increase distribution efficiency. Automated systems operating on timers or soil moisture sensors connected to a microprocessor are also being developed in an attempt to increase distribution uniformity and application efficiency (Lyle et al., 1982).

With the increasing importance of sprinkler irrigation systems on the Texas High Plains, research has focused on increasing the efficiency of sprinkler systems and, with the quadrupling of oil prices after 1972, reducing their energy requirements. One of the most promising systems developed to meet these criteria is the low-energy precision application (LEPA) system (Lyle and Bordovsky, 1980). The system operates by distributing water through drop tubes and low-pressure emitters directly to the furrow as it moves through the field in either a linear or circular fashion. Thus, traditional center-pivot or linear sprinklers can be modified to the LEPA configuration. The new cost of a LEPA system is similar to a center pivot at about $36,000 for a 136-acre system. However, an existing center pivot system can be converted to LEPA for $5,000–$10,000. The system, when combined with microbasins, is designed to maximize water application and distribution efficiency while minimizing energy costs and runoff.

In field trials of the LEPA system, measured application efficiencies averaged more than 98 percent and distribution efficiency averaged 96 percent, but runoff from both irrigation and rainfall was essentially eliminated when microbasins were included. In a two-year test on soybeans at the Texas Agricultural Experiment Station in Halfway, per-bushel costs of energy for pumping with undiked sprinkler irrigation averaged 67 percent more than for the LEPA system. Sprinkler-irrigated soybeans combined with furrow diking still required 51 percent more energy than under the LEPA system (Lyle et al., 1981).

Drip irrigation represents yet another technology for achieving high irrigation distribution efficiencies (over 90 percent). Drip irrigation systems can be placed above or below the soil surface. The technology has proven cost effective for high-value crops such as grapes, citrus, and pecans. Increasingly more acres of field crops like cotton are being irrigated with drip systems. Drip irrigation systems introduce high capital costs into row-crop farming, with the goal of greater yields and lower operating costs. A study comparing drip and LEPA on cotton for the Texas High Plains indicates no significant difference in yield (Lyle and Bordovsky, 1986). The average investment cost to implement a drip system is $500 or more per acre, compared to approximately $300 per acre for a new LEPA system (Smith, 1986).

New equipment and changes in farming practices require a higher degree of management. Further, they can be expensive. Lee, Ellis, and Lacewell (1985a) concluded that with low crop prices the new systems may not always be an economical decision. Substantial benefits can be gained by improving the efficiency of an irrigation system in place. Although shifting to more capital-intensive systems was an economic gain, it was less than that achieved by improving the old system. Thus, simply improving technical efficiency of irrigation is not a sufficient basis for a management decision to change irrigation systems. Similarly, the amount of groundwater remaining is a crucial factor. For limited groundwater areas, such as those with less than 80 feet of water-bearing sand, the amount of water remaining was insufficient to recover the initial investment. This point suggests some financial limitations for the adoption of new technology that could increase efficiency of use of the exhaustible Ogallala water supply. At issue is the role of society in providing incentives for the adoption of new technology here.

Many new technologies increase farmers' expected profits, thereby providing an incentive for adoption. New technology, along with advanced crop production practices, production systems, and innovative financial management, provides an opportunity to guide the transition from irrigated to dryland by minimizing the adverse economic and natural resource impacts. To provide estimates of the role of new technology in transitions from one level of production intensity to another, Ellis, Lacewell, and Reneau (1985) analyzed selected technologies given the level of groundwater available for the High Plains in 1980. The analysis extends from 1980 to 2020. For the study, a technology or management practice was adopted only if it increased producer profits.

Limited tillage, LEPA, and improved furrow technologies were selected as representative of the expected technological change for the region. Current use of these technologies in the region varies significantly with LEPA adoption just underway and limited tillage in use in many areas. Details of the technologies included are presented in Ellis, Lacewell, and Reneau (1985).

The analysis considers the difference in continuation of the base or current practices and simultaneous adoption of the new technologies over time. Use of center-pivot sprinkler systems, conventional furrow irrigation, and conventional tillage practices formed the base scenario.

For advanced technologies to be adopted, they must provide an economic incentive; that is, net returns to land, management, and risk must at least equal those available from use of conventional technology. Projected net returns of the base and new technologies for the 40-year period of analysis are shown in Figure 4.2. The figure shows that for the technologies, price vectors, and adoption rates used here, net returns exceed conventional technology.

Declining water supplies and increasing pumping costs resulting from greater pumping lifts for the base scenario project a downward trend in net returns if no advanced technology is adopted. The constraint employed to bound limited tillage acreages through time assumed that the practice was used on approximately 25 percent of the cropland in 1980. The shift depicted by the net returns curve for new technology reflects the benefits of this instantaneous change from zero to 25 percent adoption of minimum tillage. As adoption proceeds, the difference between net returns for new technology and the base continues to increase.

Income effects associated with the adoption of new technology are large enough to increase net returns for a time despite declining groundwater supplies. These increases proceed to the approximate end of the assumed adoption period in 1990. Then a general decline in net returns begins, even with new technology.

The 2020 projections indicate no significant reduction in total water pumped because of new technology compared to the base. Of importance, however, is the increased effectiveness in use of the groundwater

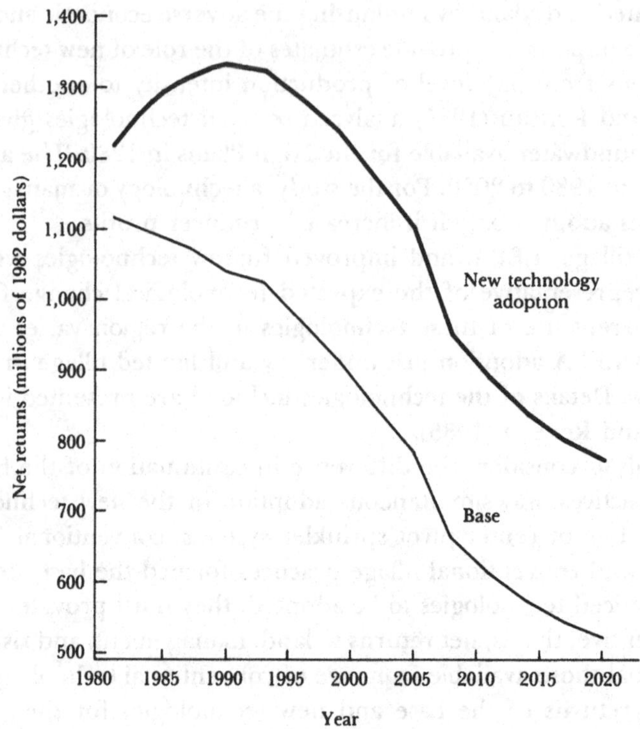

Figure 4.2. Returns to land, management, and risk over time for adoption of alternative crop production technologies, Texas High Plains, 1980–2020.

as well as greater profitability of dryland farming. For example, with only conventional technology in 2020, farmer net returns were estimated at $520 million annually. However, with all technologies, the estimated 2020 farmer net revenue value is $770 million, or 48 percent greater. Although these technologies do not significantly alter annual pumping rates, they offer great potential for minimizing the adverse effects of a major transition in intensity of crop production on the Texas High Plains.

The results indicate that the region will not follow the same path back to dryland and that dryland production will be significantly different than in the 1950s and 1960s. Farm machinery advances permit adoption of minimum tillage practices, thereby preserving crop residues on the soil surface. The value of crop rotations for long-term productivity is being demonstrated, with initial trends indicating success. This finding means a more efficient use of natural precipitation and a reduction in soil erosion except for periods of extended drought.

Cropping patterns

Implicit in the adoption of new technology and management practices is an evolution of changing cropping patterns. From 1977 to 1982, dryland crop acreage on the southern High Plains increased from 40 percent of the harvested acreage to 50 percent (Texas DWR, 1981). From a review of recent changes in cropping patterns, the response to declining water availability and high pumping costs can be identified.

Three-year moving averages of total irrigated and dryland harvested acreages of the major crops (cotton, grain sorghum, corn, and wheat) are shown for the southern High Plains between 1973 and 1982 in Figure 4.3. It does not appear that significant acreage is going out of

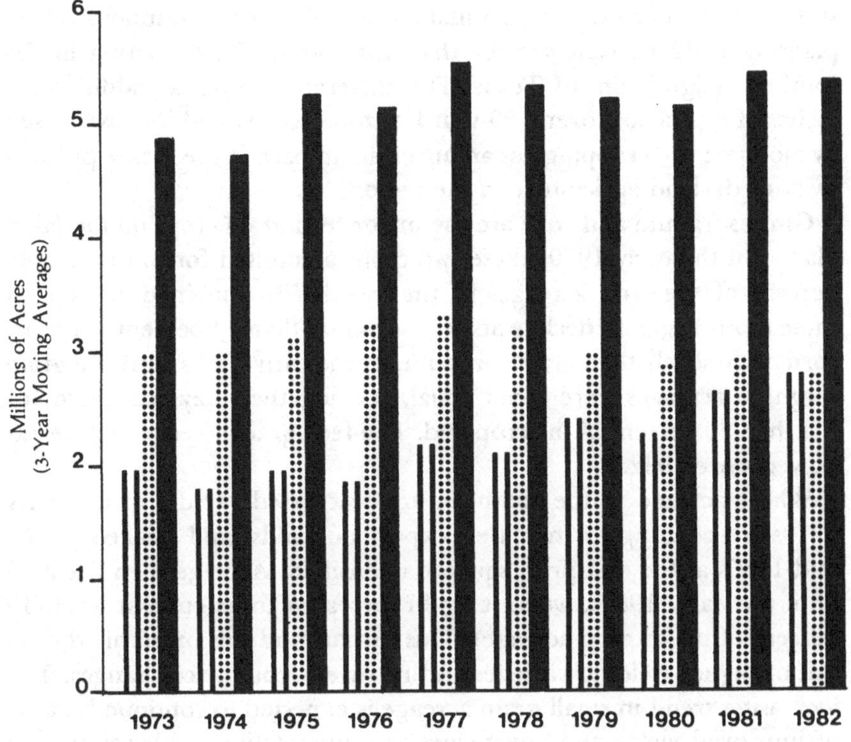

Cropland

- Dryland
- Irrigated
- Total

Figure 4.3. Acreage of major crops harvested, three-year moving averages, southern High Plains of Texas, 1973–1982 (Condra, 1984; Texas Crop and Livestock Reporting Service, 1973–1982).

production as the number of farms decreases. However, there is a clear trend in the shift of irrigated acreage to dryland on the southern Texas High Plains. Irrigated acreage has declined 2–4 percent per year since the mid-1970s (Texas DWR, 1981; Texas Crop and Livestock Reporting Service, 1973–1983).

Cotton acreages have increased steadily. However, the rate of increase in dryland cotton acreage has almost doubled that of irrigated cotton acreage. These trends are expected to continue. This point is important because cotton represents approximately 70 percent of the major crop acreage on the southern High Plains. The problems associated with monoculture crop production, particularly cotton, include pest carryover and chemical resistance, reduced soil tilth, and increased erosion susceptibility. For example, annual soil erosion from continuous cotton planting is 42 percent greater than for cotton-wheat rotation in the southern High Plains of Texas. This difference equals an additional 6 inches of topsoil loss over a 50-year horizon. Accelerated soil loss caused by monoculture cropping has an uncertain impact on the future productivity of dryland agriculture in the region.

Grain sorghum and corn are the major feed grain crops on the High Plains. In the early 1970s these two crops accounted for more than 40 percent of the crop acreage. In the mid-1970s combined acreage of these crops began to decline at a rate of more than 13 percent per year, corn more slowly than grain sorghum. By the early 1980s corn and grain sorghum acreages were about equal, and together they accounted for less than 20 percent of the cropland. The feed grain acreage decline has slowed since 1982.

Wheat acreage in the southern area increased rapidly in the early 1970s, responding to increased export demands, and peaked in the mid-1970s as the cost-price squeeze intensified. Acreage then declined until the early 1980s, when it again increased to its current level (10 percent of major crop acreage). This turnaround was primarily the result of higher-yielding varieties and relative output price changes. The increasing trend in small grain acreage is expected to continue because of improved yields and advantages in crop rotations with cotton and grain sorghum.

Less than 4 percent of the harvested acreage on the Texas High Plains is allocated to high-value crops such as vegetables and sugar beets. Although these crops offer some potential to increase farmer returns, acreage is not expected to increase significantly because of the much higher production costs, greater risks, and small specialized markets.

In projections of cropping pattern adjustments for the Texas High

Plains associated with the declining Ogallala, livestock is an important component. Livestock as a cow-calf operation is not expected to increase, but, rather, the number of stocker cattle is expected to increase in response to increases in small grain acreage (wheat and oats). Cattle feeding (feedlots) has experienced major shifts since the 1960s, with a movement toward the Central and Southern plains (Clary, Dietrich, and Farris, 1984). By 1980 about 60 percent of all finished cattle were marketed from feedlots in the Southern and Central plains. The Southern Plains include Texas, Oklahoma, and Nebraska. The outlook for west Texas, as estimated by Clary, Dietrich, and Farris (1984), is that cattle feeding and slaughter levels would not be substantially affected even if feed grain supplies (corn and sorghum) in the region were decreased 50 percent from the 1980 level. The grain needed for feeding could be shipped from surplus production areas at a lower cost than movement of cattle-feeding activities. This possibility suggests a strong competitive advantage in cattle feeding and slaughter for the Texas High Plains.

The expected adjustment in cropping patterns because of the declining Ogallala is to less-water-intensive crops. They include primarily cotton and wheat and possibly sorghum in northern areas where cotton is not grown. The acreage in corn and soybeans can be expected to decline owing to their dependence upon irrigation. The number of acres of high-value crops is expected to remain fairly constant, and they will be located in areas with ample groundwater. Most high-value crops require large quantities of water.

The evolving cropping pattern can be expected to be cotton and/or grain sorghum grown in a rotation with wheat. With the use of conservation tillage, furrow dikes, and chemical weed control, maximum use of rainfall is achieved, residues are left on the surface, and better wind erosion control results. The small grain (wheat) offers grazing to stocker cattle for several months of the year. Thus, the importance of wheat pasture for stocker cattle and wheat as a crop is expected to increase significantly. However, even with optimal cropping patterns and advanced management strategies, dryland crop production remains vulnerable to the region's frequent droughts.

Environmental issues of the transition

Owing to physical and economic conditions, agriculture on the Texas High Plains will revert back to dryland production. Because of the heterogenous nature of the Ogallala aquifer, this transition will not occur simultaneously across the region. Some areas have already made the transition, others are in advanced stages, and some have a groundwater supply that will last for several decades.

The principal focus of this case study is on the large percentage of the 5 million plus acres of irrigated cropland that will be making the transition from irrigated to dryland production in the next few decades. As the region converts back to dryland production, the incidence of wind erosion and its effects on crop productivity and health are major issues to be addressed. Blowing dust in the spring drought of 1984 in the southernmost part of the Texas High Plains was heavier than in recent years. Much of this area consists of sandyland soils, where a high percentage of groundwater mining has occurred.

Even with a cotton and wheat dryland crop rotation, average wind erosion per acre is estimated at 51 tons per year, not considering deposition (Lee et al., 1985b). For monoculture cotton, the average annual level of wind erosion is 73 tons per acre. In drought years with especially high winds, wind erosion from a cotton and wheat rotation can move as much as 99 tons of soil per acre. Thus, even though crop rotations that increase surface residue can reduce wind erosion 30 percent, the remaining amount of wind erosion is alarmingly high. In any area with several feet of topsoil, annual levels of wind erosion approaching 50 tons per acre may not drastically affect expected yield in the short run. However, for shallow topsoils the productivity effect can be significant.

Another major factor of blowing dust that affects crop yields is the impact of wind erosion on young and tender plants. Cotton is especially vulnerable to blowing dust, and exposure for a few hours destroys the plants. Jones, Unger, and Fryrear (1985) indicate that wind exposure of four-day-old cotton seedlings will cause a plant mortality rate of 34 percent or a cotton lint yield reduction of 44 percent. Thus, either there is a reduced plant population or replanting is necessary and the crop is late. Between 1947 and 1973, of all the causes of yield loss – drought, hail, excess moisture, freeze, floods, wind, insects, disease, etc. – the wind factor was particularly significant. In 14 counties in the southern High Plains of Texas, the wind yield loss for cotton averaged 10.3 percent of all yield losses on cotton and ranged from 2.8 to 27.7 percent (Crop Insurance Actuarial Association, 1947–1973). This rate is compared to an average wind yield loss of 17.7 percent of all yield losses for grain sorghum in ten counties and of 25.3 percent for wheat in five counties.

Clearly, wind damage to crops is a serious factor in reducing yields, but there are other issues. One relates to producers' electing not to plant any crop or abandoning the land entirely. With idle fields, the impact of wind erosion on adjacent fields planted with young crops increases, seriously reducing crop yields on those fields.

Off-farm damages are another consideration. If the entire 2.5 million cropped acres on the Texas High Plains in the sandyland soil classification were cropped dryland with continuous cotton, an estimated 178 million tons of soil would be moved by wind in an average year. This figure would escalate to 272 million tons at the upper limit. If 10 million acres were cropped dryland with a continuous row crop, approximately 720 million tons would be moved by wind in an average year, reaching as much as 1,088 million tons in some years. At such levels, the consequences of airborne dust would be severe. Blowing dust causes numerous problems. It "sandblasts" automobiles; impairs driving visibility, thereby increasing chances for accidents; creates medical disorders from inhalation; settles as fine particles in homes, businesses, and industry; and covers roads and ditches.

For farmers, faced with narrow profit margins, investing in long-term conservation measures could result in negative profits and potential financial failure (Brown et al., 1984). This realization seriously affects the outlook for erosion throughout the Great Plains, with its increasing acreage of dryland crop production. Further, it is suggested that conservation tillage and other conservation practices are being adopted by the more progressive and innovative farmers, who are often not the ones on marginal or erosive land. Soil productivity and nonagricultural impacts associated with high rates of wind erosion create a question for society to answer: Is the subsidy that would be required to control wind erosion less than the gains of reduced erosion?

Implications of adjustment

An overview of this section reveals the inevitable transition of the Texas High Plains from irrigated to dryland crop production. This will be a movement to less-water-intensive practices. Primary crop production will include cotton, grain sorghum, and wheat, with stocker cattle grazing the wheat in the winter and early spring.

With the transition to dryland and adoption of practices that are less water intensive, some major changes can be projected. Dryland agriculture typically returns less profit per acre than irrigated land. Thus, for development of a viable economic unit, farms will be fewer and larger. This trend is a continuation of a trend evident from 1978 to 1982, when the number of farms declined about 4 percent per year (U.S. Department of Commerce, 1978, 1982). Unlike previous trends to fewer and larger farms, producers in the Texas High Plains will be forced to expand upon a declining asset base. During the transition from irrigated to dryland production, land values are projected to decline. Because land is a major component of their total assets, producers will find it difficult to secure intermediate and long-term financing for the expansion. On the other hand, market forces will dictate an expansion of farm size to maintain the previous standard of living. Furthermore, there is evidence that the price-cost squeeze puts tenants and part-time farmers at a comparative disadvantage; thus, an increase in farms operated by full-time owners is expected.

Less-intensive agriculture will result in a decline in agriculturally related returns and a reduction in the demand for goods and services by production agriculture. This decline will impact the regional economy, which is based largely on agricultural production. Of particular significance are communities with fewer than 5,000 people. The declining Ogallala and associated adjustments are expected to double per capita annual expenditures to service bonds and to provide water, streets, hospitals, fire protection, schools, etc. (Williford, Beattie, and Lacewell, 1976), because reduced demands from production agriculture will affect the small community's tax base. There will be less need for auto mechanics to work on irrigation engines; less need for fertilizer, chemicals, and equipment; and less output moving through cotton gins and elevators. Individuals employed in these sectors will seek opportunities elsewhere. This out-migration from the region further aggravates the declining tax base for smaller communities. The socioeconomic implications for the survival of small communities in the Texas High Plains are serious.

In conclusion, ramifications and implications of the declining Ogallala aquifer are far-reaching. There are soil resource issues relating to vastly increased wind erosion. High levels of soil loss over time affect the productivity of the soil. There is also a negative effect on air quality throughout the region and beyond as dust particles from eastern New Mexico and west Texas reach Dallas and even the east coast.

Not only are natural resources vulnerable to the transition from irrigated to dryland production, but the local and regional economies are negatively affected. One result is fewer jobs, and for the small towns the transition raises a serious question about the costs of providing basic services. Thus, the declining Ogallala is associated with issues ranging from the effects on natural resources to socioeconomic implications for communities and the region. As the water becomes more scarce, pressure for policy solutions to resource problems (groundwater depletion, air quality deterioration, soil erosion, etc.) will continue to increase.

Policy alternatives

Policy alternatives relate to government regulation and incentives created to effect a farmer's decision. These policies may be at the federal, state, and local levels. The policy alternatives considered here are those that can help ameliorate the worst aspects of the transition to dryland and other program incentives that affect agricultural production in the region and thus impact on the transition. In addition, some of the policies are interrelated. For example, the requirement of establishing a cover crop on acres that must be diverted from production in order to be eligible for federal crop price programs (support and target) includes three policies: acreage diversion, cover crop establishment, and crop price support.

The policies of long-term land retirement (Conservation Reserve Program), price supports, diversion payments, and cost-sharing must be viewed in light of the federal deficit and efforts to reduce it as well as the increasing urban dominance of the legislature. The current farm program includes provisions for a reduction of the government-supported price levels. Further, subsidy levels in the 1985 farm bill may be automatically reduced to comply with the Gramm-Rudman-Hollings provision of federal deficit reduction. Government has long been a participant in agriculture, but the implications for the future suggest the use of alternatives in addition to government spending. Several of these alternatives are discussed in the following sections.

Some soil and water conservation options, including regulating pumping, imposing a water tax, well spacing, prohibiting waste, and educating farmers about conservation technologies, can be implemented at the local and regional levels by the state or by groundwater districts. Groundwater districts have been effective on the Texas High Plains for several decades. Further, according to Grubb (1981), because Texas groundwater districts are functioning well, local control and management will probably not be significantly changed by the state legislature in the near future.

Currently, groundwater districts in Texas are aggressively developing and implementing water conservation practices. The groundwater districts have regulation authority over well permits, well spacing, and prohibiting waste of water. However, a major focus of the districts is educating both irrigation and dryland farmers on water and energy conservation technologies. Their role can be particularly significant because they are involved in the transition from irrigated to dryland farming. Thus, local and regional conservation districts are in place, and expansion of their role in the transition from irrigated to dryland crop production is expected. Their role is to control water waste through strong regulatory authority and to transfer new technologies to producers making the transition from irrigated to dryland production.

Long-term land retirement

The land bank program established in 1956 offered farmers an opportunity to take cropland out of production for up to ten years and to receive an annual payment from the federal government. Under this plan, wind and water erosion can be effectively controlled. A program of long-term land retirement became law with passage of the 1985 farm bill. The Conservation Reserve Program (CRP) is designed to remove 45 million acres of highly erodible soils from crop production for ten years, including a substantial portion of the southern High Plains. It promotes both soil and water conservation. The program is administered on a competitive bid basis, and no more than 25 percent of the land in a county may be accepted without approval of the Secretary of Agriculture. An annual payment is made to the landowner no earlier than October over the period of the contract.

An important part of this program is establishing grass on the land (revegetation). As part of the package, the federal government shares costs of up to 50 percent of the total expenses for revegetation. It is expected that the cost-share payment for revegetation will be significantly greater than the annual payments. Cost sharing is a critical component of the program because establishing a grass cover is a condition for participation in the program. Cost sharing provides the needed economic incentive to encourage enrollment in long-term land retirement. A long-term program like this one prevents the loss of government farm program allotment acreage, which happens to producers involved in a one-year conservation program. However, after the contract expires, land brought back into crop production will be considered as sodbusting, and it is not eligible for government programs that might be in effect in the late 1990s.

Cotton production in the southern Texas High Plains is characterized by a high percentage of participation in federal crop programs. The fact that land coming out of the long-term land retirement program will be considered as sodbusting will present some resistance to program enrollment because federal crop programs are so important to cotton producers. Therefore, payments must offset the loss of potential benefits from federal crop programs. In view of the fact that these potential federal crop program benefits are several years off, their current value may not be significant.

The impacts of a long-term land retirement are many. In a region such as the Texas High Plains, crop production is a major component of the economy. With 25 percent of the land taken out of production, demand for inputs and the need for gins, elevators, trucks, and implements will lessen. This impact can be felt essentially upon enrollment in the program. Thus, the adjustment would not be gradual but, rather, it would be a large and immediate reduction in the base sector of the economy. This situation is analogous to the payment-in-kind program, except that the latter lasted only one year.

Further, the contract is with a landowner, and renters and tenants are left without land to farm and without benefits from the program. This program is expected to accelerate the trend of fewer tenant operators. Thus, the impact of reduced economic activity is felt primarily by the small rural communities that provide goods and services to production agriculture.

On the benefit side, there are concerns that current rates of soil erosion in the southern High Plains of Texas are creating a great new desert while exhausting the soil resource (Sheridan, 1981). Long-term land retirement is a most effective program for erosion control as well as for conservation of water. This benefit is particularly applicable given the requirement of revegetating the land.

The first round of CRP bids closed in March 1986. Preliminary expectations of acreage participation at the national level was 5 million acres in 1986, 10 million acres annually between 1987 and 1989, and 5 million acres in 1990. This level of participation is expected to reduce annual soil erosion 760 million tons and to decrease sedimentation 211 million tons. Of the 40–45 million acres eligible for the CRP nationally, Texas has the most, 11.5 million acres. The preliminary expectation for Texas for 1986 was 825,000 acres.

Results from the first bidding process indicate acceptance rates far below national and state expectations. Nationwide, the accepted bids ranged from $5 to $90 per acre, with an average bid of $41.82 per acre (U.S. Department of Agriculture, 1986). Of the approximately 5 million acres bid nationally, only 838,000 acres, or 17.4 percent, were accepted. The direct annual cost is $35 million for the 838,000 acres in the program. By contrast, only 3.6 percent of the 834,000 acres bid in Texas were accepted into the program. The average accepted bid in Texas was $30.84 per acre. Possible explanations for the high bids and low acceptance rate in Texas, particularly the High Plains, are the lucrative economics and base restrictions affiliated with the current federal crop programs, uncertainty about the time needed to establish a grass cover, and the conservation practices required over time to comply with CRP participation. However, as the federal crop program benefits are reduced in the future, participation in the CRP is expected to increase in the High Plains.

Part of the CRP is a conservation easement provision stating that the Secretary of Agriculture may acquire easements in wetland, upland, and highly erodible lands for recreational, wildlife, and conservation purposes for up to 50 years. Easement in this case would cancel part of the debt when a farmer's land is collateral for a Farmers Home Administration loan. Little to no activity is expected in the Texas High Plains relative to the conservation easement provision.

Essentially, a decision has been made that the benefits of the CRP program exceed the costs. A major issue relates to the effect of the CRP on small communities and local and regional economies owing to a possible large and rapid displacement. Relative to erosion control, this program will be effective for the enrolled acres. However, a large percentage of the land is still vulnerable to wind erosion. One drawback of the voluntary land retirement program is that federal subsidies are not targeted to the most erosive lands; therefore, it can be argued that this program may not be cost efficient.

Cost-sharing programs

Federal cost-sharing programs for water and soil conservation have existed for many years. Cost sharing is integrated into many other government programs. However, the concept and opportunities deserve separate discussion. Cost sharing can include a government agency's paying part of an investment or supplying a low-interest loan to a producer for investment in conservation.

Much of the federal soil conservation money in the southern High Plains and indeed in the Great Plains has been spent for installation of irrigation systems. Because they are more efficient, these systems may sustain moisture for a longer time, and moist soil is less vulnerable than dry soil to wind erosion (Sheridan, 1981).

Additional soil conservation cost sharing includes expenditures for terracing, irrigation ditches, drainage, and other land-shaping and structural practices. One criticism of the cost-sharing program is that federal dollars spent on structural practices are capitalized into the value of the land. As a result, income from taxpayers in general is transferred to producers who are able to acquire land-shaping assistance.

In the sandy soils of the southern High Plains, a soil-conserving practice is deep plowing. This brings up tighter soils with more clay and reduces susceptibility to wind erosion. In recent years there has been a marked reduction in deep plowing owing to diminished farm profits and limited cost-sharing funds available per farmer. Unfortunately, deep plowing is only a temporary means to reduce wind erosion (Sheridan, 1981). Thus, effective use of deep plowing for wind erosion control requires an enhanced federal program of cost sharing.

The argument for government cost sharing of deep plowing for wind erosion control extends to many farm practices. The issue relates to the effectiveness of an activity in achieving soil and water conservation and the economic incentives required to induce adoption by the producer. These opportunities can be dramatically different. Consider the following as a possible example. Monoculture cotton is the most profitable crop on the southern High Plains, and its production is associated with high wind erosion. However, a potential on-farm production practice that permits monoculture cotton while controlling wind erosion is being examined: planting wheat in the late fall or early winter after the cotton harvest. The wheat provides an effective ground cover. In the late winter or early spring, the young wheat is sprayed aerially with an EPA-labeled herbicide. Because this method has not been conventional practice over large areas, there are environmental questions about the fate and impacts of the herbicide. At any rate, the ground cover is established and is left undisturbed except to plant the cotton. While the cotton is emerging, the wheat residue provides ground cover to stop wind and water erosion, captures rainwater for the cotton, and protects the young cotton from blowing dust.

Perhaps it is not profitable for the producer to integrate the wheat into the program owing to the costs of the seed, the planting, and the herbicide application. With a small per-acre cost-sharing program, the incentive to participate could be provided and social gains realized from the wheat. This case is only an illustration because the on-farm activity has not been tested and the economics are not available. However, it clearly demonstrates the potential of cost sharing to achieve social goals.

Another form of cost sharing is providing low-interest loans to farmers to make soil or water conservation investments. This alternative is simply another way for a government agency to help pay for conservation. An example is again useful relative to the impacts of a low-interest loan program on the purchase of water-conserving equipment.

The Ogallala will be exhausted, physically or economically. This reality is a serious threat to the economics throughout the Great Plains. With this point in mind, legislation was drafted in Texas that would provide low-interest loans to farmers to purchase "water-conserving irrigation equipment" or make other water-conserving investments. The low-interest loans could be used to purchase distribution systems that are more technically efficient by improving distribution efficiency or using lower water pressure. Thus, less water and much less energy could produce the same level of crop yield. With projections of reduced water requirements for average crop yields, an extended life of the aquifer was estimated.

The legislation was based on a concept of water conservation that meant less water pumped each year, thus extending the years of irrigation. However, reducing the amount of water needed for a given crop yield (increased water-use efficiency) or increasing the yield using the same amount of water lowered the variable cost per unit of water. Repair, maintenance, and labor are approximately the same per hour, but improved distribution efficiency reduces water losses; hence, the variable costs per acre-inch of effective water (water used by the crop) delivered to the crop are essentially reduced. Further, the reduced pressure requirements simply mean less work by the engine; thus, lowering energy costs per acre-inch of water provides an incentive to use more water per year, not less. Investment in a new system, once made, becomes a fixed or sunk cost, and although important in the calculation of overall profit, it is not a factor in developing annual cropping patterns. The variable costs are the drivers in annual production decisions. Essentially, owing to limited pumping capacity in conjunction with water-critical growing periods, annual water use would be expected to remain about the same with or without the new systems. The new technology simply increases the effectiveness of the water through improved timing of application as well as through providing opportunities to shift to slightly more water-intense crops.

With the improved efficiency of water use and lower energy use, some adjustment in cropping patterns would be expected, and farmer net returns would increase substantially throughout the region (Lacewell, Ellis, and Griffin, 1985). With adoption of minimum tillage and other new technologies over time, the value of farmer net returns in 2020 is 48 percent higher than it would be with traditional technologies.

Thus, annual water use is expected to be similar to that with no subsidy, but a large increase in production would be experienced with the new technology. This benefit to the farmers increases the output from available water. However, research results do not support the contention that state-supported low-interest loans for farmers to purchase more efficient irrigation equipment will extend the life of the Ogallala. To eliminate the potential of farmers' expanding their irrigated acres with the low-interest loans, the authors considered the concept of coupling restricted irrigated acres to the loan. The notion is unpopular politically, and the potential cost of enforcement is high.

An issue beyond farmers' benefits of state-supported low-interest loans for water-conserving equipment is the source of funds for the program. If general revenue is used, then sales tax and other state revenue sources must provide the funds. Taxpayers at all income levels then share in the cost of a program that primarily benefits only one sector of the state. Further, state taxpayers would be supporting a program of subsidized loans to farmers to produce more crops that are already in surplus. Lastly, if the new irrigation systems are profitable, economic incentives exist for farmers to purchase them without subsidized interest rates.

In summary, cost sharing is among the strongest voluntary incentives to encourage soil and water conservation investments and practices. It is an effective way to accelerate use of new management practices and technologies. For the Texas High Plains' transition from irrigated to dryland agriculture, cost sharing can have a major role in encouraging investments and practices that are most effective in controlling erosion and that maximize use of rainwater. The principal issue is discerning which investments and practices are effective in controlling wind erosion, how much cost sharing is necessary to achieve adoption, and whether that cost is acceptable relative to the erosion control obtained. This requires an analysis on each type of investment and practice.

The federal crop program

The federal commodity program is an important factor in crop production on the southern High Plains of Texas. Essentially all cotton producers are participants in the program. Major components to be considered are crop price supports, acreage diversion, loans, and crop disaster insurance.

Farm programs make crop production activities more attractive. They raise commodity prices and reduce risk and production costs (Osteen, 1985), and they provide economic incentives to crop the more sensitive soils such as the sandylands in the southern High Plains. Cotton and other program crops are more erosive than grassland; hence, to the extent that farmers produce cotton because of program benefits, wind erosion is increased. Further, because commodity programs deny base acreage status, which is the basis of the farm program, to land not recently used to produce a crop, the farmer who adopts long-term conservation measures is penalized by the ineligibility of this land for commodity benefits (Reichelderfer, 1985).

Acreage diversion is a part of the requirement for government program eligibility. Diverted cropland requires a cover crop. This land is thereby protected from erosion and captures the rainwater; it may be more productive after diversion. However, diverted acres are only a small percentage of total cropland.

Policies to stabilize crop prices and assure adequate farm income may result in crop production on the more erosive soils of the southern High Plains. Price and income policies are sometimes inconsistent with soil and water conservation policies (Reichelderfer, 1985). This conflict can be resolved by modifying provisions of the crop programs to make them consistent with conservation goals; for example, crops grown on erosive soils without conservation protection would not be eligible for farm program benefits. Conservation protection may include deep plowing; wheat cover on cotton land, as described in the cost-sharing section; and other activities. The 1985 farm program provides for cross-compliance or consistency between crop programs and conservation policies. Cross-compliance between soil conservation and crop program benefits are a greater incentive for practicing soil conservation (Grano, 1985). Given the provisions of the 1985 farm program requiring target price and support price of crops to be systematically reduced, the continuing debate on reducing farmer deficiency payments from $50,000 to $25,000, Gramm-Rudman-Hollings federal deficit provisions, and cross-compliance, much of the economic incentive provided by the farm program may be reduced. These potential changes in farm programs over the next few years will have dramatic implications for dryland cotton production in the southern High Plains of Texas. Input levels will be reduced, cropland will be left idle in many of the dry years, and perhaps some land will even revert to grassland.

Cropping-pattern restrictions

Dryland crop production, particularly cotton on sandyland soils of the Texas High Plains, results in soil exposure to the elements for several months of the year. High levels of wind erosion are widely recognized. The dust from wind erosion is damaging long-term productivity of the land, is a threat to young field crops, and causes both nuisance and damage to others through air pollution.

Field dust is a form of nonpoint pollution and is subject to governmental regulation. Possible controls could be deletion of row crops and required revegetation, mandatory crop rotations that include high-residue crops such as wheat and oats, mandatory cropping and management practices, and limited frequency of cropping.

Deletion of field crops by government regulation would be effective in controlling wind erosion. However, it would also cause the loss of the cash crop that is the basis of the economy of most of the southern High Plains rural communities. This option is not politically feasible.

Mandatory crop rotation could be somewhat effective in controlling wind erosion. However, an analysis suggests that erosion is still about 50 tons per acre per year in a cotton-wheat rotation, compared to 70 tons for monoculture cotton. Thus, the soil loss would still be large, and the regulation would be difficult to enforce and not politically popular.

Mandatory cropping and management practices are aligned closely with the current cross-compliance provisions in the farm bill. It is a blend of the stick-and-carrot approach. For this discussion, a more stringent regulation encouraging adoption of cropping and management practices that result in significant conservation is assumed. It is possible that as the cross-compliance provision of the farm program begins to take effect, applicable conservation practices will be carefully defined and expanded. In addition, it is possible that compliance will be monitored, and with noncompliance crop program benefits will stop. Both are economically and politically feasible. The major issue relates to developing, defining, and promoting effective conservation practices for the region.

The last option relates to restrictions on how often fragile land in a semiarid region may be cropped. Australia and some other countries with a similar climate permit only one-third of the land to be cropped in any given year (Wendt, 1986). The land in noncropped years must be in a cover crop. Implications of this type of program are that soil erosion is reduced; the land is permitted to rest two out of three years, which can sustain productivity; and the land is better able to withstand frequent and lengthy droughts.

This policy option cannot be considered without strong evidence of its effects. Currently, it is not a viable policy option because its impacts on erosion and productivity are not known. The need to investigate is suggested, and if it satisfies the goals of significant erosion reduction, sustained productivity, and drought tolerance, policies can be developed to encourage adoption. Mandatory regulation of how often land may be cropped would not be politically popular. However, integrated into either the crop program or a long-term land retirement program, it could be an attractive alternative; it does not eliminate crops and thereby leaves a large part of the economic base of the region intact.

Regulating pumping

The policy options thus far have emphasized soil and water (rainfall) conservation. Continuing with regulation as a policy but adjusting to the pumping of the Ogallala, pumping restrictions or quotas represent an extreme. Regulating annual pumping rates could be accomplished by prohibiting new irrigation wells, establishing water quotas, limiting hours or days of pumping, prohibiting irrigation on sandy soils, and apportioning available water (Kromm and White, 1985).

If reduced pumping is an objective, the issue is what the appropriate or optimal annual rate of pumping is. The concept of balancing withdrawals to recharge rates is most popular for aquifers, that is, to control pumping so that there are sustained well yields and the aquifer is in equilibrium. In all cases, an aquifer will eventually reach equilibrium. The question is whether to mine groundwater for some period. Mining groundwater increases lift and reduces well yield, increasing the cost of pumping. Optimal use of groundwater over time is basically an economic question. It can be complicated by factors such as deterioration of quality and subsidence. However, the decision can be placed in an economic decision framework.

Consider, for example, a region of the Ogallala where recharge is negligible. Annual water withdrawals could be limited to the rate of recharge and that quantity of water would theoretically be available into perpetuity. Alternatively, the water can be mined over some period and withdrawals reduced to the rate of recharge. With mining, however, although the same annual supply is available after equilibrium is reached, it must be lifted from a greater depth by smaller wells. The optimal rate of water use or rate of mining from an aquifer then depends upon the value of the water in use in each time period, the cost to pump, interest rates or the time preference for money, impacts on the quality of the water in the aquifer, and the effects on neighbors or externalities. Given all these factors, some aquifers are now at equilibrium or are pumped at the rate of recharge. However, others justify some mining. Only with a zero interest rate (time preference for money) could an argument be made in absolute terms for pumping only at the rate of recharge. This statement is not to minimize the complexities surrounding the pumping of groundwater, including impact on neighbors, legal and institutional barriers, and environmental impacts.

Farmers using groundwater for irrigation must pay the cost of pumping and distributing the water across a field. In most cases groundwater users pay more for their water than surface water users do, and consequently they have an economic incentive to use less water per unit of land (Frederick and Hanson, 1982). Of course, water rights and other institutional factors are involved. For the Northern and Southern plains (Ogallala region) and the Mountain and Pacific regions of the West, there are gross estimates of water application rates per acre of irrigated land. Based on 1975 data, about 1.4 acre-feet of water was applied to each acre of irrigated land across the Great Plains (Ogallala region). In the Mountain and Pacific regions, which rely primarily on surface water, the figure was 3.8 acre-feet (Frederick and Hanson, 1982). Although these values are somewhat gross, they are an indication of the effects on water use when farmers pay the full cost of pumping and distributing groundwater and when surface water irrigators receive water at a much lower cost.

If the groundwater is in a common pool used by many farmers and belongs to no one until it is withdrawn for use, it may be used at an inefficiently rapid rate. This statement is true only if the farmer does not incur the total cost of using the groundwater; that is, the loss is shared with neighbors and future users. An accelerated and rapid use of water, in addition to exceeding the socially desirable rate of use from an aquifer, can result in serious long-term quality impacts. Thus, maintenance of groundwater quality is also a critical issue.

The arguments for controlling pumping from the Ogallala are typically based on the concept of a common property resource. Nieswiadomy (1985) investigated the common property issue and the concept of inefficient overutilization for a seven-county area of the southern Texas High Plains. The Gisser-Sanchez rule states that if the natural recharge and the slope of the demand curve for groundwater are small relative to the area of the aquifer times storability, the welfare losses from the intertemporal misallocation of pumping effort are negligible. The analysis indicates that an optimal temporal use of water from the Ogallala differs little from the common property use in the Texas High Plains.

Another analysis of the Ogallala concludes that if aquifer withdrawals are arbitrarily reduced, reaching a 30 percent reduction by the year 2000, the farmers' returns to land and management would be reduced 3 percent annually in the early years and 8 percent by 2020 compared to no mandatory controls (High Plains Associates, 1982). These two studies provide evidence that the Ogallala is not being exploited as a common property resource and that controlling annual pumping is an unwise economic decision. Further, groundwater pumping regulations are politically unpopular for the Texas government. Unquestionably, controlling annual pumping rates below those that farmers would select with no regulation would extend the economic life of the aquifer. However, the evidence indicates that this action is not justified economically.

Because there is evidence that the Ogallala for the most part does not act as a common property resource, market mechanisms with minimal interference are preferred (Beattie, 1981; Nieswiadomy, 1985). Groundwater in Texas is the property of the landowner and may be bought and sold. Cities have purchased the water beneath the land; hence, water is transferred from irrigation to municipal use.

There are strong justifications for not interfering with the market system that exists for Ogallala water on the Texas High Plains. Regardless of the above limitations and reasons, a water market has not emerged for many areas of the Ogallala; economists have no doubt that an organized way of trading or exchanging rights to water would increase the net social value of water. Without a market, a governmental entity assumes water allocation responsibilities. Priorities for use of limited water supplies are then established by the political or legislative system rather than the market system. It can be argued that, historically, government intervention has failed, particularly when trying to correct a market failure. Thus, the market system should be given an opportunity to provide signals for water allocation and transfers while certainly including safeguards to protect third parties and considering the unique characteristics of water (Office of Technology Assessment, 1983). Because the market system already exists in Texas, the argument raises a serious question about changing allocation of water from the Ogallala to one of governmental regulation.

Water tax

If water is being pumped at a rate exceeding the socially desirable rate, the farmer is not incurring the full cost associated with mining the aquifer. Thus, another alternative that would reduce annual withdrawals is to estimate the social costs of each acre-inch of water pumped and add a tax that would rectify the externality. Applied to groundwater withdrawals, the tax provides an economic incentive to use less water. Limitations of the tax option are quantifying an appropriate tax and policing action.

Ellis et al. (1983) evaluated the expected effect of a groundwater tax. A per acre-foot tax on water pumped of zero to $20 per acre-foot would not significantly reduce annual pumping in the southern High Plains of Texas.

Owing to past years of groundwater mining in the southern High Plains, remaining groundwater is quite limited. Thus, water is the most limiting resource in this area. With the relatively small amount of groundwater that can be pumped over several days and the large areas that need it, not all acres are irrigated during the water-critical periods of the crop. For these limited irrigation situations, the marginal value product of water is higher than the cost to pump owing to the limited period available for irrigating that corresponds to the plants' critical water period. However, net returns to irrigated farming were dramatically impacted. For the northern High Plains of Texas, a $20 per acre-foot tax on water pumped reduced withdrawals over a 40-year period approximately 5 percent. However, the present value of returns to water was reduced $1.7 billion in the region. For the farmer, the cost is about $400 for each acre-foot of water *not* pumped because of the tax. The study suggests that a tax to provide water conservation incentives is not economically justified. In addition, tax revenue collection and allocation have not been addressed.

Research and education

The government encourages soil conservation practices. Many soil conservation programs were discussed with the grass-establishment and deep-plowing programs. Other soil conservation practices include minimum tillage; crop rotation, including high organic matter crops like barley and wheat; strip-cropping; and windbreaks. Often these practices can work together, for example, minimum tillage incorporated with crop rotation. Crop rotation with small grains produces the residue that is desirable in minimum tillage systems. This residue can reduce wind erosion and conserve soil moisture, enhancing yield of the crop following the small grain. In addition, furrow dikes or row dams can capture the rainwater in the field and hold it for crop production.

Typically in the southern High Plains, government farm programs and inexperience with crop rotation and minimum tillage practices do not produce profits sufficient to compete with monoculture cotton. Developing new cropping systems that are economically competitive is the challenge for simultaneous wind erosion control and economically viable crop production. The need for research, development, and technology transfer is clear.

Research and education need to be strong components of regulation and government policy. In published literature the need for research and education is often expressed. For example, Strange (1983) indicates that: "soil conserving and drought-resistant grain crops should be explored," and "agricultural research should seek a sustainable future."

In his policy alternatives for the Ogallala, Beattie (1981) lists two of the most productive public-policy investment opportunities. The first was education, and the second was a stepped-up research-extension effort in developing and disseminating water- and energy-conserving technologies for both dryland and irrigated production systems. Grubb (1981) reiterated the need to inform irrigation and dryland farmers about existing and emerging water conservation technology. He suggests more emphasis on soil moisture monitoring, irrigation scheduling, testing system efficiency, and field demonstrations.

Similarly, an Office of Technology Assessment study (1983) identifies as a major issue the need for an interdisciplinary program of basic and applied research on arid and semiarid water resources, including work on irrigated and dryland production. The Texas Agricultural Experiment Station considers the economic and resource problems of the Texas High Plains a major issue and has initiated a major research program in the area of "farming systems" to address these problems. According to Clarke, director of the Experiment Station (1980), "It is not business as usual for the Texas High Plains." Major programs in drought-resistant cultivars, new innovative and effective crop rotations, new irrigation technology, minimum tillage opportunities, and integration of all sciences into irrigated and dryland farming systems have been initiated. Thus, the partnership of research and development of new technology and optimal farming systems with policies that economically and effectively promote water and soil conservation provide both adjustments and alternatives for the producer and preservation of resources.

Summary

The southern High Plains of Texas is in a transition from irrigated to dryland agriculture. Serious resource and economic implications are associated with both the transition and the final dryland production system. A smooth transition over many years permits orderly adjustments in local and regional economies, providing an opportunity for approaching dryland production slowly while identifying and adopting optional management strategies. However, the farm commodity program provides strong incentives to continue irrigation at rates sufficient to maintain crop yields under established cropping patterns; that is, it encourages water use, not water conservation. For much of the region, monoculture cotton, which is associated with high rates of wind erosion, is the primary crop and has high enrollment rates in the farm program. Further, the farm program is so lucrative that there is only limited interest in the conservation reserve program.

Government policy has not been designed to minimize the adverse effects of a transition from irrigated to dryland production but, rather, will perhaps enhance these negative effects. Benefits of the commodity program may be reduced each year of the program; abiding with the deficit reduction provisions of Gramm-Rudman-Hollings may cause an even faster reduction in benefits. The result could be farmers' moving from a significant level of irrigation to essentially dryland production in a few short years. Furthermore, with commodity program benefits reduced, the conservation reserve program becomes an attractive alternative. It takes land from irrigated production to an idle state. Satisfying the provisions of the conservation reserve program assures both water and soil conservation. However, early indications suggest that the cost to the federal treasury may be large. Further, repercussions on economies of local communities can be expected to be severe because a gradual transition has been changed to an erratic and sharp one.

The commodity program needs to provide for an orderly transition to dryland and to allow for needed changes in management and cropping practices. An effective dryland production system would provide for wind erosion control yet be economically viable for the farmer. Thus, viable short-term alternatives for soil and water conservation on the Texas High Plains relate to modifications in the commodity program to permit cropping pattern changes without severe penalties and adoption of new irrigation technology and improved management practices.

There are regulatory options that would certainly promote soil and water conservation. For example, Australia permits only one-third of the land in the semiarid region to be cropped in any specific year. This option provides an opportunity to maintain soil productivity and control erosion. Again, there are serious local and regional economic issues related to reducing cropped acres nearly two-thirds. Water conservation in the form of reduced annual pumping from the Ogallala could be accomplished by pumping allocations or a water tax. The justification for such regulation would be that the Ogallala is a common property resource. Research suggests that regulation will not increase the present value of the water supply and may indeed reduce it.

These findings imply that policy needs for resource conservation and orderly economic adjustment on the Texas High Plains include flexibility in commodity programs for cropping pattern changes, fine tuning of cost-sharing programs, targeting the conservation reserve program to maximize erosion control per dollar of federal expenditure, and research and education to develop new technology and improve farming systems and to transfer information to the farmers.

References

Banks, Harvey O., Williams, Jean O., and Harris, Joe B. 1984. "Developing New Water Supplies." In Ernest A. Engelbert and Ann Foley Sheuring, eds., *Water Scarcity Impacts on Western Agriculture*. Berkeley, Los Angeles, and London: University of California Press.

Beattie, Bruce R. 1981. "Irrigated Agriculture and the Great Plains: Problems and Policy Alternatives." *Western Journal of Agricultural Economics*, *7*(2), 289–299.

Beattie, Bruce R., Frank, Michael D., and Lacewell, Ronald D. 1978. "The Economic Value of Water in the Western United States." Proceedings of a conference on Legal, Institutional, and Social Aspects of Irrigation and Drainage and Water Resource Planning and Management. New York: American Society of Civil Engineers.

Brown, Lester R., et al. 1984. *State of the World*. New York: W.W. Norton and Company.

Clark, R.N. 1979. "Furrow Dams for Conserving Rainwater." Presented at the Crop Production and Utilization Symposium, Texas A&M Research and Extension Center, Amarillo, Texas, Feb. 22.

Clarke, Neville P. 1980. *Texas Agriculture in the 80's: The Critical Decade*. Texas Agricultural Experiment Station B-1341. Texas A&M University.

Clary, G.M., Dietrich, R.A., and Farris, D.E. 1984. "Interregional Competition in the U.S. Cattle Feeding Fed Beef Economy with Emphasis on the Southern Plains." Texas Agricultural Experiment Station B-1487. Texas A&M University.

Condra, Gary D. 1984. "The Transition from Irrigated to Dryland on the Southern High Plains of Texas." Unpublished final contract report to the Texas Agricultural Experiment Station.

Cordonier, Melissa. 1985. "Are Dust Bowl Days a Threat Now?" *Southwest Farm Press*, *12*(20).

Crop Insurance Actuarial Association. 1947–1973. "Crop Insurance Analysis Sheet: Cause of Damage." Selected counties, Texas High Plains.

Elliott, Dennis L. 1977. "Synthesis of National Wind Energy Assessments." Battelle Pacific Northwest Laboratories BNWL-2220, WIND-5, Richland, Washington.

Ellis, John R., Lacewell, Ronald D., and Reneau, Duane R. 1985. "Economic Implications of Water-Related Technologies for Agriculture: Texas High Plains." Texas Agricultural Experiment Station MP-1577. Texas A&M University.

Ellis, John R., et al. 1983. "Pricing and Conservation of Irrigation Water in Texas and New Mexico." Texas Water Resources Institute TR-125.

Frederick, Kenneth D., and Hanson, James C. 1982. *Water for Western Agriculture.* Washington, D.C.: Resources for the Future.

Godfrey, C.L., Carter, C.R., and McKee, G.S. 1967. "Land Resource Areas of Texas." Texas Agricultural Experiment Station B-1070. Texas A&M University.

Grano, Anthony. 1985. "Analysis of Policies to Conserve Soil and Reduce Surplus Crop Production." U.S. Department of Agriculture, Economic Research Service, Agricultural Economics Report 534. Washington, D.C.

Greb, B.W., Smika, D.E., and Black, A.L. 1967. "Effect of Straw Mulch Rates on Soil Water Storage During Summer Fallow in the Great Plains." *Soil Science Society of America, 31*, 556–559.

Green, Donald E. 1973. *Land of the Underground Rain.* Austin and London: University of Texas Press.

Grubb, Herbert W. 1981. "Water Issues in Texas Agriculture." Texas Agricultural Experiment Station MP-1485.

Heimes, Frederick J., and Luckey, Richard R. 1982. "Method for Estimating Historical Irrigation Requirements from Groundwater in the High Plains in Parts of Colorado, Kansas, Nebraska, New Mexico, Oklahoma, South Dakota, Texas, and Wyoming." U.S. Geological Survey, Water Resources Investigations 82–40.

High Plains Associates. 1982. "Six State High Plains Ogallala Aquifer Regional Resources Study." Final report to the U.S. Department of Commerce.

Hughes, William F., and Magee, A.C. 1960. "Some Economic Effects of Adjusting to a Changing Water Supply, Texas High Plains." Texas Agricultural Experiment Station B-966.

Huszar, Paul C. 1985. "Dusting Off the Sodbuster Issue." Unpublished. Colorado State University.

Huszar, Paul C., and Young, John E. 1984. "Why the Great Colorado Plowout." *Journal of Soil and Water Conservation, 39*(4), 232–234.

Johnson, W.C., and Davis, R.G. 1972. "Research on Stubble-Mulch Farming of Winter Wheat." U.S. Department of Agriculture, Conservation Research Report 16.

Jones, O.R., Unger, P.W., and Fryrear, D.W. 1985. "Agricultural Technology and Conservation in the Southern High Plains." *Journal of Soil and Water Conservation, 40*(2), 195–198.

Knowles, Tommy R. 1984. "Assessment of the Ground-Water Resources of the Texas High Plains." In George A. Whetstone, ed., Proceedings of the Ogallala Aquifer Symposium II, Texas Tech University, Lubbock, Texas.

Kromm, David E., and White, Stephen E. 1985. "Conserving the Ogallala: What Next?" Kansas State University, Department of Geography, Manhattan, Kansas.

Lacewell, Ronald D., and McGrann, James M. 1982. "Research and Extension Issues in Production Economics." *Southern Journal of Agricultural Economics, 14*, 65–74.

Lacewell, Ronald D., Ellis, John R., and Griffin, Ronald C. 1985. "Economic Efficiency Implications of Changing Groundwater Use Patterns." In E.T. Smerdon and W.R. Jordan, eds., *Issues in Groundwater Management.* University of Texas, Center for Research in Water Resources, Austin.

Lee, John G., Ellis, John R., and Lacewell, Ronald D. 1985a. "Valuation of Improved Irrigation Efficiency from an Exhaustible Groundwater Source." *Water Resources Bulletin, 21*(3), 441–447.

Lee, John G., et al. 1985b. "Implications of Dryland Crop Rotations on Profits and Wind Erosion in a Semi-Arid Environment." Presented at the Western Agricultural Economics Association annual meeting, Saskatoon, Saskatchewan, Canada.

Luckey, Richard R., Gutentag, Edwin D., and Weeks, John B. 1981. "Water-Level and Saturated Thickness Changes, Predevelopment to 1980, in the High Plains Aquifer in Parts of Colorado, Kansas, Nebraska, New Mexico, Oklahoma, Texas, and Wyoming." U.S. Geological Survey, Hydrologic Investigation Atlas HA-652.

Lyle, William M., and Bordovsky, James P. 1980. "New Irrigation System Design for Maximizing Irrigation Efficiency and Increased Rainfall Utilization." Texas Water Resources Institute TR-105, Austin, Texas.

Lyle, William M., and Bordovsky, James P. 1986. "LEPA vs. Drip Irrigation on the Texas High Plains." In Proceedings of Drip Irrigation Cotton Symposium, Midland, Texas, February 18–19. College Station, Texas: Texas Agricultural Extension Service.

Lyle, W.M., and Dixon, D.R. 1977. "Basin Tillage for Rainfall Retention." *Transactions of A.S.A.E., 20*, 1013–1021.

Lyle, W.M., et al. 1981. "Evaluation of Low Energy, Precision Application (LEPA) Irrigation Method at the Texas Agricultural Experiment Station, Halfway, Texas." Annual progress report.

Lyle, W.M., et al. 1982. "Water-Related Technologies for Sustained Agriculture in the U.S. Arid and Semi-Arid Lands." U.S. Congress, Office of Technology Assessment, draft.

Nieswiadomy, Michael L. 1985. "The Demand for Irrigation Water in the High Plains of Texas: 1957–1980." *American Journal of Agricultural Economics, 67*(3), 619–626.

Osteen, Craig D. 1985. "The Impacts of Farm Policies on Soil Erosion: A Problem Definition Paper." U.S. Department of Agriculture, Economic Research Service, Natural Resource Economics Division, AGE584109.

Reichelderfer, Katherine H. 1985. "Do U.S.D.A. Farm Program Participants Contribute to Soil Erosion?" U.S. Department of Agriculture, Economic Research Service, Agricultural Economic Report 532.

Runkles, J.R. 1980. "Critical Issues in the 80's: The Critical Decade." Texas Agricultural Experiment Station B-1341. Texas A&M University.

Sheridan, David. 1981. *Desertification of the United States*. Washington, D.C.: U.S. Government Printing Office.

Smith, Jackie. 1986. "A Brief Examination of the Donald Love Cotton Drip Irrigation Project." In Proceedings on Drip Irrigation Cotton Symposium, Midland, Texas, February 18–19. College Station, Texas: Texas Agricultural Extension Service.

Stewart, B.A., and Harman, W.L. 1984. "Environmental Impacts." In Ernest A. Engelbert and Ann Foley Sheuring, eds., *Water Scarcity: Impacts on Western Agriculture*. Berkeley, Los Angeles, and London: University of California Press.

Strange, Marty. 1983. "Of Whooping Cranes and Family Farms: Another Look at the High Plains Study." *Journal of Soil and Water Conservation, 38*(1), 28–32.

Texas Almanac and State Industrial Guide 1982–1983. 1983. Dallas, Texas: A.H. Belo Corporation.

Texas Crop and Livestock Reporting Service. 1971–1983. "Texas Agricultural Crop Acreages, Production, Cash Receipts, Prices Received and Paid by Farmers." U.S. Department of Agriculture, Austin, Texas.

Texas Department of Water Resources. 1981. "Inventories of Irrigation in Texas 1958, 1964, 1969, 1974, and 1979." Report 263.

U.S. Congress, Office of Technology Assessment. 1983. "Water-Related Technologies for Sustainable Agriculture in U.S. Arid/Semiarid Lands." Washington, D.C.: U.S. Government Printing Office.

U.S. Department of Agriculture. 1986. "News – Office of Information, USDA Accepts 838,356 Acres for Conservation Reserve Program." Washington, D.C., Mar. 28.

U.S. Department of Commerce, Bureau of the Census. 1949–1982, selected issues. *Census of Agriculture*.

Unger, Paul W., and Box, John. 1972. "Technical Problems Associated with Minimum Tillage." Presented at the Cooperative Conservation Workshop, College Station, Texas, July 11.

Unger, Paul W., and McCalla, T.M. 1980. "Conservation Tillage Systems." *Advances in Agronomy, 33*, 1–58.

Unger, Paul W., and Weise, Allen F. 1979. "Managing Irrigated Wheat Residues for Water Storage and Subsequent Dryland Grain Sorghum Production." *Soil Science Society of America Journal, 43*(3).

Wendt, Charles. 1986. Texas Agricultural Experiment Station, Lubbock, Texas. Personal communication.

Williford, George H., Beattie, Bruce, and Lacewell, Ronald D. 1976. "Effects of a Declining Groundwater Supply in the Northern Plains of Oklahoma and Texas on Community Service Expenditures." Texas Water Resources Institute TR-71. Texas A&M University.

Wyatt, A. Wayne. 1984. "Secondary Recovery of Capillary Water." In George A. Whetstone, ed., Proceedings of the Ogallala Aquifer Symposium II. Texas Tech University, Lubbock, Texas.

Zingg, A.W., and Whitfield, C.J. 1957. U.S. Department of Agriculture. Technical Bulletin 116.

5 Water resources of the Upper Colorado River Basin: problems and policy alternatives

CHARLES W. HOWE AND
W. ASHLEY AHRENS

Background*

The Colorado River is the major surface water resource of the Southwest. In spite of John Wesley Powell's forecast that the region would never be useful or inhabited, the river basin has been fought over and romanticized more than any western river. Certainly, it has been the subject of more writing – from geomorphology to politics – than any other western river. Nadeau (1950) wrote of the Owens Valley controversy and of the heroic attempts to conquer the Lower Colorado River. Terrell (1965) described in detail the political battle between Arizona and California over the waters of the Lower Basin, and Fradkin (1968) touchingly described the great changes in the river that have come about through human attempts to control and harness its forces. Throughout this history, the influence of the financial power and the concentration of political power on water issues have dominated the policy scene, and since passage of the Reclamation Act of 1902, the federal government through the Bureau of Reclamation has been the agent of project construction and water supply provision.

The Upper Basin has grown more slowly than the Lower Basin – a typical pattern for development for river basins – generally owing to the superior climate, soils, and accessibility of the Lower Basin. Because the Colorado River's waters are produced primarily by snowmelt in the mountains of Colorado, Wyoming, and Utah, this uneven growth pattern has created a pattern of mutual fears between Upper and Lower basins, the former fearing that the latter would establish title through

*The physical description of the basin is taken mostly from Howe, 1977. The institutional aspects are drawn from Howe and Murphy, 1981.

early use, the latter fearing the former might eventually develop to a point at which it would be using much of the river's water. This competition is still present, sometimes explicit, sometimes under the surface, and it helps explain much of the current political and legal maneuvering in the basin.

The Upper Basin is perceived as a water-short area, and it appears to be in terms of low rainfall and runoff: 10 inches per year and 124 acre-feet per square mile per year. Yet its water supply is not really lowland precipitation and runoff, but the melting mountain snowpack, a supply that is widely distributed through the region in natural streams and man-made distribution systems. According to the U.S. Geological Survey, more than 70 percent of all man-made diversions of this supply are directed toward agriculture, and more than 90 percent of all consumptive uses occur in agriculture. The prospects of large-scale energy development that dominated water and environmental concerns a decade ago (for example, Spofford, Parker, and Kneese, 1980) have now largely faded from view. The projections of water use and population have fallen in keeping with the slower growth. Projected agricultural water use needs to be reduced to reflect the oversupply of agricultural commodities nationally and internationally, combined with an increasing reluctance of the federal government to subsidize new irrigation water supply.

It is a major contention of this study that the Upper Basin is not and will not be short of water if the states of the basin use their supplies in an economically reasonable way. Changing values call for greater protection of instream flows, and the high costs of developing new supplies for municipal and industrial uses indicate the desirability of transferring water from agriculture to urban areas. Only extreme shortsightedness and a thoughtless scramble to put all their water to use quickly could lead to a future water crisis.

The methods of water allocation used and the development of new uses take place within an institutional framework consisting of interstate compacts, state constitutional and legal provisions, public water agencies, and supervised water markets. This institutional framework that has developed over the past century is now outdated in some important respects that interfere with economically reasonable use of water in the face of changing values and demands. Water rights and other forms of claims to water can be exchanged, but often only within a limited geographical area and at high transactions costs. Institutional reform, perhaps assisted by new information technologies, can greatly improve the effectiveness with which water is used in the Upper Basin. Water markets, both intrastate and interstate, can play a much larger role in this process, though there will be conflicts between what individual water owners want to do and what their state is willing to approve.

Water management in the Upper Basin cannot be separated from land use and management. Forest areas constitute the main snowsheds of the Upper Basin, so forest management practices affect water yields, erosion, and water quality. Agricultural practices affect soil erosion and siltation of streams, and saline irrigation return flows resulting from excessive irrigation applications constitute the major water quality problem of the Upper Basin. Unfortunately, water quantity is administered separately from water quality, with the result that water quality (salinity) improvement programs in the basin are needlessly costly. Again, institutional changes are needed.

The Upper Colorado River Basin has an area of approximately 102,000 square miles, located in southwestern Wyoming, western Colorado, eastern Utah, northwestern New Mexico, and northeastern Arizona. Figure 5.1 shows the entire Colorado River Basin, of which the Green, Upper Main Stem, and San Juan river basins comprise what is called the Upper Basin.

The Upper Basin is sparsely populated, with about 537,000 persons (1980 census), for an average of 5.3 persons per square mile, compared with a national average density of 57.4. The low density is primarily due to the mountainous terrain and the arid to semiarid climate of much of the remainder of the region. The population is expected to grow at a rate of approximately 1.4 percent per year (Spofford et al., 1980, table 4, scenario A, p. 227), thus rising to about 706,500 in the year 2000. There are no major cities in the basin; the main towns are Green River, Wyoming (population 8,000); Grand Junction (30,000) and Durango, Colorado (12,000); and Farmington, New Mexico (22,000). However, water is exported to the Denver metropolitan area in the east and is likely to be exported to the Salt Lake City area in the future.

The climate ranges from continuous snow cover and heavy precipitation on the western slopes of the Rocky Mountains to desert conditions in the south. Most of the moisture falls as winter snow; spring and summer rainfall is localized infrequent storm activity. Thus, water supply in the basin is dependent upon construction of dams to capture spring runoff and distribution systems to carry the water to points of use. There is little transfer of water among subbasins of the Upper Basin.

The major economic activities of the basin have traditionally been agriculture, cattle and sheep raising, and the mining of metallic miner-

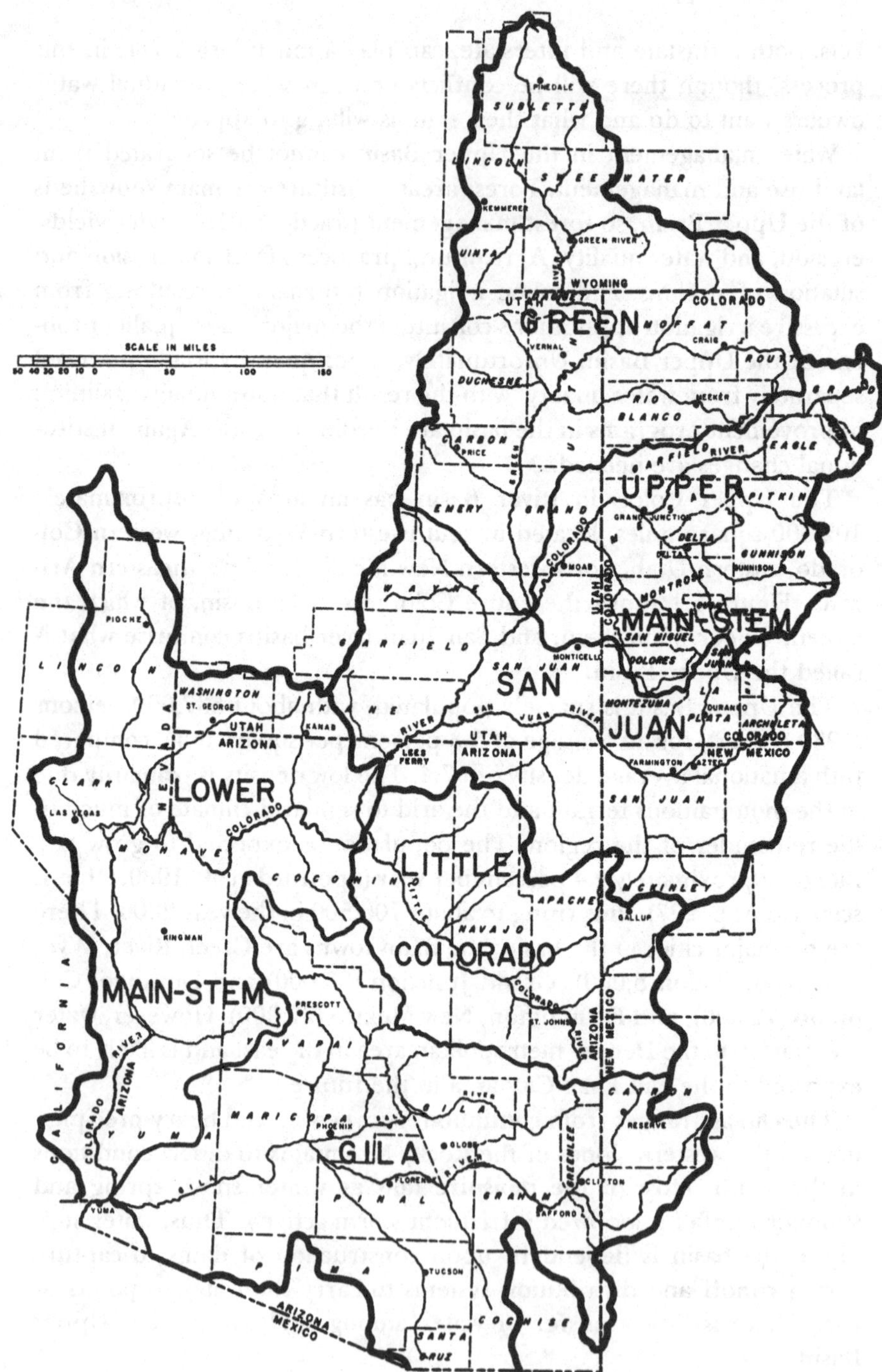

Figure 5.1. Major subbasins of the Colorado River Basin (U.S. Department of the Interior).

als. In the post-World War II period, recreational use of all parts of the basin has expanded tremendously, ranging from exclusive international skiing resorts like Aspen and Vail to mountain hiking, fishing, open river and reservoir boating, and desert area exploration. This broad recreational use is facilitated by the extensive federal land holdings that constitute 70 percent of the basin's area.

Throughout the 1970's, the huge coal and oil shale deposits that underlie much of the basin attracted development interest. From the mid-1970s until 1981, the federal government was intent upon development of these energy resources. The state governments of the area were less certain about the desirability of large-scale strip mining of coal, with associated power plants and the development of oil shale refining. Environmental problems could be critical, water requirements would be large, and interference with the growth of recreation and tourism could be severe. All energy development other than coal-fired thermal electric generation have now been abandoned.

The availability of water for further economic expansion of the basin is a major issue. At the time of the Colorado River Compact (ratified in 1929), average annual water availability was thought to be 15 million acre-feet (maf) per year. Since that time, estimates of long-term average flows have been decreasing; it is now felt that the average annual availability may be as low as 13.5 maf. In addition, a U.S.-Mexican treaty of 1944 calls for a guaranteed delivery of 1.5 maf per year to Mexico, and it is not clear how this obligation is to be divided between the Upper and Lower basins, although it has been declared a national obligation by more recent legislation. The Lake Powell Research Project estimated water available to the Upper Basin at 5.25 maf, and the U.S. Department of the Interior has used 5.8 maf, both figures dependent on the assumed long-term average virgin flow. It is probable that about 6.0 maf per year is legally available for consumptive use in the Upper Basin, with 2.5–3.0 maf now in consumptive use.

Certainly, Upper Basin demands will grow. Figure 5.2 presents demand projections commonly used during the energy boom of the mid- to late 1970s. The future energy use indicated on the demand side of Figure 5.2 has nearly disappeared with the collapse of the federal synthetic fuels program. It also seems unlikely that food and fiber consumption will increase at all, and it is likely to decrease. Annual exports from the basin to Colorado's Eastern Slope and the Wasatch Front in Utah will probably increase 400,000 acre-feet (Spofford et al., 1980, table 12, p. 393). Projected total consumptive uses for the year 2000 are thus likely to be around 4.2 maf, well short of either estimate of availability in

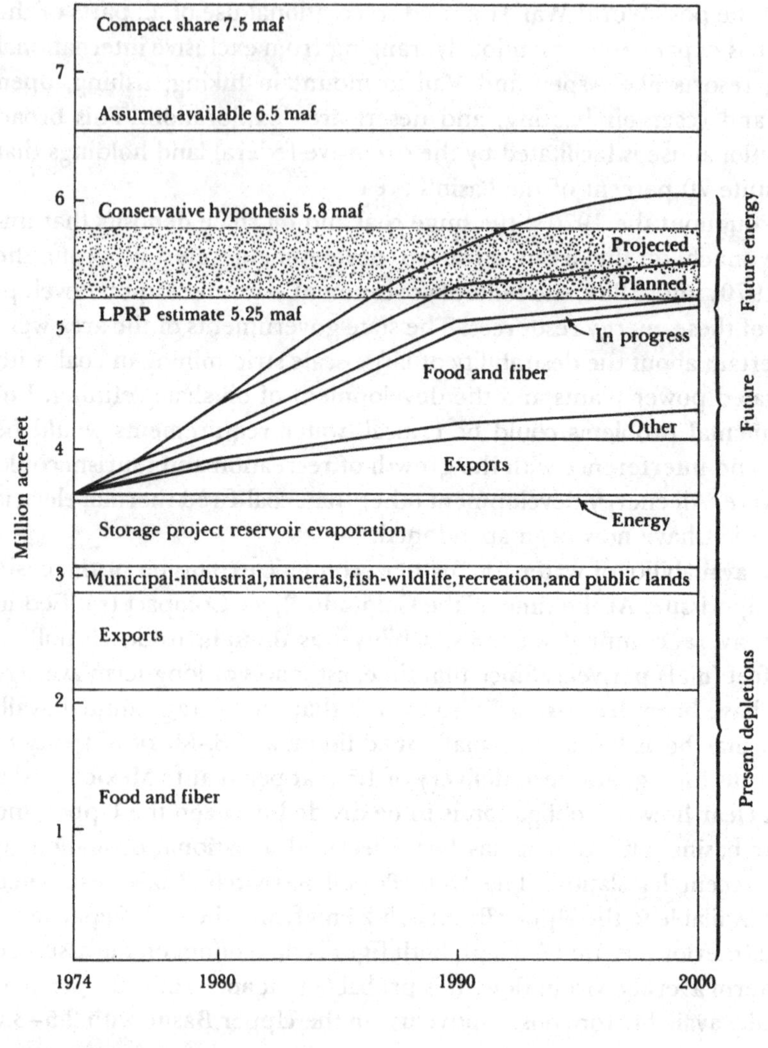

Figure 5.2. Surface water available for consumptive use, Upper Colorado River Basin (Weatherford and Jacoby, 1975, fig. 1, p. 186).

Figure 5.2. Naturally, these average figures do not preclude shortages during extreme drought or in local drainage basins.

The fact that the Upper Basin is not fully consuming the available water does not mean that the water is unused. The river's average flow is fully used, and indeed in a typical year no flow whatsoever reaches the river's original terminus, the Gulf of California.

As further development takes place in the Upper Basin, some of the current Lower Basin uses of water will have to be foregone. Southern California is a prime candidate to give up water use, because the state has a legal right to only 4.4 maf per year although it currently uses 5.2 maf per year. These reductions will not be without cost. It has been calculated that the direct opportunity cost of marginal withdrawals of water from agriculture in southern California ranges from $6.50 to $30 per acre-foot (Vaux and Howitt, 1984); regional income effects might range up to $200 per acre-foot, depending upon the availability of substitute commodities as inputs into the agricultural and food-processing industries and the mobility of resources outside agriculture (Howe and Easter, 1971, updated). Additional water demands in the Upper Basin could also be met through reductions in current water uses in the Upper Basin itself.

We now know that the virgin flows of the Colorado River have been highly variable. Figure 5.3 shows Stockton's reconstruction of 400 years of annual runoff measured at Lees Ferry, Arizona. The filtered series in Figure 5.3 exhibits significant persistence, that is, sequences of years of positive deviation from the long-term mean (about 13.5 maf per year) followed by sequences of negative deviations. Table 5.1 gives estimated average virgin flows at Lees Ferry for various periods.

It is clear from these data that above-average or below-average flow can persist for 10 to 20 years, that is, for significant parts of the intended lifetime of a large water project. Given the current impossibility of long-term climate prediction, it is difficult to specify any one number as the average flow to use for planning purposes. Rather, attention must be paid to the nature and range of climatological variability likely to be faced and to the flexibility of the system being planned.

Accompanying the growth of water use in the Upper Basin has been a deterioration of water quality in the form of a rising trend of total dissolved solids (TDS). Current TDS levels at Imperial Dam in the Lower Basin average 850 parts per million, and they are predicted to rise to 1,100 parts per million by the year 2000 in the absence of a more effective salinity control program (Bureau of Reclamation, 1983a). Current TDS levels are 59 percent attributable to natural point and diffuse sources, 30 percent to agricultural return flows, and 11 percent to municipal-industrial uses and out-of-basin transfers. A major program of salinity reduction is being planned and carried out by the Bureau of Reclamation. The program involves control of several major natural point sources and increased efficiency of water use in the agricultural sector in the Upper Basin. Some agricultural areas are shallowly under-

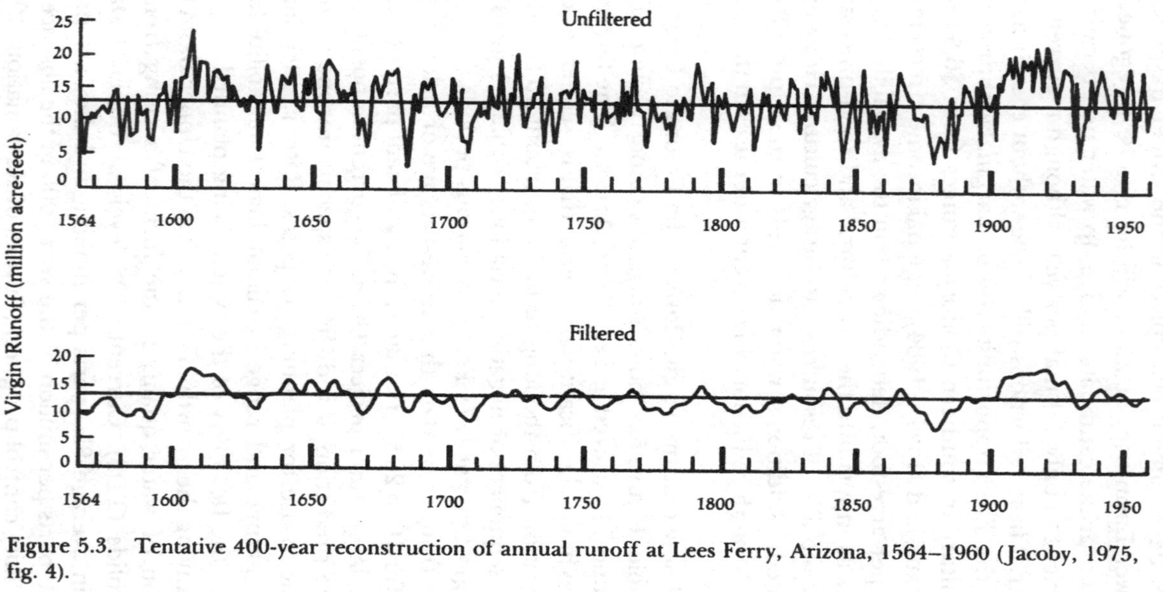

Figure 5.3. Tentative 400-year reconstruction of annual runoff at Lees Ferry, Arizona, 1564–1960 (Jacoby, 1975, fig. 4).

Table 5.1. *Estimated average annual virgin flows at Lees Ferry, Arizona*

Period	Flow (maf)	Remarks
1896–1968	14.8	Federal estimates
1896–1929	16.8	34-year wet period
1930–1968	13.0	38-year dry period
1914–1923	18.8	10-year wettest period
1931–1940	11.8	10-year driest period
1917	24.0	Greatest 1-year flow
1934	5.6	Smallest 1-year flow

Source: Dracup, 1977, p. 121.

lain by salt deposits and are estimated to contribute as much as 10 tons of salt per acre per year. Damages from increased salinity concentrations in the Lower Basin are estimated at approximately $493,000 per milligram per liter per year (Gardner, 1983). It takes approximately 10,000 tons of TDS in the Upper Basin to raise the concentration at Imperial Dam 1 milligram per liter per year, but salinity concentrations are also raised by consumptive uses and water exports. Thus, it is clear that further Upper Basin development will have significant water quality and quantity effects downstream.

Relationships among land forms, land uses, and water systems

Natural land forms, geological structures, land cover, and patterns of land use affect hydrologic patterns, including runoff, infiltration, wind and water erosion, sedimentation, dissolved solids, and other chemical properties of water. Land management measures can be utilized to improve hydrologic characteristics of the basin: agricultural practices can reduce erosion and salinity in return flows, forestry practices can affect snowpack and runoff, reseeding and reforestation of denuded areas can increase infiltration and reduce erosion, grazing controls can help maintain a healthy turf and reduce soil compaction, and controls over recreational activities can prevent excessive disruption of land cover. Management programs in the Upper Colorado Basin are facilitated by the fact that roughly 75 percent of the land area is public land (Upper Colorado Region Group, 1971). Figure 5.4 shows the approximate patterns of land ownership, administration, and vegetal cover, and Table 5.2 gives estimates of the total runoff, sediment, and salt loads from the public lands in Colorado, Wyoming, and Utah. Clearly, land management policies on the public lands are important factors in determining water quality and quantity in the Upper Basin.

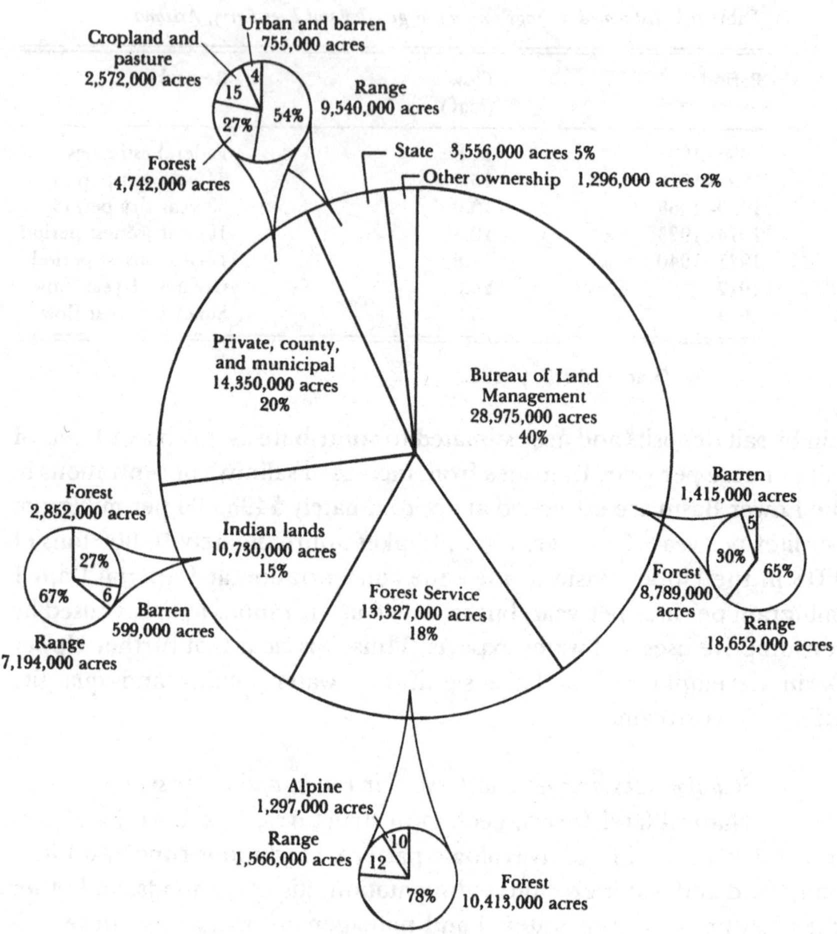

Figure 5.4. Land ownership, administration, and major vegetal cover types, Upper Colorado region (Upper Colorado Region Group, 1971, app. VIII).

Legal and institutional history of the Colorado Basin

The institutional framework for decision making in the Upper Basin consists of state law and a sequence of compacts and national laws that have evolved over the past 60 years to deal with perceived problems and development opportunities. The key compacts and federal laws, each of which significantly affects the management of Upper Basin water today, are listed and described below. Elaboration on this history is found in Hundley (1986).

The Colorado River Compact, 1922 (Meyers, 1966). The compact was ratified in 1922 by all states except Arizona. The major provi-

Table 5.2. *Summary of estimated total runoff, sediment, and salt produced from public lands in the Upper Basin, by state*

	Salinity class			
	Highly saline	Moderately saline	Slightly saline	Total
Colorado				
Runoff (acre-feet)	7,600	17,900	287,000	312,500
Sediment (tons/yr)	897,000	986,000	8,099,000	9,982,000
Salt (tons/yr)	34,400	19,400	113,000	166,800
Utah				
Runoff (acre-feet)	36,900	40,000	445,000	521,900
Sediment (tons/yr)	4,363,000	2,210,000	12,550,000	19,123,000
Salt (tons/yr)	167,000	43,600	176,000	386,600
Wyoming				
Runoff (acre-feet)	7,000	21,300	201,000	229,300
Sediment (tons/yr)	831,000	2,449,000	5,658,000	8,938,000
Salt (tons/yr)	31,900	33,300	79,300	144,500
Total of three states				
Runoff (acre-feet)	51,500	79,200	933,000	1,063,700
Sediment (tons/yr)	6,091,000	5,645,000	26,307,000	38,043,000
Salt (tons/yr)	233,300	96,300	368,300	697,900

Source: Bureau of Land Management, 1977, p. 14.

sions are to (1) define Lees Ferry, Arizona, as the dividing point between the Upper and Lower basins; (2) limit the Upper Basin to 7.5 maf of beneficial consumption use per year; (3) limit the Lower Basin to 8.5 maf of beneficial consumptive use per year; (4) require the release from the Upper Basin of at least 75 maf over every ten-year interval; (5) require the two basins to share equally any future Mexican delivery requirement not met by surplus waters; and (6) forbid the Upper Basin from withholding any water that could not reasonably be applied to domestic and agricultural use.

The Boulder Canyon Project Act, 1928. This act provided for the construction of Boulder (later Hoover) Dam for Lower Basin water supply, flood control, and electric generation. As *quid pro quo* for the Upper Basin, the act provided for the study of the development of Upper Basin water. The result was the *Krug Report* of 1946, which identified the projects included in the 1954 Colorado River Storage Project.

The treaty with Mexico, 1944. To resolve long-standing conflicts with Mexico and to effect President Roosevelt's Good Neighbor

Policy, a treaty was signed in 1944 that guaranteed Mexico a minimum of 1.5 maf annually. Significantly, the treaty did not cover water quality.

The Upper Basin Compact of 1948. As noted in the *Krug Report*, the federal government felt it important that interstate divisions be clarified to facilitate long-term planning. Although this order of events seems backwards, it appeared unlikely that substantial federal aid for further water development would be forthcoming until basin waters were divided. The states agreed to a percentage allocation of annual available water: Colorado, 41.75 percent; Utah, 23 percent; Wyoming, 14 percent; New Mexico, 11.25 percent; and Arizona, a fixed 50,000 acre-feet per year.

The Colorado River Storage Project Act, 1956. This act was intended to provide for development of the Upper Basin waters in the way that Boulder Dam had controlled Lower Basin waters. Its passage involved the first major environmental fight over a dam proposed for Echo Park in Dinosaur National Monument. That dam was deleted from the final authorization that included Flaming Gorge in Wyoming; Blue Mesa, Morrow Point, and Crystal in Colorado; Navajo in New Mexico; and Glen Canyon in Arizona. (See Figure 5.1.)

The Colorado River Basin Project Act, 1968. This act authorized the Central Arizona Project (CAP), long sought by Arizona as a way of transferring water from the Colorado to central Arizona, where groundwater was being overdrawn some 5 maf annually. Although such a project had been studied for decades, the economics was so poor that only a huge federal subsidy could ever pay for the project. Major environmental fights occurred over proposed power dams in Bridge and Marble canyons, revenues from which would presumably (in a bookkeeping sense) help to pay for the CAP. A large thermal power plant was finally included for this purpose.

In addition to authorizing the CAP, this act included the following steps that further defined or constrained development of the river: (1) assigning priority to California's 4.4 maf, so that Arizona would have to absorb any shortages that might occur from shortfalls of Upper Basin deliveries; (2) authorizing various Upper Basin projects and Hooker Dam on the Gila; (3) declaring the Mexican treaty obligation a national obligation to be satisfied (at federal expense) from any *future* supply augmentation plans; (4) forbidding any federal studies of importation of water from other river basins (to placate the fears of Columbia River

Basin interests); (5) authorizing Upper Basin retention of waters not needed to satisfy compact and Mexican obligations to build up reservoir stocks sufficient to give reasonable protection to the Upper Basin's established consumptive uses; and (6) requiring approximate equality in the volumes of water in storage in Lake Powell and Lake Mead (Glen Canyon and Hoover dams).

To obtain aid from national programs in competition with other regions, a consensus among basin states was necessary. The potential magnitude of federal aid outweighed any gains likely from one state's taking advantage of its neighbor. Federal aid changed a zero-sum game into a positive-valued game for the Colorado Basin states. Obtaining consensus meant agreement on policies and projects, such as rules for distributing the river's waters and locating major storage projects. The effectiveness of the policies and projects chosen much depended on the true climatological and hydrologic regimes of the region, about which little was known at the time of many key decisions. Yet, the consensual process had to continue once it began, even when the scientific data base and desired study results were not at hand. The political costs of failure were perceived to be greater than any likely economic or physical inefficiencies that might result from decisions based on inadequate data.

In spite of the complex legislative history summarized above, no river basin agency has management responsibility for the entire basin. Water has been legally allotted to individual states. This institutional setting has the following effects: (1) states with claims to water in excess of their current uses are eager to put this water to use regardless of efficiency considerations lest some change of law or adverse political alignments deprive them of their unused water, (2) it appears doubtful that water can be reallocated among states or between Upper and Lower basins without substantial changes in existing laws and compacts, and (3) states tend not to be concerned with the downstream quantity and quality effects of their actions.

Hydrology, water use patterns, and the value of water in the subbasins of the Upper Colorado River Basin

The first objective of this section is to characterize the surface hydrology of the eight principal subbasins of the Upper Basin by looking at the average surface outflows and the average net surface outflows, the latter defined as water originating in that subbasin less that subbasin's consumptive use. Each subbasin generates substantial net outflows that, on the average, total 11,145,000 acre-feet per year for the entire Upper Basin. If the obligation to the Lower Basin of 7.5 maf annually and half Mexico's obligation, 0.75 maf, are subtracted, the result is an average annual Upper Basin surplus of 2,894,000 acre-feet. Of course, these outflows are variable year by year, and this variability and its relation to the risks facing potential users are an important issue.

The second objective of this section is to describe patterns of consumptive water use in the Upper Basin. Because more than 90 percent of all water consumption is in agriculture and 65 percent of agricultural consumption is in the growing of low-value feed grain and forage crops, attention is directed to those low-value agricultural uses, involving an average of 1.6 maf per year.

A third objective is to estimate the values of this water to the farmer and to the state where the farming operation is located. From a slightly different viewpoint, the farmer's income per acre-foot consumed represents the price at which a farmer is likely to consider selling water. Among the low-valued crops considered, the highest net farm income per acre-foot is $72, with an average of about $25 per acre-foot. However, state income generated directly and indirectly ranges from $75 per acre-foot consumed in Subbasin 1 (Wyoming) to $160 per acre-foot consumed in Basins 5 (Colorado) and 8 (Utah) assuming a permanent shutdown of directly and indirectly linked activities – a worst case scenario. Thus, the effects of reallocating water from agricultural to nonagricultural uses may be seen quite differently by the individual farmer and the state government.

Further, water has quite significant values when left in the stream. The purposes served by increasing Upper Basin instream flows include better water quality, recreation, fish and wildlife values beyond their recreational values, hydroelectric generation, and more agricultural production in the Lower Basin. Estimates of these values make it clear that instream values today exceed the income-producing values to the farmer, and in some cases they significantly exceed the total income-producing value to the state. These findings have important implications for the desirable and likely patterns of future water use.

Hydrology of the subbasins

Wyoming, Colorado, Utah, and New Mexico all lie partly within the Upper Colorado River Basin. The three main subbasins, the Green, the Upper Main Stem of the Colorado, and the San Juan, are divided into smaller subbasins for analytical purposes in this paper. These subbasins are described in Table 5.3.

Table 5.3. *Characteristics of the subbasins of the Upper Colorado River Basin*

Subbasin	Name	Primary state	Counties included	Outlet gauging station, USGS[a]	Major reservoirs
1	Green River to Flaming Gorge	Wyoming	Sublette, Lincoln, Uinta, Sweetwater	9-2345	Fontenelle, Flaming Gorge
2a	Yampa River	Colorado	Routt, Moffat	9-2510	None
2b	White River	Colorado	Rio Blanco	9-3065	None
3	Green River above Colorado River	Utah	Carbon, Daggett, Duchesne, Emery, Uintah	9-3070	None
4	Gunnison	Colorado	Delta, Hinsdale, Gunnison, Ouray	9-1525	Blue Mesa, Morrow Point, Crystal
5	Upper Main Stem, Colorado	Colorado	Garfield, Grand, Eagle, Mesa, Pitkin, Summit	9-1635	None
6	Dolores	Colorado	Grand (Utah) Dolores, Montrose, San Miguel	9-1800	McPhee
7	San Juan	Colorado	Archuleta, La Plata, San Juan (New Mexico), San Juan, Montezuma	9-3795	Navajo
8	Colorado above Lees Ferry	Utah	Garfield, Kane, San Juan, Wayne	9-3800	Glen Canyon

[a]USGS = U.S. Geological Survey.

Table 5.4. *Monthly median discharges of the subbasins of the Upper Colorado River Basin*[a] *(thousand acre-feet)*

	Subbasin								
	1	2a	2b	3	4	5	6	7	8
Oct.	926	217	320	1,597	809	2,570	98	845	5,935
Nov.	969	217	287	1,604	865	2,681	99	669	5,387
Dec.	925	194	252	1,316	712	2,418	97	542	4,581
Jan.	843	177	252	1,255	645	2,069	103	508	4,110
Feb.	936	208	273	1,708	658	2,273	133	805	5,102
Mar.	1,012	407	383	2,627	797	2,397	181	1,023	6,629
Apr.	1,293	1,815	427	4,761	1,820	3,492	979	2,046	11,528
May	1,436	4,571	1,096	8,949	4,923	8,942	1,557	3,399	19,905
June	2,310	3,993	1,347	12,355	4,957	12,123	978	3,308	28,847
July	1,768	867	435	4,005	1,486	3,580	239	1,294	11,644
Aug.	1,392	246	327	1,897	746	2,226	160	968	8,115
Sept.	1,145	143	286	1,347	697	2,154	94	779	6,123

[a]The years between 1963 and 1983 are not reported for Subbasin 8 because filling Lake Powell (begun in 1963, with full capacity reached in 1983) caused the measured flows at Lees Ferry to diminish significantly. Observations for Subbasin 3 cover only 1951–1966 because the gauging station at Green River near Ouray, Utah, was discontinued in 1967.

Source: U.S. Geological Survey, Water Resources Division, Colorado District.

The hydrology of each of the eight subbasins can be partially represented by the median monthly outflows in acre-feet from the subbasin. (See Table 5.4.) The monthly discharges represent the flow that is met or exceeded 50 percent of the time in the water year. The basin outlet gauging stations for each subbasin are identified in Table 5.3. Computations are based on all daily values available, and sample sizes range from 18 to 84 years.

In a quantification of the water that might still be available for new uses, it is useful to estimate the amount of water that originates in and flows unused from each of the eight subbasins. This is water that would be available for out-of-basin uses without diminishing existing uses. Subbasins 1, 2, 4, and 7 are headwater basins that receive no water from other basins. Subbasin 6 is drained primarily by the Dolores River and can thus be considered a headwater basin, although a 30-mile segment of the Colorado main stream runs through the basin in Grand County, Utah. Subbasins 3, 5, and 8 receive water from Basins 1 and 2; 4; and 3, 5, 6, and 7, respectively. These inflows are netted out to compute the net increment to the flow from within the basin. Mean annual net discharges and their standard deviations for the eight subbasins are presented in Table 5.5. (These means and standard deviations are calculated over different periods of record, so their sum is not the true time average of outflows.)

The sum of 11,144,926 acre-feet indicates that the states of the Upper Basin are far from fully utilizing their compact-apportioned waters. It also indicates that a potentially sizable amount of water is available for out-of-basin use without impairing existing uses. Subtracting required Lower Basin deliveries indicates excess outflows that the Upper Basin is entitled to consume. Potential out-of-basin users of these excess waters would surely be interested in the reliability of the supplies: more reliable supplies are certainly more valuable and would command a higher price in a water market. Those who need only supplemental water would not be as concerned about the reliability of the flow and would be willing to pay only a lower price for the water. The reliability of these subbasin net discharges can be pictured with a frequency distribution like Figure 5.5 (for Subbasin 1) or by statistical measures like the standard deviation and coefficient of variation.

Agricultural water uses and values

Agriculture and agriculturally related activities are the largest consumers of water in the Upper Basin. Livestock production is a dominant economic activity in the basin, to which agriculture plays a supportive role. Here the water consumed is quantified by the large-volume, low-value feed and forage crops of the basin. Alfalfa, barley, wheat, oats, corn grain and silage, potatoes, and pasture represent the predominant

Table 5.5. *Mean annual net outflows and standard deviations for the subbasins of the Upper Colorado River Basin (thousand acre-feet)*

Subbasin	Mean	Standard deviation	Coefficient of variation
1	1,512	569	0.38
2a	1,094	350	0.32
2b	503	150	0.30
3	930	476	0.51
4	1,843	670	0.36
5	2,615	893	0.34
6	581	373	0.64
7	1,657	821	0.50
8	405	217	0.54
Total mean outflow	11,140		
Less required deliveries	−8,250		
Excess water	2,890		

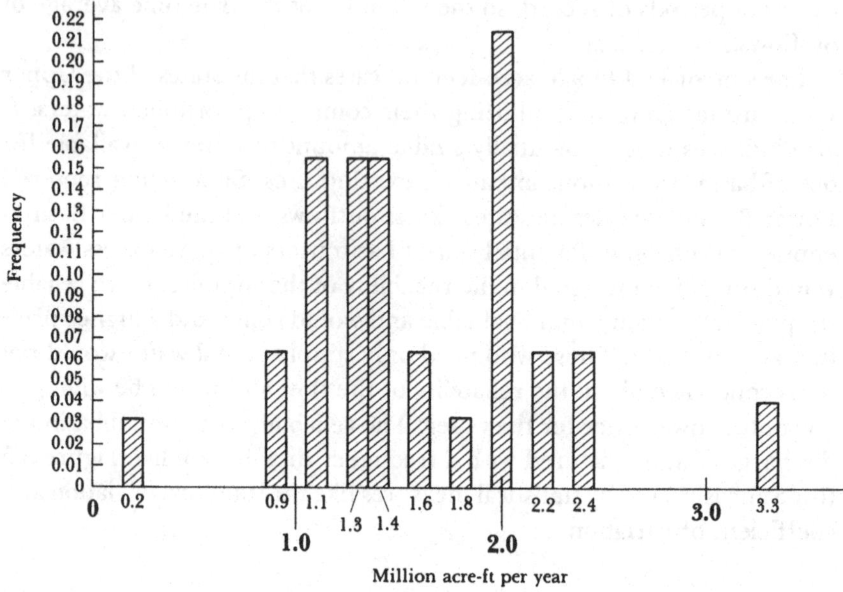

Figure 5.5. Frequency distribution of net discharges, Subbasin 1.

Table 5.6. *Irrigated acreages by crop, Upper Colorado River Basin*

Sub-basin	Alfalfa	Barley	Wheat	Oats	Corn Grain	Corn Silage	Potatoes	Pasture	Total
1	6,139	14,778	0	934	0	0	0	107,985	129,836
2	16,573	18	633	470	0	0	0	29,478	47,172
3	68,215	6,060	1,981	3,537	1,803	10,871	117	102,547	195,131
4	23,885	1,790	975	1,015	3,930	3,693	0	45,809	81,097
5	66,414	376	5,397	2,160	10,000	4,646	83	81,534	170,610
6	27,552	4,903	2,513	1,299	9,495	7,946	56	36,870	90,634
7	58,828	601	1,716	1,738	256	2,309	10	65,200	130,658
8	22,354	2,238	35	941	0	94	21	15,565	41,248
Total	289,960	30,764	13,250	12,094	25,484	29,559	287	484,988	886,386

uses of irrigation water in the basin. The crops are primarily forage and feed for livestock. Irrigated acres by crop are given in Table 5.6.

Total consumptive uses for each crop in each basin have been computed using Narayanan, Padungchai, and Bishop's (1979) data on consumptive use (acre-feet per acre per year; see Table 5.7) and the acres irrigated (U.S. Department of Commerce, 1982). The crops shown in

Table 5.7. *Consumptive water use, Upper Colorado River Basin (acre-feet per acre per year)*

					Corn			
Subbasin	Alfalfa	Barley	Wheat	Oats	Grain	Silage	Potatoes	Pasture
1	1.77	1.2	0.00	1.6	0.00	0.00	0.00	1.75
2	1.61	1.2	1.67	1.6	0.00	0.00	0.00	1.70
3	1.77	1.2	1.67	1.6	2.08	1.4	1.75	1.80
4	1.67	1.2	1.67	1.6	2.08	1.3	0.00	1.70
5	1.67	1.2	1.67	1.6	2.08	1.3	1.83	1.70
6	2.50	1.4	1.67	1.6	2.08	1.8	1.83	2.20
7	1.57	1.3	1.67	1.6	2.08	1.8	1.83	2.00
8	1.57	1.3	1.67	1.6	0.00	2.08	1.83	2.00

Table 5.8. *Consumptive water use by crop, Upper Colorado River Basin (thousand acre-feet per year)*

					Corn				
Subbasin	Alfalfa	Barley	Wheat	Oats	Grain	Silage	Potatoes	Pasture	Total
1	11	18	0	2	0	0	0	189	219
2	27	I	1	1	0	0	0	50	79
3	121	7	3	6	4	15	I	185	340
4	40	2	2	2	8	5	0	78	136
5	111	1	9	4	21	6	I	139	289
6	69	7	4	2	20	14	I	81	197
7	92	1	3	3	1	4	I	130	234
8	35	3	I	2	0	I	I	31	71
Total	506	38	22	19	53	45	1	883	1,566

I = insignificant volume.

Table 5.8 represent a total of 886,386 irrigated acres and consumptive use of 1,566,205 acre-feet of water per year, roughly 25 percent of the Upper Basin's compact-apportioned water.

What is the value of this water? For the farmer, the value is the net return to water per acre-foot consumptively used. Consumptive use is the relevant measure because water transfers are generally limited by the water courts to that volume. Farm budget data (Narayanan et al., 1979), data on consumptive use (Table 5.7), crop yields per acre for the eight selected crops (Table 5.9), crop prices (Table 5.10), and production costs (Table 5.11) permit the calculation of the net return to water in dollars per acre-foot consumed. The data have been updated to 1982 through the use of U.S. Department of Agriculture (USDA) price and cost indices.

Table 5.9. *Annual crop yield per acre of irrigated land, Upper Colorado River Basin*

Subbasin	Alfalfa (ton)	Barley (bushel)	Wheat (bushel)	Oats (bushel)	Corn Grain (bushel)	Corn Silage (ton)	Potatoes (hundred weight)	Pasture (animal unit months)
1	3.25	50.0	0.00	50	0.00	0.00	0.00	4.5
2	3.10	50.0	50	50	0.00	0.00	0.00	6.8
3	3.35	62.5	50	62	55.43	12.50	106.30	6.8
4	3.35	55.0	50	50	99.80	16.44	0.00	6.8
5	3.35	57.0	50	50	97.58	15.38	145.70	6.8
6	3.85	62.0	50	50	87.64	17.72	212.38	6.8
7	3.08	50.0	50	50	87.64	11.80	90.25	6.8
8	3.08	62.5	50	62	0.00	10.75	156.25	6.8

Table 5.10. *Crop prices, Upper Colorado River Basin, 1982 (dollars per unit)*

	Alfalfa (ton)	Barley (bushel)	Wheat (bushel)	Oats (bushel)	Corn Grain (bushel)	Corn Silage (ton)	Potatoes (hundred weight)	Pasture (animal unit months)
Colorado	62.08	2.46	3.40	1.70	2.64	19.75	3.2	12.27
New Mexico	81.08	2.06	3.58	NA	2.87	19.75	3.2	12.27
Utah	66.67	2.06	3.55	1.70	3.27	19.75	3.2	12.27
Wyoming	55.92	2.42	3.55	1.62	3.04	19.75	3.2	12.27

NA = not applicable.

Table 5.11. *Estimated annual cost of production, Upper Colorado River Basin (1982 dollars per acre)*

Subbasin	Alfalfa	Barley	Wheat	Oats	Corn Grain	Corn Silage	Potatoes	Pasture
1	131.10	225.01	0.00	69.35	0.00	0.00	0.00	21.06
2	134.00	119.03	138.72	69.35	0.00	0.00	0.00	24.58
3	127.73	144.32	138.72	69.35	155.03	168.31	333.71	24.58
4	138.84	119.03	138.72	69.35	180.27	219.73	0.00	24.58
5	138.84	119.03	138.72	69.35	176.24	219.73	457.50	24.58
6	185.25	230.70	138.72	69.35	158.30	219.73	666.90	24.58
7	143.74	119.03	138.72	69.35	158.30	219.73	283.36	24.58
8	113.03	176.56	138.72	69.35	0.00	168.31	490.63	24.58

Table 5.12. *Net return on water consumptively used, Upper Colorado River Basin (dollars per acre-foot)*

Subbasin	Alfalfa	Barley[a]	Wheat	Oats	Corn Grain	Corn Silage	Potatoes	Pasture
1	28.67	−86.68	0.00	7.28	0.00	0.00	0.00	21.95
2	36.30	3.31	18.73	9.78	0.00	0.00	0.00	36.65
3	54.02	9.92	23.22	22.53	12.61	56.08	3.69	34.61
4	41.40	13.56	23.22	9.78	40.00	3.82	0.00	36.65
5	41.40	17.66	23.22	9.78	39.12	64.64	4.78	36.65
6	10.61	−55.84	23.22	9.78	41.76	72.36	6.95	28.32
7	48.87	3.05	23.22	9.78	35.13	7.40	2.97	31.15
8	58.80	−15.62	18.73	22.53	0.00	21.15	5.12	31.15

[a]Negative returns to barley occur when it is used as a nurse crop for alfalfa.

The cost data include all relevant variable and fixed production costs, including an allowance for family farm labor and the opportunity costs of management. Thus, the estimated net return per acre of irrigated land for each of the eight crops represents a pure return to water. Dividing the estimated net return by the amount of water consumptively used, the average net return per acre-foot of water consumed is derived (Table 5.12). Several biases affect this figure. First, some cropping operations are integrated with cattle operations. The integrated operation should be budgeted as a unit. Doing so would probably increase some values per acre-foot of consumption. Unfortunately, no information on farm structure is available for the region. Second, conservation measures such as reduced water application initially affect yields very little. As conservation steps are increased in intensity, their costs increase. Therefore, our figures *overstate* the values of initial quantities of water that might be withdrawn from crop irrigation. Further, if water were to be partially withdrawn from some of these cropping operations, a rational response by the farmer would be to change cropping patterns. All these steps could be included in a programming approach to derive more accurate value figures (for example, see Gisser et al., 1979). A programming approach was not feasible for this study.

In Table 5.13, consumptive uses in acre-feet are ranked according to the net crop values of Table 5.12. Crop 1, crop 2, etc., are different for each basin. In Subbasin 4, for example, water exhibits its lowest net value when used in the cultivation of corn for silage. If farmers in this area were offered more than $4 per acre-foot for their water, it is likely that they would stop raising corn for silage, and each year about 5,000 acre-feet of water would be available for other uses. If the offer price for

Table 5.13. *Direct farm income per acre-foot of consumptive use, by crop, in ascending order, with cumulative thousand acre-feet*

					Corn			
Subbasin	Alfalfa	Barley	Wheat	Oats	Grain	Silage	Potatoes	Pasture
1	$I	$7	$22	$29				
	18	20	208	219				
2	$3	$10	$19	$36	$37			
	I	1	2	28	79			
3	$4	$10	$13	$23	$23	$35	$54	$56
	I	8	11	17	20	205	326	341
4	$4	$10	$14	$23	$37	$40	$41	
	5	6	9	10	88	96	136	
5	$5	$10	$18	$23	$37	$39	$41	$65
	I	4	10	11	149	170	281	289
6	$I	$7	$10	$11	$23	$28	$42	$72
	7	8	9	78	82	163	183	197
7	$3	$3	$7	$20	$23	$31	$35	$49
	I	1	5	8	11	141	142	234
8	$I	$5	$19	$21	$23	$31	$59	
	3	3	3	3	5	36	71	

Total consumptive use = 1,566

I = negative value or insignificant quantity.

water were to rise to approximately $10 per acre-foot, farmers might be induced to stop producing oats, providing up to an additional 1,624 acre-feet per year for a total of 6,425, etc. This schedule constitutes a crude supply curve for water taken out of existing uses.

For the Upper Basin as a whole, combining all eight subbasins, offer prices in excess of $15 per acre-foot would lead to cessation of the production of barley, oats, and potatoes, in the longer term freeing 139,118 acre-feet per year for other uses. An offer price of $25 per acre-foot would be likely to induce farmers to drop the production of wheat, making available a total of 348,671 acre-feet per year. Further, if the offer price for water were higher than $72 per acre-foot, none of the crops presented here would likely be produced in the long run, and 1,566,200 acre-feet per year would be available for other uses.

State income effects of agricultural water use

There are sectors in the states' economies that are tied to agriculture either as suppliers (backward linkages) or as processors of agricultural outputs (forward linkages). In Colorado, for example, the food-processing sector is one of the largest sectors in the state owing primarily to the processing of meat provided by the livestock sector, which in turn is largely fed by outputs of irrigated agriculture (Gray and McKean, 1975). Taking crops out of production may affect these sectors to some extent. To what extent would the withdrawal of irrigation water in the Upper Basin affect the various state economies? State governments and the state water agencies can be expected to consider the overall effect on state income and employment, not simply the direct loss of income to the farmer, even though from a national accounting stance, these secondary income impacts may largely cancel out, with gains to other areas.

What are the determinants of the extent to which state income losses might exceed the direct loss of farm income? The following factors would be important:

- The extent to which the agricultural crops being phased out have been exported without further value added or the extent to which they have been used as inputs into other state sectors, for example, livestock;
- The extent to which substitute agricultural inputs are available through imports to sustain the agriculturally linked activities;
- Whether the water released from agriculture substitutes for costly new supply projects to service continuing state growth; and
- Whether the money from the sale of agricultural water is reinvested within the state.

If one can answer "to a large extent" to the first two items and "yes" to the remaining two, then one would expect few income losses beyond the loss of direct farm income. If the answers are "to a very small extent" and "no," one would expect fairly severe income losses beyond the farm level.

In Colorado, 92 percent of irrigated agriculture's output is used in the livestock sector, food processing, and consumption or export (Gray and McKean, 1975). The crops in western Colorado that would be phased out are primarily feed grains and forage crops, so the relevant linkage is to the livestock sector, because the rest is exported. The livestock sector, in turn, is linked to food processing. Regarding substitutes, it seems unlikely that *no* agricultural substitutes would be available from in-state or out-of-state sources. It is assumed here that half the reduction in agricultural outputs could be substituted from such sources.

Western Colorado has a surfeit of water overall (although some localities experience shortages), but eastern Colorado will provide a ready market for agricultural water from the Colorado Basin through sequences of exchanges, if not by direct transmission. It seems likely that waters withdrawn from agriculture are likely to substitute for expensive

Table 5.14. *State income lost on the average per acre-foot of reduced consumptive use: worst case scenario with forward linkage to feeding livestock and processing food*

Subbasin	Income lost
1	$ 75
2	90
3	149
4	151
5	160
6	145
7	137
8	160

new in-state projects. On the other hand, it seems unlikely that the money from water sales will be reinvested in Colorado, certainly not in western Colorado.

Thus, state income impacts beyond direct loss of farm income would range from moderate for in-state sales of water to fairly severe for out-of-state sales. The statewide income impacts are estimated for a worst case scenario.

The methodology for estimating these direct and indirect income losses is complicated by the absence of appropriate models. As water is progressively withdrawn *from* agriculture, farmers substitute other inputs (labor, capital) for water, and they adjust their cropping patterns and their total acreage. These adjustments result in a nonlinear response of farm income to water withdrawal. A detailed programming model is needed to approximate the kinds of decisions that would be made by a profit-maximizing farmer.

The models that are available are state input-output (I-O) models for Colorado and Utah. (None is available for Wyoming.) I-O models are fixed coefficient models that imply no substitution possibilities. However, they do present the historical interrelationships among the sectors of the state economy. If used with good judgment and an acquaintance with the workings of the state economy, the I-O models can provide reasonable approximations of statewide effects.

The sequence of steps in the analysis is as follows: (1) a reduction in irrigated output is postulated, (2) the reduction is divided into reductions in deliveries to final demand (consumption or export) and reductions in inputs into the livestock sector, (3) reductions in livestock output are calculated and treated as reductions in inputs into the food-processing sector, (4) all reductions in food-processing output are treated as reductions in deliveries to final demand, and (5) all reductions in deliveries to final demand are "run through" the I-O model to obtain the total reduction in payments to households, insurance, real estate, rent, interest, and profits. The average result for Colorado and Utah is that a $1 reduction in irrigated agricultural output leads to a $1.80 reduction in state incomes. The results are given in Table 5.14. These state income losses should be compared with the direct on-farm income losses shown in Table 5.13. The higher state income losses are likely to lead to state resistance to private out-of-state water sales.

The value of water in instream uses

Values of water consumed in agriculture to the farmer and to the state economy have been estimated. Naturally, some of this water will be bid away from agriculture for other uses, either inside or outside the Upper Basin itself. However, the water would generate quite substantial values if left in the stream: in hydroelectric power generation, in recreation, in maintaining higher water quality, and in downstream consumption. In many cases, these values far exceed the value to the farmer, and in some cases they exceed state income losses that would follow from withdrawing the water from agriculture.

The doctrine of beneficial use in western water law generally does not allow the appropriation of water rights for instream uses of water. However, hydroelectric power generation is treated as a nonconsumptive diversion use for which water can be appropriated. Federal installations have not filed for such flow rights. The value of water in hydroelectric production is the value of the marginal electric power produced by an acre-foot of water. Reservoir releases are generally managed so that most of the water passing through the reservoir is used in hydroelectric production.

The Colorado River Storage Project (CRSP) provides for developing the Upper Basin's compact-apportioned waters while still meeting its flow obligations at Lees Ferry, Arizona. Four storage units, Flaming Gorge, Wayne Aspinall (formerly Curecanti, consisting of Blue Mesa, Morrow Point, and Crystal), Glen Canyon, and Navajo, were authorized to provide long-term regulatory storage. The reservoirs created by the CRSP have a combined storage capacity of 34 maf (Water and Power Resources Service, 1981, p. 355). All but one of the units, Navajo, create hydroelectric power. The power-producing units are located in Subbasins 1, 4, and 8. A complex transmission system has been provided for the electricity created by the CRSP. Data on the energy generated and the revenues from sale to electric utilities were obtained from the Bureau of Reclamation's *Annual Report* (1981, app. III). Average revenue from electricity sales for the entire CRSP is found by dividing total revenue by the kilowatt hours sold. The average revenue from electricity sales in the Upper Basin is $0.0106 or 10.6 mills per kilowatt hour (based on 1981 data). This figure might be compared with the Bonneville Power Authority's minimum surplus power rate of 18 mills or Idaho Power's avoided cost rate for cogeneration of 44 mills (Gardner, 1985). New thermal power-generating and distribution costs are around 8.5 cents per kilowatt hour (Hamilton and Lyman, 1983). It is quite clear that Bureau of Reclamation power is substantially underpriced.

Flaming Gorge Dam is located on the Green River approximately six miles south of the Utah-Wyoming border, thus placing it in Subbasin 1. The facility's net electricity production in 1981 was 360,789,000 kilowatt hours (Water and Power Resources Service, 1981, p. 357). Total revenue accruing from electricity sales was $3,834,430. The flows available for hydroelectric production amounted to 1,042,753.6 acre-feet in 1981 (flows measured at Greendale, Utah). Thus, the marginal value of an acre-foot of water allowed to pass through the turbines in the power plant is $3.68 if priced at 10.6 mills or, more realistically, $14.72 if valued at short-run avoided cost rates.

The Aspinall Unit (consisting of Blue Mesa, Morrow Point, and Crystal dams) is located on a 40-mile segment of the Gunnison River in Subbasin 4. It is assumed that the 1981 flow of 743,600 acre-feet on the Gunnison above the North Fork (but before the substantial diversions to the Uncompaghre Valley) passed through the three dams. Blue Mesa Dam is approximately 30 miles southwest of Gunnison, Colorado. Total revenue from electricity sales of 245,859,300 kilowatt hours in 1981 was $2,625,145. Consequently, the value of an additional acre-foot of water allowed to pass through the turbines is $3.53, or $13.20 at avoided cost rates. Net electrical generation of 314,808,000 kilowatt hours at Morrow Point produced revenues of $3,362,849 in 1981. The marginal value product of water at Morrow Point Dam is thus $4.53, or $18.12 avoided cost. Crystal Dam is six miles downstream from Morrow Point Dam, near the town of Montrose, Colorado. Active capacity of the reservoir is 13,000 acre-feet (Water and Power Resources Service, 1981, p. 360). Net hydroelectric production in 1981 amounted to 159,604,400 kilowatt hours, bringing revenues of $1,808,699. The value of an acre-foot of water passing through the power plant is thus $2.43, or $9.72 avoided cost.

Glen Canyon Dam and Lake Powell are located on the mainstream of the Colorado River (Subbasin 8) approximately four miles south of the Arizona-Utah boundary and 15 miles upstream from Lees Ferry, Ari-

Table 5.15. *Hydroelectric power values foregone through consumptive use of 1 acre-foot in the subbasins of the Upper Colorado River*

Subbasin	Power generating stations	Value per acre-foot at: 10.6 mills/kWh	44 mills/kWh
1	Flaming Gorge	$ 3.68	$14.72
	Glen Canyon	4.97	19.88
	Hoover	2.76	11.48
		$11.41	$46.08
2, 3, 5–8	Glen Canyon	$ 4.97	$19.88
	Hoover	2.76	11.48
		$ 7.73	$31.36
4	Blue Mesa	$ 3.53	$13.20
	Morrow Point	4.53	18.12
	Crystal	2.43	9.72
	Glen Canyon	4.97	19.88
	Hoover	2.76	11.48
		$18.22	$72.40

zona. The lake provides the largest storage for the Upper Basin, with an active capacity of 20,876,000 acre-feet (Water and Power Resources Service, 1981, p. 355). Net hydroelectric production in 1981 was 3.8 billion kilowatt hours, and revenue was $41.3 million. Based on 8.3 maf of unused water flowing through Lees Ferry in 1981, the marginal value product of an acre-foot of water in electrical production is $4.97, or $19.88 avoided costs.

Water that is released from Lake Powell proceeds downstream to Lake Mead in back of Hoover Dam. Enroute and during storage in Lake Mead, water is lost by evaporation and seepage, so an acre-foot released is roughly equal to 0.9 acre-feet available for release from Hoover Dam. Historically, Hoover Dam has generated about 290 kilowatt hours per acre-foot of water released (Bureau of Reclamation, 1981). Thus, the value of an acre-foot released upstream, allowed to enter Lake Mead and then released, would be $290 \times 0.0106 \times 0.9 = \$2.76$ if priced at 10.6 mills, or $11.48 valued at avoided cost. Table 5.15 summarizes the hydroelectric values calculated above as they relate to water released in each subbasin.

In addition to the hydropower values of water left in or released to the stream, recreation values can be enhanced for both stream-related and reservoir-based recreation. For stream-related recreation, the stabilization of flow at levels that enhance aesthetics and fish life is important, but both extremely high and low flows are detrimental. For reservoir-

Table 5.16. *Willingness to pay for increased flows for recreational activities on the Cache La Poudre River, Colorado, 1978*

Flow rate (cubic feet per second)	Marginal values (dollars per cubic foot per second per day)		
	Whitewater	Fishing	Shoreline
100	—	23	16
200	0.95	17	14
300	0.95	11	11
400	0.95	4	8
500	0.95	−2	6
600	0.95	−8	3
700	0.95	−15	0
800	0.95	−21	−2

Source: Daubert and Young, 1980, table 4.

related recreation, maintenance of the reservoir level, that is, avoidance of a large drawdown, is most important.

Many studies of instream flow values for fishing and shoreline activities have been carried out, but Daubert andYoung's (1980) measurement of marginal willingness to pay for increased streamflow on the Cache La Poudre River is closest to what is needed for estimating the recreational opportunity cost of water consumed in agricultural and other uses. Table 5.16 summarizes their results.

These figures show that, under some conditions, higher streamflows are highly valued for recreation, but these values depend on timing. For example, an increase above 200 cubic feet per second (cfs) during the late summer would be valued at $23 per cfs per day. Because 1 cfs maintained over a day totals 2 acre-feet, a release of 2 acre-feet per day would generate recreation values of $11.47 per acre-foot. On the other hand, added releases during spring high flows would reduce recreational values. Thus, such values are quite specific to certain river reaches, and little can be said about such values in the absence of specialized studies.

Reservoir recreation is also valuable. Howe et al. (1982) estimated the worth of reservoir recreation along the Denver–Fort Collins corridor at $18.75 per person per day. Values at a unique site such as Lake Powell may well be even higher in spite of the lake's remote location. Again, such values are highly site specific. However, the issue for this study is the *marginal* effect of added streamflow. In dry years, increased flows would reduce reservoir drawdown, adding to recreational values, but in wet years, the water would simply be passed through the reservoir. On the average, however, the additional water would have some value, but no specific studies are available for the Upper Basin.

The other quite important instream value stems from the increases in water quality that would result from reductions in Upper Basin consumptive uses. As explained in detail in the following section on water quality issues, the removal of high-quality (low total dissolved solids) water in the Upper Basin increases the TDS concentration downstream in two ways: the removal of low TDS water reduces the dilution factor for the lower-quality waters downstream, and the return flows from the water, if any, may be high in TDS, thus increasing the overall TDS concentration.

The average salinity of the river at its source is 50 milligrams (mg) per liter; at Imperial Dam, the last major diversion point before it reaches Mexico, it is more than 800 mg per liter. The Bureau of Reclamation estimates that 59 percent of the salinity concentration at Hoover Dam emanates from natural sources (saline springs, erosion of sediments, and the concentrating effects of evaporation and transpiration), and 41 percent comes from man-made causes (irrigation applications, municipalities and industry, and out-of-basin transfers) (Bureau of Reclamation, 1983a). Of the man-made sources of salinity, roughly 72 percent (or 30 percent of the total) can be attributed to the extensive irrigated agriculture within the basin.

Most of the damages attributable to high salinity concentrations are borne by the water users in the Lower Colorado River basin. The extremely high salt load of 9 million tons annually entering Lake Mead adversely affects more than 12 million people and about 1 million acres of irrigated farmland (Bureau of Reclamation, 1983a). Total (direct plus indirect) agricultural damages in the 875–1,100 mg per liter range (the hypothesized salinity values with and without control) are estimated at $121,969 per mg per liter per year; municipal impacts have been estimated at $371,000 per mg per liter per year. Thus, total damages to the Lower Basin are $492,969 per mg per liter (Gardner, 1983). The Bureau of Reclamation estimates total damages (agricultural, municipal, and industrial) at $561,000 (1984 dollars) per mg per liter increase at Imperial Dam. Annual municipal damages, 70 percent of total damages, are allocated as follows: Metropolitan Water District, 54 percent; Central Arizona Project, 8 percent; and lower main stem users, 8 percent (U.S. Department of the Interior, 1985, p. 15). Damages to agriculture account for the remaining 30 percent.

Consider first the effects of consumptively using (or exporting) water in (from) the Upper Basin at some point where the TDS concentration is

Table 5.17. *Instream values per acre-foot of reduced consumptive use in the Upper Colorado River Basin*

Subbasin	Water opportunity cost in Lower Basin	Salinity damages averted	Power value at 44 mills	Total
1	$30	$ 38	$46	$114
2, 3, 6–8	30	38	31	99
4	30	38	72	140
5	30	280	31	341

100 mg per liter. One acre-foot contains approximately 0.136 tons of dissolved solids. The approximate *change* in TDS concentration at Imperial Dam is then 0.000078 mg per liter per acre-foot. Using Gardner's damage figures, the removal of 1 acre-foot would cause $38.50 in Lower Basin damages.

Now consider the effects of consumptively using an acre-foot in crop irrigation in the Grand Valley of Colorado, on the Colorado River main stem. With a consumptive fraction of about 50 percent in that area, this consumptive use would be accompanied by a 1 acre-foot return flow. Return flows in the area average a TDS concentration of 4,200 mg per liter (Soil Conservation Service, 1977, p. 49), or 5.7 tons per acre-foot of return flow. The resulting change in the TDS concentration at Imperial Dam is approximately 0.00057, or added damage of $281 to Lower Basin users.

In addition to the instream values mentioned above, there is the opportunity cost of the water to downstream users in the Lower Basin. Vaux and Howitt (1984) estimated these costs as in the $8–$30 per acre-foot range; Howe and Young (1978) estimated a $30 value. The sum of all these foregone values is shown by subbasin in Table 5.17. Thus, it seems clear that the values generated by leaving water in the stream considerably exceed many of the values generated on-farm in agriculture. (See Table 5.13.)

Water quality issues in the Colorado Basin

The previous section discussed costs to the Lower Basin caused by increased salinity concentrations. Consumptive uses, evaporation, and exports in the Upper Basin cause these increased TDS levels. Thus, Upper Basin uses cannot be fully evaluated without quantifying the consequent water quality impacts.

An ideal river basin management scheme would simultaneously optimize both the allocation of water quantities and the control of water quality. A property right in water is not fully specified unless the dimensions of water quality are specified. Howe, Shurmeier, and Shaw (1986) have shown that economically efficient water allocation can be achieved only by joint quantity-quality optimization. In this practice, this jointness has been finessed by specifying ambient water quality standards. A water right is then defined in terms of quantities that are better than or equal to the standard quality.

In the absence of laws and agencies that might provide joint optimization of water quantities and quality, it is important that the quality standards be reached at minimum cost. This section identifies the various steps that can be taken to improve water quality, estimates their costs, and arranges them in increasing order of cost.

The Environmental Protection Agency (1971) has called salinity the major water quality problem in the Colorado River Basin. Since 1949, the general trend in salinity concentrations at Imperial Dam has been upward, reaching a high of over 900 mg per liter in the mid-1950s. Concentrations at Imperial Dam have decreased since 1970 owing primarily to the filling of Lake Powell behind Glen Canyon Dam. However, without an accelerated salinity control program, concentrations are expected to reach 1,100 mg per liter by the year 2010 (Bureau of Reclamation, 1983a).

The United States–Mexico Treaty for Utilization of Waters of the Colorado and Tijuana Rivers and of the Rio Grande, signed in 1944, guarantees Mexico an annual delivery of 1.5 maf from the Colorado River below Imperial Dam. Between 1951 and 1960, Mexico received an average of 4 maf per year, the quality of which was near that of the water used in California and Arizona. However, Glen Canyon Dam was completed in 1961, and the need for storage in Lake Powell caused the flows to Mexico to fall to the compact limit. At the same time, the Wellton-Mohawk Irrigation Project came into operation, discharging large volumes of brine into the river. The saline concentration of the water delivered to Mexico rose to approximately 1,500 mg per liter, causing extensive damage to irrigated agriculture in the Mexicali Valley (Oyarzabal-Tamargo and Young, 1978).

With passage of the Colorado River Storage Project Act in 1956, the Secretary of the Interior was directed by Congress to study the quality of the Colorado River and its tributaries and to investigate possible means by which the quality of water could be improved. In 1971, the Colorado River Water Quality Improvement Program (CRWQIP) was initiated; its purpose was to analyze methods by which salinity control objectives can be set and achieved. Title II of the 1974 Salinity Control Act instructs the Secretary of the Interior to expedite the salinity control program outlined by the CRWQIP (Colorado River Basin Salinity Control Forum, 1984). Numerical salinity criteria established for Hoover, Parker, and Imperial dams were 723 mg, 747 mg, and 879 mg per liter, respectively. It was estimated that 2.2–2.8 million tons of salt per year would have to be removed from the river system by 2010 in order to meet the criteria. Title II authorized the construction, operation, and maintenance of four salinity control units (Grand Valley, Paradox Valley, Las Vegas Wash, and Crystal Geyser) and the completion of preliminary reports on 12 other projects. The 1984 amendments to the Salinity Control Act direct the secretaries of the interior and agriculture to give preference to projects that reduce salinity at the least cost per unit of reduction, instruct the Secretary of the Interior to submit final implementation reports to Congress and basin states prior to spending construction funds, and direct the Secretary of the Interior to undertake feasibility studies on the use of saline or brackish wastewaters in industrial production.

Subbasin 5, the Grand Valley, encompasses 126,000 acres of land in west central Colorado along the mainstream river. Agricultural activity covers roughly 50,000 acres, mostly irrigated from unlined canals and laterals. The valley itself is cut into the Mancos shale formation, which is a high salt-bearing shale. The average salinity of the water delivered to farms is 500 mg per liter. The Soil Conservation Service (1977, p. 49) has tested the return flows at numerous locations and found concentrations ranging from 1,600 to 9,000 mg per liter and averaging 4,200 mg per liter. The Bureau of Reclamation estimates the Grand Valley salt load contribution at 580,000 tons (U.S. Department of the Interior, 1983, p. 52), that is, 11.6 tons per irrigated acre, raising the concentration at Imperial Dam 59 mg per liter. The salts emanate from deep percolation from irrigation applications and seepage from conveyance systems coming into contact with the Mancos shale.

The Bureau of Reclamation's Grand Valley Unit was built to increase irrigation efficiency by improving conveyance systems and irrigation techniques. Stage I involves lining 6.8 miles of the Government Highline Canal and associated laterals in order to reduce conveyance seepage. Monitoring of Stage I to date indicates a reduced salt load of 14,200 tons (a reduction of 14.2 mg per liter at Imperial Dam) (Colorado River Basin Salinity Control Forum, 1984). The ultimate objective of Stage I is to remove 28,000 tons of salt annually. Stage II entails lining the west, middle, and east reaches of the Government Highline Canal and associated laterals. An estimated 164,000 tons of salt will eventually be removed from the river at an annual cost of $719,000 and $766,000 per mg per liter reduction at Imperial Dam for stages I and II, respectively (U.S. Department of the Interior, 1985, p. 63).

While the Bureau of Reclamation has been establishing large capital-intensive projects, the Department of Agriculture (USDA) has been experimenting with various on-farm measures for reducing the salt load of the river. The USDA Salinity Laboratory is experimenting with the use of saline water in irrigating certain salt-tolerant crops; the Soil Conservation Service has been evaluating automated irrigation systems in Colorado.

The USDA's beneficial saline strategy involves intercepting irrigation drainage return flows before they are mixed with the river (Bureau of Reclamation, 1984). This saline water in turn is used for irrigation at certain periods during the irrigation season of some crops. When the drainage water is no longer useful for irrigation, the water is discharged to evaporation ponds. This strategy is intended to reduce diversions and the salt loading of the river by irrigating salt-sensitive crops (lettuce, alfalfa, corn, etc.) in rotation with river water, while salt-tolerant crops (wheat, cotton, sugar beets, barley, oats, etc.) are irrigated with drainage water. The switch to drainage water would occur after seedling establishment, and preplant and initial irrigations would be done with river water. Long-term feasibility has not yet been established.

There is growing evidence that light, frequent irrigations are beneficial to plant growth and that they reduce the salt load entering the river (Bureau of Reclamation, 1982). With a "cablegation" automated irrigation system, soil water contents generally do not reach the extreme lows and highs of typical flood irrigation. Through avoidance of the lows, water stress and associated crop yield reductions are lessened. Eliminating extremely high soil water avoids deep percolation, thereby reducing the return of saline water to the river.

A major project of the Salinity Control Program is the Paradox Valley Project. The Paradox Valley is a collapsed salt anticline in west central Colorado along the Dolores River. Numerous brine seeps enter the river along a 1.2-mile stretch. With an average concentration of 260,000 mg per liter, the brine contributes about 205,000 tons of salt to the river system each year (Colorado River Basin Salinity Control Forum, 1984). The salinity control proposal calls for lowering the freshwater-brine interface below the river channel by groundwater pumping, injecting the brackish water into deep wells within the valley. The project is to remove 180,000 tons of salt annually from the river system at a cost of $11–$28 per ton, or $107,000–$266,000 per mg per liter (Bureau of Reclamation, 1983b).

Another major project involves Las Vegas Wash in Nevada, a natural drainage channel, the lower reach of which is now a perennial stream owing to sewage discharges from the Las Vegas metropolitan area and other wastewaters. Nearly 230,000 tons of salt were discharged in 1982 (Bureau of Reclamation, 1982). The high salt load is caused by the disposal of wastewater and the consequent leaching of salt from the underlying saline deposits. Reducing groundwater flow has been proposed to reduce salinity. The Las Vegas Wash Unit is to remove 71,000 tons of salt from the river system at an annual cost of $10.30–$11.50 per ton, or $102,000–$114,000 per mg per liter (Bureau of Reclamation, 1983b).

Economic analysis of salinity control alternatives indicates that salinity control should be undertaken to the point at which the marginal benefits of salinity reduction (that is, damages avoided) equal the marginal cost of control. We have seen (Gardner, 1983) that these damages approximate $492,969 per mg per liter in the 875–1,100 mg per liter range. Both the Paradox Valley and Las Vegas Wash Units appear to be cost-effective means for salinity control because benefits appear to exceed costs per ton of salt removed. For the Grand Valley Unit, however, costs of abatement are greater than the damages avoided. The other CRWQIP projects authorized by Congress (LaVerkin Springs, Lower Gunnison Basin, Unita Basin, McElmo Creek Basin, Glenwood-Dotsero Springs, and Big Sandy River) have benefit-cost ratios of less than one (Gardner, 1983, p. 187). A summary of Bureau of Reclamation and Department of Agriculture salinity projects is given in Table 5.18 and Figures 5.6 and 5.7.

In addition to the poor cost-effectiveness of most of the projects, annual salt removal is significantly below what is needed to ensure meeting the criteria in 2005. The Bureau of Reclamation, in cooperation with private firms, is therefore looking at new ways of either disposing of or putting to beneficial use saline and brackish water. The most promising prospects for wastewater use at this point are in energy development, slurry lines, and disposal in dry lakes (for example, Sevier Dry Lake and dry lakes in the desert of southeastern California). Roughly 610,000 acre-feet per year of saline water containing 2.6 million tons of salt could be collected for disposal or use in energy development and in slurry lines (Colorado River Basin Salinity Control Forum, 1984, p. 40). Desalination or evaporation of the saline waters could cost $4–8 billion. The same degree of salinity control through beneficial consumptive use of saline waters may cost much less (Bureau of Reclamation, 1983a). One possibility is the use of highly saline wastewater in power plant cooling. With

Table 5.18. *Salinity control program summary, Upper Colorado River Basin*

Unit	Potential salt reduction[a] (1,000 tons/yr)	Estimated salt reduction to date (1,000 tons/yr)	Annual cost[b] ($/ton)	Effect at Imperial Dam: TDS reduction (mg/liter)	Effect at Imperial Dam: Annual cost[b] ($/mg/liter)
U.S. Department of the Interior					
Authorized for construction and/or completed					
Grand Valley, Stage I	28	17.7	72	2.8	719,000
Grand Valley, Stage II	136		77	13.6	766,000
Las Vegas Wash	92		10	9.2	102,000
Lower Gunnison Basin	141		71	14.1	712,000
McElmo Creek	24		50	2.4	500,000
Meeker Dome	57	48	15	4.8	152,000
Paradox Valley	180		25	18	250,000
Authorized for planning					
Big Sandy River	78		69	7.8	691,000
Dirty Devil River	20		74	2	740,000
Glenwood–Dotsero Springs	284		121	28.4	1,210,000
LaVerkin Springs	53		190	5.3	1,900,000
Lower Gunnison Basin, North Fork	NA		NA	NA	NA
Lower Virgin River	NA		NA	NA	NA
Palo Verde Irrigation District	11		28	1.1	280,000
Price–San Rafael Rivers	30		35	3	350,000
Saline Water use	160		NA	NA	NA
San Juan River	NA		NA	NA	NA
Sinbad Valley (BLM)	7		75	0.7	751,000
Uinta Basin	26		90	2.6	903,000
U.S. Department of Agriculture[c]					
Authorized for construction					
Big Sandy River	35		30	3.5	300,000
Grand Valley[a]	130	23.3	24	13.0	240,000
Lower Gunnison Basin	335		56	33.5	560,000
Mancos Valley (preliminary)	20		89	2.0	890,000
McElmo Creek	38		79	3.3	790,000
Moapa Valley	20		38	2.0	380,000
Price River (preliminary)	62		NA	6.2	NA
San Rafael River (preliminary)	62		NA	6.2	NA
Uinta Basin	77	12.8	96	7.6	960,000
Virgin Valley	37		9	3.7	90,000

NA = not available.

[a]Reflects values presently included in the Colorado River Salinity Study data base.

[b]The estimates represent, at best, either appraisal- or feasibility-level costs. Caution must be used in drawing comparative conclusions for placing priorities on projects based on these cost-effectiveness values.

[c]Indexed to 1982 prices.

Source: U.S. Department of the Interior, 1983.

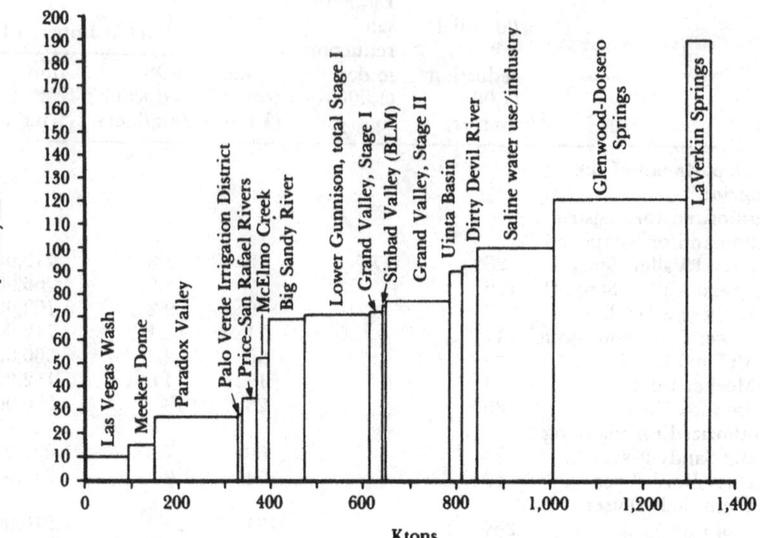

Figure 5.6. Cost-effectiveness and salt reductions for Department of the Interior projects at Imperial Dam (U.S. Department of the Interior, 1985).

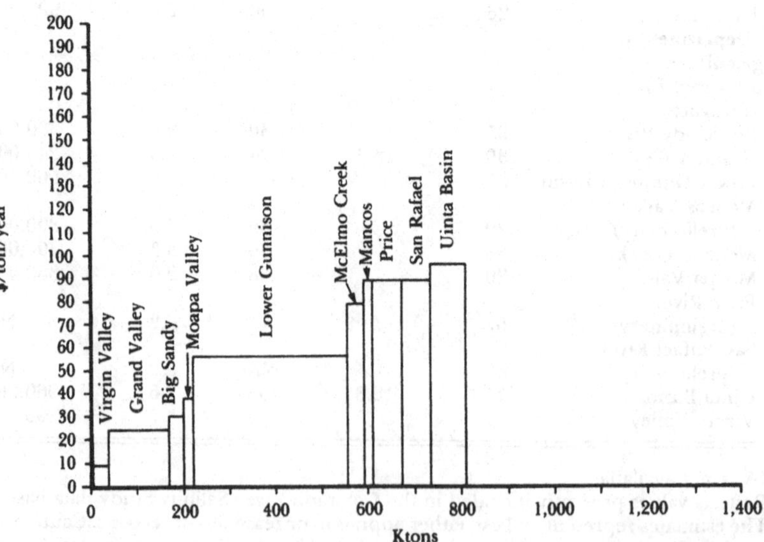

Figure 5.7. Cost-effectiveness and salt reductions for Department of Agriculture projects at Imperial Dam (U.S. Department of the Interior, 1985).

minimal pretreatment, water from 2,100 to 4,000 mg per liter TDS can be used in some cooling processes (Water and Power Resources Service, 1980, p. 3). California, Wyoming, and Utah are active in this pursuit.

The Big Sandy River rises in the Wind River Mountains of southwestern Wyoming, and the water is of good quality. By the time irrigation return flows from the Eden project are mixed with the river, 19,565 acre-feet of highly saline water (1,000–6,000 mg per liter) and 164,000 tons of salt are discharged into the Green River (U.S. Department of the Interior, 1985). The Bureau of Reclamation has recommended piping saline water to the Jim Bridger Power Plant for use in power plant cooling (Bureau of Reclamation, 1985a).

In May 1985, the Bureau of Reclamation issued its final report on using Big Sandy River Unit water at the Jim Bridger Power Plant. One option would remove 50,250 tons of salt from the river per year at a cost of roughly $70 per ton (Bureau of Reclamation, 1985a, tables ES-4 and ES-5). This option would reduce the salinity concentration at Imperial Dam 4.57 mg per liter at an annual cost of $70,000 per mg per liter. The most cost-effective process option would remove 25,125 tons of salt per year at an annual cost of $45 per ton, for reduced salinity concentrations at Imperial Dam of 2.54 mg per liter at an annual cost of $450,000 per mg per liter. Thus, the latter option may be economically marginal. The Bureau of Reclamation has studied available technology for saline water use at Hunter Powerplant in Utah. Findings indicate that the binary cooling tower is not cost effective compared to other saline water use equipment and that other processes involving off-the-shelf hardware are efficient in using saline water in cooling applications (Colorado River Basin Salinity Control Forum, 1984).

Although interest in using saline water for cooling is widespread, the process is acceptable only for new generating capacity. The use of highly saline water in other forms of energy development (oil shale, tar sands, coal gasification, etc.) holds promise if those resources are ever developed.

The use of saline water in slurry pipelines for transporting coal, potash, trona, and other marketable minerals from western fields to market areas is under consideration in a number of instances. The most ambitious program is the Aquatrain program. All the aforementioned uses of saline wastewater would be tied together by a pipeline carrying highly saline water to points of beneficial consumptive use in the western states. In its original form, the project envisioned a saline water pipeline carrying plastic capsules of clean, dry coal to the West Coast. In 1983, this proposal was dropped in favor of a double-barrel pipeline, one carrying saline water and the other carrying a liquid carbon dioxide and coal slurry. A report completed in 1984 identified various input and output points (that is, coal mines, carbon dioxide, saline water sources, power plants, and export sites) in southwestern Wyoming, western Colorado, Utah, northern Arizona, central and southern Nevada, and southern California. Probable uses of the saline water include power plant cooling, oil shale development, solution mining, tar sand development, and hydraulic mining. If all potential sources of saline water are used, 160,000 acre-feet per year of water could be transported to users and 900,000 tons of salt could be removed annually (Bureau of Reclamation, 1985b, p. 88). The Bureau of Reclamation will attempt to determine the potential benefits and costs of the project.

The disposal alternative (transporting water to dry lake beds and evaporating the water) for removing highly saline water from the river system is probably the least viable. It will most likely involve an inter-basin transfer of water because the proposed dry lake beds lie in areas outside the basin of origin. Consequently, a plethora of legal and institutional constraints are brought into the picture. The Bureau of Reclamation proposed to Wyoming the piping of saline water into the Great Divide Basin, and the state rejected the proposal. Colorado law, in addition to requiring compensatory storage for the basin of origin in trans-basin diversions, does not recognize evaporation as beneficial use of water.

Regulation of the Colorado's natural flow has significantly altered the seasonal and annual variations in flow and salinity concentrations. Between 1963 and 1980, massive net amounts of water were stored in the basin. Storage capacities reached 50 maf in 1980 (U.S. Department of the Interior, 1985, p. 25). Between 1976 and 1980 the average yearly reservoir evaporation in the Colorado River Basin was 2,114,000 acre-feet (U.S. Department of Interior, 1985, p. 14), leaving lower-quality water to pass through the turbines or spillways for downstream uses. With the initial filling of Flaming Gorge, Reudi Reservoir, and Lake Mead, significant leaching of calcium sulfate (gypsum) occurred. Long-term salt leaching at Flaming Gorge Reservoir is being studied. There is strong evidence that between 1965 and 1980, Flaming Gorge Reservoir and Lake Powell stored high TDS water and routed lower TDS spring runoff downstream (U.S. Department of the Interior, 1985, p. 25). Bank storage, chemical precipitation, ion exchange, and oxidation reduction are thought to have prevented high TDS water from being released from these reservoirs. TDS may also be influenced by sedimentation in reservoirs. In contrast to a riverine environment, where suspended sediment may continue to release salts and exchange ions, sediment once settled out in a reservoir may limit salt and ion exchange capabilities (U.S. Department of the Interior, 1985, p. 25).

Unconventional approaches to salinity control

The alternative approaches discussed above include both large capital-intensive projects and some on-farm measures by the Soil Conservation Service (USDA). The latter involve installing gated pipe and other improved mechanical systems for improving irrigation efficiency. In addition, Young and Leathers (1978) have studied other on-farm options that primarily involve more careful management of irrigation water application and modified cropping patterns.

Howe and Young (1978) and Howe and Orr (1974) studied the savings in consumptive use of water and salt in the return flow that would be involved in phasing out marginal agricultural lands in the Grand and Uncompaghre Valleys of Colorado. The former study considered a phase-out of 8,800 acres in the Grand Valley and 10,200 in the Uncompaghre, calculating the direct and indirect losses that might ensue. Income losses were about $16.30 per ton of salt reduction, but for every ton of salt reduction, 0.17 acre-feet of consumptive use was also avoided. Valuing this water saved at the hydroelectric opportunity cost of $31.36 per acre-foot (Table 5.16) plus the direct Lower Basin agriculture opportunity cost of $30 per acre-foot, the net cost is reduced to about $6.50 per ton from the viewpoint of the entire Colorado Basin.

Taking into account both the salt reductions and changed consumptive uses (positive and negative), the net costs per ton of salt reduced are compared in Table 5.19. The activities have been ranked by net cost per ton of salt reduction from a basinwide viewpoint. It is clear that on-farm management and irrigation system changes and irrigated acreage constitute the most cost-effective approaches to salinity reduction. The larger projects are substantially more costly.

The problem is to motivate use of the low-cost alternatives. The on-farm measures of Young and Leathers (1978) would be paid for by the farmer, and all the federal programs are likely to be paid for by the federal government. Acreage retirements could be made attractive to farmers by having the state government or a federal agency offer to buy either the land or the irrigation water. Given the low direct income per acre (or per acre-foot), the offering price would not have to be high.

Table 5.19. *Calculation of net cost per ton of salt removed by various methods, Upper Colorado River Basin*

Activity and source	Independent salt-saving potential (tons/yr)	Salt-saving potential with Grand Valley acreage retirement (tons/yr)	Cost per ton of salt reduction ($/ton)	Consumptive water use reduction (acre-feet/yr)	Value of water saved per ton of salt reduction[a] ($/ton)	Net cost per ton of salt for: Entire basin	Net cost per ton of salt for: Upper Basin only
On-farm practices (Young and Leathers, 1978)	110,000	93,500	$ 3.33	44,700	$29.33	–$25.33	$ 3.33
Grand Valley acreage retirement (Howe and Young, 1978)	88,000	88,000	16.30	14,800	10.30	6.00	16.30
Uncompaghre acreage retirement (Howe and Young, 1978)	102,000	102,000	16.30	16,000	9.63	6.67	16.30
U.S. Department of Agriculture on-farm Grand Valley program	130,000	110,500	24.00	15,850	8.80	15.20	24.00[b]
Paradox Valley (Bureau of Reclamation, 1982)	180,000	180,000	25.00	–10,000	–3.40	28.40	25.00[b]
Modified crops (Young and Leathers, 1978)	25,000	21,250	67.27[c]	7,870	22.72	44.55	67.27
Bureau of Reclamation Grand Valley I and II	164,000	139,400	76.00	20,000	8.80	67.27	76.00[b]
Bureau of Reclamation Big Sandy	50,250	50,250	70.00	–8,000	–12.10[d]	82.10	70.00[b]

[a]Value per acre-foot = $30 Lower Basin agricultural uses plus $31.36 hydropower value, except Big Sandy.

[b]Federal projects whose costs will probably be paid largely by the federal government.

[c]Measured by loss in net farm income.

[d]Value per acre-foot = $30 plus $46 hydropower value. (See Table 5.17.)

The potential for intrastate and interstate water markets involving the Upper Basin

The recent literature on privatization of resources emphasizes the likely advantages of private ownership and market exchange over bureaucratic control and allocation. Stroup and Baden (1983) emphasized the rigidities of nontransferable public rights in resources and the inefficiencies of bureaucratic resource management in energy, groundwater, and timber. Anderson (1983a, 1983b) has analyzed the transition of western water rights from early (mining) appropriations doctrine to a more restrained system based on the common law concept of beneficial use. The thrust of these arguments is a return to the market, away from centralized, bureaucratic control.

It has always been recognized, however, that the establishment of markets is severely inhibited and the outcome likely to be inefficient in the presence of pervasive externalities that result either from the user's consumption of the resource (for example, water pollution) or directly from the buyer-to-seller transaction (for example, through changes in return flows). Water resources are a case in point, in which the sale of water nearly always has positive and/or negative direct impacts on third parties. Yet, fairly extensive markets have developed for water. These markets, although sometimes involving rather high transactions costs, have been successful in transferring water from lower- to higher-valued uses over time in many locations of the West.

This section identifies characteristics that would be generally desirable for resource allocation mechanisms and argues that, for water, markets possess more of these characteristics than their alternatives. Some shortcomings of water markets are also described.

Desirable characteristics of resource allocation mechanisms and their implications for markets

Increased flexibility in the allocation of existing water supplies is vital, so that water can be shifted from use to use and place to place as climatic, demographic, and economic conditions change over time. Both short- and long-term flexibility is needed. Within a given season, rainfall and temperature conditions frequently vary from expected values, opening up possibilities for beneficial water exchanges among agricultural users and between the agricultural sector and municipalities. In the longer term, regional population patterns and economic structures change, increasing the relative marginal value of water in some uses and decreasing it in others.

A second desirable characteristic is security of tenure for established users. Only if the water user can be assured of continued use will the user invest in and maintain water-using systems. Security of tenure need not be inconsistent with flexibility as long as users can voluntarily respond to incentives for reallocating water (the third characteristic).

Fourth, it is important that water allocation mechanisms confront the user with the real opportunity cost of the water being used. Only then will the economically efficient technology of water use be chosen. Naturally, pursuit of this goal must be subject to constraints imposed by other social goals – for example, equity and environmental protection.

Fifth, a water allocation process should be perceived by the public as equitable and fair. Water users should not impose uncompensated costs on other parties, and parties giving up water should be compensated, as should those injured by changes in points of diversion or return flows. This goal, plus the security characteristics mentioned above, again imply that most changes in water allocation should be noncompulsory.

Sixth, a socially responsible water allocation process must reflect public values that may not be adequately considered by individual water users. For example, water quality and instream flow maintenance may generate large public values but may be of little concern to individual water users.

Seventh, it is desirable that the outcome of the allocation process be predictable in particular applications. If a buyer and seller identify a particular transaction, it is desirable that they be able to predict how much water they can transfer. Applications to a water control board may be subject to adjustments by the state engineer or the water court to protect third parties. All allocation mechanisms will incorporate some degree of hydrologic uncertainty.

The characteristics mentioned above are found to a large degree in market processes for the allocation of scarce resources. First, of course, for a market to exist and function well, there must be well-defined property rights in the resource being allocated. Such definition is particularly difficult in the case of water. Property rights and the existence of markets guarantee flexibility in response to changing values. They also guarantee security of tenure in that no one has to sell water unless it is advantageous to do so. Market prices will confront potential users with the appropriate opportunity costs if the geographical extent of the market incorporates all physical impacts of water transactions. Fair compensation is guaranteed between buyers and sellers in a market almost by definition, because transactions are voluntary and payment does take place (in contrast to the Hicks-Kaldor criterion, in which beneficiaries must be able only to compensate the losers). Various public values could be reflected in market-determined allocation if units of local government and/or consortia of private interests could buy and sell in the market to protect instream flows and water quality (this practice is certainly not widespread). If such purposes are not possible, the market environment can sometimes be appropriately conditioned through taxes, subsidies, or regulations to reflect externalities or inadequately priced opportunity costs. Further, under some state systems of market allocation, especially those involving a count process, there is some uncertainty concerning the outcome of the process.

Thus, the market is not a perfect mechanism, but it fares well compared with other arrangements on most of the criteria above.

Water market failures and their mitigation

The main problem is the imposition of externalities on parties other than the buyer and seller. Such third-party effects take the form of changed return flows that are being used by others, groundwater level changes, and water quality changes. The return flow problem was first discussed by Hartman and Seastone (1970), and water quality externalities have been extensively discussed in the water quality literature. When third-party effects result from market transactions, there is no guarantee that the transactions will result in an overall increase in net benefits unless the third-party effects can be clearly quantified and are required to be compensated.

However, in determining the comparative desirability of markets versus other allocative mechanisms, the relevant comparison is not between markets that can function without externalities and water markets with their inherent third-party problems; rather, it is between water markets and other rational allocative mechanisms. All rational mechanisms would require the determination of third-party effects and their consideration in the allocation process. In this regard, the costs associated with markets are no greater than the costs of other rational mechanisms.

A second problem in many water market settings is that of adequate communication among potential buyers and sellers. Although large water conservancy districts with professional staffs may be able to communicate desires to buy or sell easily and cheaply, individual farmers or small towns often have a hard time finding buyers or sellers, or even knowing where to look. Matching buyers and sellers across greater distances would be more difficult.

A third problem is the maintenance of instream flows and adequate water quality. Motivation for provision of adequate instream flow and adequate water quality is lacking because of the public-good nature of their benefits (beauty of the river, sports fishing, wildlife maintenance, etc.) and because market boundaries do not extend as far as reduced flow, increased salinity, and other physical effects.

We have seen clearly in Table 5.16 that large instream power and water quality values, plus the opportunity cost of water in Lower Basin applications, are not reflected in current market prices for water because present water markets are limited to state boundaries. In addition, the limitations placed on instream uses as beneficial uses make it difficult to protect those values. Howe and Lee (1983) have argued that the best way to protect instream values within a state is to permit units of local government to acquire water rights and to hold them for purposes of instream flow maintenance.

A fourth problem is the asymmetric treatment of return flows in a market context. Although third parties are protected from damage (or are guaranteed adequate compensation) resulting from reduced return flows, third parties that gain from increased return flows are not made to pay for these benefits. Consequently, the costs to some third parties (or the costs of protecting those third parties) are reflected in markets, but the benefits to others are not. And some transactions that should take place do not. Third-party effects must be taken into account, and, from an equity viewpoint, compensation should be paid to losing third parties so that no one is worse off as a result of the transaction. The main issue in making markets work more efficiently is to identify and quantify these effects accurately and quickly and to reach agreement on their magnitudes so that compensation or adjustment to the original property rights can be made without excessive transactions costs.

In the United States, systems of state water law provide safeguards for potentially damaged third parties. Prospective transactions must be advertised, and parties that perceive themselves to be damaged have recourse to administrative actions or court hearings that will result either in compensation for damages from the buyer or in modification of the water right being transferred (for example, the amount and timing of permitted diversions). In New Mexico, the state engineer's office identifies third-party effects and also proposes modifications to the right being transferred (for example, the amount and timing of permitted diversions) that should make the transfer acceptable to all parties. The buyer has the option of paying compensation to damaged third parties to effect their agreement to the transfer (see Hartman and Seastone, 1970, chap. 3). In Colorado, proposed transfers of water rights are advertised, and third parties are allowed to register their objections with the district water court. The buyer can then either negotiate with these parties to get their consent, agree to a court hearing on the transfer, or drop the transfer. In many cases, negotiation over compensation to third parties is successful (see Howe, Alexander, and Moses, 1982).

The court trial process is costly and time consuming, and it fails to produce the best analysis of the case. Because the process is adversarial, each party brings its own experts, who have strong motivation to bias their analyses in favor of their clients. The court must select a compromise position, so the outcome of a case is hard to predict. In one case, the city of Boulder was permitted to divert 18 percent of the entitlement of a ditch company share that the city purchased and was also restricted to diverting water at the same times as the ditch company. When buyer, seller, and third parties cannot agree, court modification of rights to prevent third-party damages can result in a large reduction in benefits to buyer and seller in order to avoid only small losses to third parties.

Communications problems between potential buyers and sellers may require a state agency to help make a market by establishing procedures for the publication of desire to buy or sell. Although such services could be provided by private brokers, the market may be too thin to support an institution that does not already exist for other purposes.

Instream flow values can be reflected in the purchase and holding of water rights by units of government or consortia of private interests that are concerned about the maintenance of these values. The obstacles, however, are legal and institutional. State laws frequently do not recognize instream uses as beneficial uses. In Colorado, state law permits the state government to file or buy water rights and to devote that water to instream flow maintenance, although lower levels of government and private parties are not allowed to do so. Institutionally, it may be costly to organize the large number of individuals, each of whom has a small interest in these values. However, in a dynamic setting where town, county, and state governments could buy water rights on behalf of their citizens, these problems would not be as severe as implied in the literature (Howe and Lee, 1983).

To date, under appropriations doctrine, water sales have been confined within a state because state law governs such transactions. Under state appropriations doctrine, river basins may sell water to each other, subject only to existing or compact conditions. Most sales, however, take place on the same river. Water conservancy districts (WCDs) are typically established to contract with state and federal governments for raw water from large projects. In most cases, district constitutions permit transfers among users, but in many cases transfers of such state and federally provided water outside the district are not permitted. Because localized markets within WCDs and small river basins have been active for many years in the western United States, some of the largest opportunities for increased efficiency lie in interdistrict markets.

Many WCDs persist in following inefficient allocation methods. Thus, it is worthwhile to describe and evaluate any market arrangements that appear to be working efficiently. One that could well be adopted by other WCDs is that found in the Northern Colorado Water Conservancy District (NCWCD). The federal Colorado–Big Thompson Project was begun in 1937 and completed in 1957 to bring supplemental irrigation water from the western side of the Rocky Mountains to northeastern Colorado. The NCWCD was established to contract with the federal government for purchase of the water, repayment of project costs, and distribution of the water to final users (see Howe, Schurmeier, and Shaw, 1982; Maass and Anderson, 1978). The Colorado–Big Thompson Project has provided a historical average of 230,000 acre-feet, or about 17 percent of the total water supply of the region. Although this supply is primarily for supplemental irrigation, towns and a growing number of industries use the project as a raw water supply.

The NCWCD was established in 1937 to contract with the Bureau of Reclamation for Colorado–Big Thompson water and to devise methods for allocating the water available to it among potential users. The NCWCD includes areas of quite different natural water supplies in relation to the amount of arable land. As a result, potential users did not want a mandatory uniform assignment of water to the land. These sentiments finally led, in 1957, to a clear-cut definition of an NCWCD share (allotment) as a freely transferable contract between the district and the holder. Each share entitles the holder to $^1/_{310,000}$ of the water available – approximately 0.7 acre-feet of water per year. The transferability of the allotments stimulated creation of a market in which they could be traded.

Much of the water needed for urban and nonagricultural industrial growth is provided by the sale of NCWCD allotments from agriculture. These nonagricultural users often "rent" excess water to irrigation on a short-term (annual) basis. In the early years, many irrigators gave away their allotments because they did not want to pay the $1.50 annual charge. Allotment prices increased from $30 in 1960 to $291 in 1973. This trend accelerated sharply in 1974, with average prices reaching a peak of $2,161 in 1980. Since 1980, prices have fallen to about $900 because of the completion of a new transmountain water supply project that will supply water to some towns in the region.

Rentals are transfers of water among users for one season only without transfer of the allotment titles. The NCWCD office facilitates communication among prospective buyers and sellers, so that rentals are easily effected in response to relatively small discrepancies in water values among users. About 30 percent of the Colorado–Big Thompson water is involved in rental transactions each year; towns are the main renters of water to agriculture.

Third-party and instream flow problems have not been solved in the NCWCD. The complexities and high transactions costs imposed on most water transfers by possible third-party intervention have been circumvented because the district has retained title to all return flows. This practice is permitted under Colorado law for waters newly imported to a basin. Third-party effects still do occur, leaving open the possibility that some sales are economically inefficient. Regarding instream flows, Greenly, Walsh, and Young (1982) believe that improved water quality in the South Platte River and its tributaries would be valued highly by recreationists, residents, and citizens expressing strong option values. Their estimates indicate a high value for instream flows in the NCWCD, but no specific arrangements have been made to guarantee such flows.

Although the NCWCD market arrangements have shortcomings from an efficiency viewpoint, they appear to be much more efficient than alternative water distribution mechanisms. Transaction costs are much lower than with the transfer of water rights under state laws. Flexibility is infinitely greater than under many Bureau of Reclamation contracts that prohibit water transfers from specific land parcels. Security of tenure is greater than that found under administrative procedures, such as those in the Southeastern Colorado Water Conservancy District, where water is reallocated annually by the board of directors (see Hartman and Seastone, 1970, chap. 6). The possibility of easily and advantageously replicating these NCWCD market structures in other project areas warrants serious consideration.

The potential for interstate water markets

The major device for allocating the waters of interstate rivers among the riparian states has been the congressionally approved interstate compact. Major examples are the Delaware and Potomac compacts in the eastern United States and the Yellowstone, Colorado, Rio Grande, and Columbia (international) compacts in the West. They were negotiated among the riparian states (and with Mexico) and approved by Congress. Any revisions would require unanimous agreement among the signatories.

Another device for allocating interstate streams is litigation between states under the doctrine of equitable apportionment. In suits between states, the Supreme Court has original and exclusive jurisdiction. The doctrine is quite complex and is sharply tailored to individual situations, generally following neither appropriations nor riparian doctrine.

Western states continue to express concern that water allocated to them by interstate compact may, in fact, be effectively taken away by other states through nonuse. There appears to be no legal basis for such loss, but political leaders apparently fear that "use it or lose it" will somehow apply between states as it does within states, perhaps through political alliances against financing upstream water projects. Some erosion of the ability to effect the utilization of compact-granted water has occurred because of salinity control programs and the Endangered Species Act. As a result of these fears, political pressures have risen for the construction of new water projects to tie down this water. Some projects proposed largely for this purpose (for example, the Animas–La Plata Project in Colorado) are grossly inefficient from an economic viewpoint and would certainly tie the water to uses of little value in the long run, rather than protecting it for important future developments.

It is clear that the status of potential interstate water sales by either private appropriators or public bodies is in a state of legal flux. Although it seems clear from *Sporhase* v. *Nebraska* and *City of El Paso* v. *Reynolds* that blanket prohibitions of interstate transfers are unconstitutional, necessary conditions for legal sales have not become clear. An interesting recent proposal is that of the Galloway Group, Ltd., a Colorado corporation that wants to sell surface water apportioned to the state of Colorado but purportedly claimed under Colorado water rights by entities in Arizona and Southern California. (For an excellent analysis, see Gross, 1985.) Galloway claims to have water rights to 1.3 maf of water per year on the Yampa and White rivers in western Colorado, in the Upper Colorado Basin. Galloway intends to raise more than $200 million of private capital to build dams on the two rivers to generate electric power and store water.

In August 1984, the San Diego Water Authority paid Galloway $10,000 for an option to lease 300,000–500,000 acre-feet for 40 years. Many questions remain unanswered. Gross (1985) has concluded that the Colorado River Compact and the Upper Colorado River Compact preclude the Galloway proposal, mainly through implied territorial use limitations. Gross further concludes that the compacts, as federal law, are immune from Commerce Clause attack.

On the economic side, there are questions of the price it would be reasonable for San Diego to pay, given the alternatives, and the effects that a clearing of legal barriers would have on the total supply coming from the Upper Basin. At a time when the western power market is overbuilt, when Colorado has excess storage capacity in existing Western Slope reservoirs, and when the Colorado Basin's total storage is so large that total basin yields fall with added storage, it seems to make little sense from the financial and economic viewpoints to make such large investments in storage. If the proposal is someday permitted, it should be without the waste of added storage.

The answers to many questions are yet unknown. Must the water be confined to a pipeline? Is it sufficient that it be made part of a larger product (chicken soup or coal slurry)? Can water sold be allowed to remain in the stream to be abstracted downstream by the buyer? Can a state government lease part of the water allocated to it under interstate compact but not currently used (for example, waters unappropriated under state water law or held by the state for state uses)?

Would interstate water leases or sales help affirm the titles to such waters? Would there be a market for such water? Against which state's compact allotment would such transactions be counted? Would California, which has been using waters unused in Colorado and Arizona for many decades, be willing to pay something for a longer-term lease that would assure continued delivery for a known period? Would such an arrangement eliminate the pressure for nonsensical "use it or lose it" projects? The status of water allocated to western Indian tribes under the federal reservation doctrine and the *Winters* decision could be a much larger issue in the next decade.

In a river basin context, supplies of water for transfers out of state or from the Upper Basin to the Lower Basin can come from unused water in excess of deliveries that may be required by compact and from water withdrawn from current consumptive uses. Regarding unused water, it is not clear that conditions for exchange exist, because there is no practical way to withhold the water; nor is it clear that the Upper Basin can legally claim water that it cannot use consumptively. The water cannot be stored unless there is a consumptive use, and it will continue flowing downstream anyway. The only motivation for making a contract on such water would be to guarantee that consumptive uses will not be developed over a specified time so that continued downstream availability could be guaranteed.

Table 5.5 showed the mean annual discharge of unused water originating in each subbasin of the Upper Colorado Basin. Out-of-basin parties willing to pay the Upper Basin not to develop this excess water (averaging 2,894,000 acre-feet per year) would naturally be concerned about the reliability of this supply. However, there is so much storage on the main-stem Colorado and its major tributaries (55.6 maf of active storage, approximately 4.3 years' average flow) that water deliverable below Hoover Dam could be made quite reliable through Bureau of Reclamation storage and release arrangements.

Water could also be transferred out of the Upper Basin through transfer of established water rights that are currently being used. Only agricultural rights are relevant to potential transfers because of their relatively low value, and they represent 70 percent of total diversions and more than 90 percent of total consumptive use. The annual consumptive use of water by crop in each subbasin and the associated net return to the farm enterprise per acre-foot of consumptive use were presented in

Table 5.20. *Supply curve for water trading, Upper Colorado River Basin*

Offering price per acre-foot	Incremental offering (acre-feet)	Cumulative acre-feet offered
$ 5	36,452	36,452
10	30,565	67,017
15	65,803	132,820
20	7,591	140,411
25	208,079	348,490
30	92,171	440,661
35	346,646	787,307
40	322,256	1,109,563
45	170,548	1,280,111
50	92,361	1,372,427
55	120,741	1,493,213
60	15,219	1,508,432
65	18,400	1,516,832
72	49,373	1,566,205

Table 5.13. These data permit construction of a crude supply curve of water from existing agricultural uses, assuming that offer prices in excess of the net returns experienced per acre-foot consumed will, sooner or later, induce farmers to sell that water. For each subbasin, such a supply curve can be constructed by arranging the crops in increasing order of net return and cumulating the quantities of water that would be forthcoming at that net return (or lower).

Based on the data of Table 5.13, a supply, or offer, curve is constructed for the Upper Basin, using $5 offering price intervals and cumulating the amounts of water that might be forthcoming at each offering price (see Table 5.20).

The conclusion is that lots of water is likely to be forthcoming from the agricultural sector at relatively low prices if the process is not subject to state control. The above data represent a private agricultural sector point of view of accounting stance, that is, private profitability as a criterion for giving up irrigation water. However, from a state or overall Upper Basin viewpoint, things look quite different, as shown earlier in Table 5.15. In the various subbasins, average state income losses per acre-foot consumed range from $74 to $160 per acre-foot. A regional official looking at Table 5.15 would be concerned that the loss of regional income would not be made up by new water-using activities (that might be out-of-state) or that the proceeds to the seller might not be reinvested in the state. Transfers that are highly beneficial from a national stance and are modestly beneficial from a private stance can be perceived as highly harmful to the economy of the exporting region.

This difference will continue to be a major point of contention regarding out-of-state or out-of-basin sales.

Principal findings and policy recommendations

Principal findings

The Upper Colorado River Basin is not and under foreseeable circumstances will not be short of water for consumptive uses. Estimated current consumptive uses per year total 2.7 maf (Spofford et al., 1980, chap. 6, modified), including those associated with publicly supplied waters, rural domestic and livestock supplies, irrigation, self-supplied industrial uses, thermal-electric generation, tributary groundwater use, and export. Table 5.21 indicates the Upper Basin and state availabilities for consumptive use under three assumed values for average virgin flow. The 13.5 maf estimate is the lowest in current use, but because there are periods of persistent low flow, the effects of a repetition of the lowest ten-year flow in this century (from 1931 to 1940) are considered. Only in the latter case could there be a shortage. Then, the annual shortage could be met by net releases from the 50 maf of active storage on the river and the reallocation of water from agriculture to other uses in order to avoid serious damage to the nonagricultural sectors.

It should also be noted that the average excess outflows from the Upper Basin of 2,890,000 acre-feet (Table 5.5) exhibit high year-to-year

Table 5.21. *Upper Colorado River Basin and state water availabilities for consumptive use under three virgin flow assumptions (million acre-feet per year)*

	Assumed virgin flows		
	13.50[a] −8.30[d]	14.05[b] −8.30	11.80[c] −8.30
Upper Basin availability	5.20	5.75	3.50
Colorado (51.75%)	2.691	2.976	1.811
New Mexico (11.25%)	0.585	0.647	0.394
Utah (23.00%)	1.196	1.322	0.805
Wyoming (14.00%)	0.728	0.805	0.490

[a]Lake Powell research project estimate (Jacoby, 1975).

[b]Used by U.S. Department of the Interior Water for Energy Management Team.

[c]Lowest ten-year flow in the twentieth century (Dracup, 1977, p. 121).

[d]Release required by Colorado River Compact plus half the Mexican obligation plus 50,000 acre-feet per year for Arizona.

Source: Modified from Spofford et al., 1980, table 10, p. 387.

variability as a source of supply for the Upper Basin, capable of being regulated only at high cost. However, these same supplies can be regulated and made available to the Lower Basin at no additional cost (as they are today) through the vast amount of storage on the river system.

Low-valued agricultural uses consume approximately 31 percent of the water available to the Upper Basin (that is, 1.6 maf, Table 5.8, out of 5.2 maf, Table 5.21). The values of these waters in terms of net farm income per acre-foot consumed range up to $72 but average $25 per acre-foot. On the other hand, if some of this water were to be transferred out of the Upper Basin, there would be somewhat larger impacts on state incomes, ranging from $75 to $160 per acre-foot (Table 5.15). These estimated state income effects stem from a "worst case" scenario and do not take into account the positive income effects of new in-basin uses or the possibility that agricultural sellers might reinvest their sales proceeds in Upper Basin activities. As more market incentives are felt to transfer agricultural water, this conflict between individual willingness to sell and state concern will escalate, with the states (rightly or wrongly) increasingly opposing transfers that are privately profitable, especially out of state.

Instream value of waters currently consumed in the Upper Basin, when viewed from the standpoint of the entire Colorado Basin, are quite high, often surpassing even the state income values that may be associated with the consumptive uses. These values arise from the effects on water quality, water-based recreation, fish and wildlife values (beyond direct recreational values), hydroelectric power generation, and Lower Basin irrigation uses. Some of these values have not been quantified in monetary terms, but it is possible to place values on the hydroelectric power effects, the value of the water to Lower Basin irrigators, and the water quality effects. The hydroelectric values were shown to range from $31 to $46 per acre-foot, depending on the subbasin of origin (Table 5.15). The likely value of irrigation water at the margin of application in the Lower Basin is about $30 per acre-foot, allowing for evaporative losses. (See the preceding section on hydrology, water use patterns, and the value of water.) The effect of consumptive use on the concentration of dissolved solids and the consequent damages to the municipal and agricultural sectors of the Lower Basin range from about $38.50 per acre-foot of consumption for water exported from the headwaters areas of the Upper Basin to about $280 per acre-foot for water applied in the Mancos shale areas of the Grand Valley (in Colorado). Thus, not counting recreational and fish and wildlife values, the values of water left in stream range from approximately $100 to $350. Surely, some rethinking of water allocation among uses is called for by the differences between these values and the private and state income values mentioned earlier.

Bureau of Reclamation electric power generated within the Colorado River Storage Project sells at extremely low prices, an average of only 11 mills per kilowatt hour. Compare this figure with the cost-avoided prices being paid for cogenerated power of around 44 mills or, in the extreme, the approximate full cost of electric energy from newly constructed coal-fired thermal electric plants (being built in the Southwest) of 8.5 cents (85 mills) per kilowatt hour. Such underpricing leads to misallocation of energy resources and energy-related investments, and it shortchanges the regions that provide the water for the hydroelectric generation.

A basic motivational problem is created by the fact that most of these instream values accrue not to the Upper Basin but to the Lower Basin and wider areas that use the power.

The high levels of dissolved solids in the Lower Basin have been seen to cause quite significant damage, approximately $0.5 million per mg per liter of water (Gardner, 1983). Yet, the Colorado River Salinity Control Program is failing to make adequate use of the most cost-effective methods of reducing salinity: changes in on-farm irrigation water management and the retirement of irrigated land in areas that contribute huge amounts of salt through their return flows (Table 5.19). The reasons for this failure are again motivational: the Bureau of Reclamation and the Soil Conservation Service find administering on-farm programs for large numbers of farmers difficult compared to constructing large point-source projects, although the farmers themselves prefer projects whose costs they do not share.

There exists no basinwide agency that is concerned with or is able to study and influence the pattern of public values and negative externalities noted above, that is, the impacts of changes in Upper Basin water use on Lower Basin users. Results of the absence of such oversight include increased Upper Basin fear of Lower Basin political power over the use of water, a consequent "hurry up, build any kind of project" attitude, lack of concern with substantial Lower Basin losses and power losses, and fear of considering ways for reallocating water over the short and long terms to the mutual benefit of all parts of the basin.

Further, it seems clear that there exists an unexploited potential in the Upper Basin for an increased role for water markets. Although water markets cannot solve all problems, they can provide the flexibility in water allocation that economic and demographic change necessitates. State water agencies can facilitate an expanded market function by providing information (for example, where there are excess water and shortages); there is also a need for public monitoring of the water market process in order to ensure important public values. Cases of successful markets need to be studied with an eye to replication.

Colorado lags behind the other Upper Basin states in relying on the water court system to deal with water reallocation and in having no mechanism for guarding public values. All the Upper Basin states lack the tools for reflecting instream values adequately. The water laws of all the states fail to recognize conservation as a beneficial use; they in fact encourage inefficient uses.

Policy recommendations: basinwide

1. Because there exists *no* river basin agency with interest in and responsibility for monitoring and overseeing the entire Colorado River Basin, and in light of the potential benefits to be gained from the existence of such an agency as noted above, it is recommended that the Colorado River Basin states that are signatory to the Colorado River Compact consider establishment of an interstate river basin commission along the lines of the Potomac and the Delaware river basin commissions, to act as a focal point for study, exchange of information, continuing dialogue, and enforcement and monitoring of agreed-upon policies. Basinwide management implies the need for basinwide compensatory arrangements so that all parties can benefit from both water planning and water transfers.

2. In light of the excess water supply in the Upper Basin that will be costly to develop for Upper Basin purposes but is now regulated through storage for Lower Basin use at almost no cost, and given the Upper Basin fears that are leading to the costly development of inefficient consumptive uses, it is recommended that thought be given to mechanisms needed to negotiate a long-term agreement with the Lower Basin states, especially California, by which the Upper Basin would be paid to agree *not* to develop new uses for a portion of the water during a specified time.

3. In light of the low private values generated by much of the consumptive water use in the Upper Basin and the likely opposition of state governments to transfers out of agriculture under current institutional arrangements, it is recommended that:

a. studies be undertaken to quantify the direct and indirect recreational values generated *in* the Upper Basin by added streamflows;
b. hydroelectric prices for power from the Colorado River Storage Project be raised toward market levels; and
c. revenues from CRSP power sales be shared proportionally among the

Upper Basin states, thereby providing them with badly needed revenues and motivating them to recognize instream values.

Policy recommendations: state level

4. Because appropriations doctrine fails to recognize conservation as a beneficial use of water, thereby denying the owner any reward for increased efficiency in use, it is recommended that the Upper Basin states consider legislation such as the Katz-Bates bill in California (1983) that so recognizes water conservation. This change would be doubly effective in motivating on-farm water management change that also reduces return flows and their dissolved solids loads. It would also motivate retirement of unproductive acreage.

5. The salinity management program is unable to motivate on-farm measures and acreage retirement sufficiently. Cost sharing as practiced by the Soil Conservation Service is not adequate because farmers resist any increase in cost that does not provide directly offsetting benefits. Although Recommendation 3 above will help, it is recommended that further steps be taken to redefine "beneficial use" in a way that reflects the availability of modern water management methods available at moderate costs. Beneficial use should require reasonable water control methods and should be differentially defined among areas to reflect special attributes of each area, especially where return flows are extremely saline.

6. To facilitate market transactions but simultaneously to give weight to those public values not reflected fully in private values (water quality, aesthetics, species preservation, public recreation, etc.), it is recommended that the riparian states consider changing to the New Mexico system under which the office of the state engineer, rather than the water courts, monitors water transfers, carries out needed hydrologic studies, and imposes public interest criteria. This change would decrease transactions costs of transfers while protecting public values.

7. In all Upper Basin states, it is recommended that the state engineer's office facilitate water transfers by providing information on local water availability and shortage. Systems such as Colorado's satellite-linked water monitoring system can provide valuable information for this purpose.

8. It is recommended that an agency of state government stand ready to buy water rights at stated prices from designated low productivity-high salinity lands to facilitate the retirement of those lands. The water would be either sold or retained by the state for instream flow maintenance.

9. It is recommended that efficiently working markets such as the Northern Colorado Water Conservancy District be carefully studied with an eye to their extension to all conservancy and irrigation districts.

The real issues confronting the Colorado Basin are primarily institutional, not technical. More research, including actual experimentation, should be devoted to institutional-motivational design. A basinwide institution is needed to identify and negotiate "win-win" changes in water allocation and management for all parts of the basin. These are exciting challenges.

Appendix: Detailed derivation of state income impacts

Use of a state input-output (I-O) model allows analysis of the forward and backward linkages from irrigated agriculture to other sectors of the state economy. An analysis of the effects of the withdrawal of water currently consumed in agriculture in Colorado and Utah has been undertaken. Because of economic proximity, it is assumed then that Grand County, Utah, and all the San Juan River Basin are part of Colorado. No I-O table from Wyoming is available.

Colorado has a well-developed economy characterized by a high degree of interdependence among the various producing sectors. The 28-sector Colorado I-O model based on 1970 data is taken from Gray and McKean (1975). The flows of five sectors are shown in Table 5.22. The household sector was included so that wage and salary income changes could be estimated and consumer multiplier effects included.

Table 5.22. *Gross flows, 1970 (thousand dollars)*

	Livestock	Irrigated agriculture	Food processing	Households	Other	Total intermediate demand
Livestock	$265,585	$ 0	$ 585,110	$ 12,454	$ 45,931	$ 909,080
Irrigated agriculture	192,276	0	77,127	144	27,047	296,594
Food processing	0	34,529	54,599	172,410	672,557	934,095
Households	151,330	61,586	161,333	850	3,017,818	3,392,917
All other	156,660	142,868	104,061	3,592,523	15,361,357	19,357,469
Total interindustry	765,851	238,983	982,230	3,778,381	19,124,710	24,890,155
Primary inputs	185,405	81,999	805,030	4,306,453	8,159,536	13,538,423
Total	$951,256	$320,982	$1,787,260	$8,084,834	$27,284,246	$38,428,578

To estimate the state income effect of withdrawing an acre-foot of water from agriculture in Colorado, one needs to use both the input-output model and certain judgments. To illustrate our calculations, we work through a typical sequence of assumptions and calculations for a reduction of irrigation output in western Colorado. First we observe that 92 percent of irrigated agriculture's output is used for livestock, for food processing, and for consumption and export. Because the irrigated crops on the Western Slope are primarily forage crops and feed grains, it appears reasonable to eliminate the direct linkage to food processing.

Because the technical coefficient for the input from irrigated agriculture per dollar of livestock output is 0.2, a $1 reduction in irrigated agriculture deliveries to livestock may cause as much as a $5 (1/0.2) reduction in livestock output if no substitutes are available. However, complete absence of substitutes seems unlikely, so we assume that one-half the reduction of irrigated agriculture inputs into the livestock sector will be substituted by imported supplies. The reduction of $1 of Colorado irrigated inputs into the livestock sector would then cause a $2.50 reduction in livestock output.

Livestock output is mainly for the food-processing sector (62 percent); the remainder is largely inputs to itself (cow-calf outputs into range livestock, etc.). Thus, each $1 reduction in livestock output would result in a $0.62 reduction in deliveries to food processing. Because the input coefficient for livestock into food processing is 0.33, the reduction in food processing output would be $1.89 (0.62/0.33).

In summary, as a consequence of our assumptions about the structure of the regional economy of western Colorado, a $1 reduction in irrigated agricultural output would exhibit the following consequences: (1) $0.11 represents a reduction in deliveries in final uses; (2) $0.89 comes from livestock, causing a $2.23 reduction in livestock output; (3) the $2.23 reduction in livestock output causes a $1.38 ($0.62 × $2.23) reduction in livestock deliveries to food processing; (4) the $2.23 reduction in livestock output also causes a $0.85 ($0.38 × $2.23) reduction in deliveries to final uses; and (5) the $1.38 reduction in livestock deliveries to food processing may lead to a $4.18 (1.38/0.33) reduction in food-processing output that then represents a reduction in deliveries to final demand because processing output is mostly exported.

The direct and indirect state income effects of these reductions in deliveries to final demand depend upon the average reductions in outputs by the various sectors and the related reductions in wage and salary payments and financial payments. In particular, the reduction in wage and salary payments to the household sector for each dollar reduction in

Table 5.23. *Cumulative reduced consumptive use, state income lost per acre-foot of reduced consumptive use, incremental state income lost (thousand dollars), and cumulative state income lost (thousand dollars)*[a]

	Crop 1	Crop 2	Crop 3	Crop 4	Crop 5	Crop 6	Crop 7	Crop 8	Average income loss/acre-foot[b]
Subbasin 1									
Water used[c]	17,734	19,928	208,020	219,067					$\bar{v}$ = \$75
Income loss (\$/acre-foot)[d]	182	91	57	185					
Total loss (\$1,000s)[e]	3,228	136	10,772	2,010					
Cumulative loss (\$1,000s)[f]	3,228	3,364	14,136	16,146					
Subbasin 2									
Water used	22	774	1,831	28,348	78,629				
Income loss (\$/acre-foot)	185	96	184	216	89				$\bar{v}$ = \$90
Total loss (\$1,000s)	4	72	194	2,347	4,460				
Cumulative loss (\$1,000s)	4	76	270	2,617	7,077				
Subbasin 3									
Water used	205	7,477	11,227	16,886	20,195	204,779	325,520	340,739	
Income loss (\$/acre-foot)	349	193	156	118	191	83	227	317	$\bar{v}$ = \$149
Total loss (\$1,000s)	72	1,403	585	668	632	15,321	27,408	4,824	
Cumulative loss (\$1,000s)	72	1,475	2,060	2,728	3,360	18,681	46,089	50,913	
Subbasin 4									
Water used	4,801	6,425	8,573	10,201	88,077	96,251	136,139		
Income loss (\$/acre-foot)	450	96	203	184	86	228	225		$\bar{v}$ = \$151
Total loss (\$1,000s)	2,160	156	436	300	6,697	1,864	8,975		
Cumulative loss (\$1,000s)	2,160	2,316	2,752	3,052	9,749	11,613	20,588		

Subbasin 5									
Water used	152	3,608	10,084	10,712	149,320	170,120	281,031	289,431	
Income loss ($/acre-foot)	459	96	210	183	88	223	224	421	$\bar{v}$ = $160
Total loss ($1,000s)	70	332	95	1,649	12,198	4,648	24,844	2,543	
Cumulative loss ($1,000s)	70	402	497	2,146	14,344	18,932	43,826	46,369	
Subbasin 6									
Water used	6,864	6,967	9,145	78,025	82,222	163,336	183,085	197,371	
Income loss ($/acre-foot)	196	670	95	172	183	68	200	350	$\bar{v}$ = $145
Total loss ($1,000s)	1,345	68	197	11,862	768	5,516	3,950	5,006	
Cumulative loss ($1,000s)	1,345	1,413	1,610	13,472	14,240	19,756	23,706	28,712	
Subbasin 7									
Water used	18	800	4,956	7,737	10,602	141,002	141,534	233,895	
Income loss ($/acre-foot)	284	170	234	96	183	75	201	219	$\bar{v}$ = $137
Total loss ($1,000s)	5	133	973	267	524	9,780	107	20,227	
Cumulative loss ($1,000s)	5	138	1,111	1,378	1,902	11,682	11,789	32,016	
Subbasin 8									
Water used	2,909	2,948	3,006	3,202	4,707	35,837	70,933		
Income loss ($/acre-foot)	178	490	190	182	118	75	235		$\bar{v}$ = $160
Total loss ($1,000s)	518	19	11	36	178	2,335	8,248		
Cumulative loss ($1,000s)	518	537	548	584	762	3,097	11,345		

[a]The phasing out of crops in each subbasin is assumed to be in ascending order of private profitability, as in Table 5.12.
[b]Average loss of basin income per acre-foot of consumptive use.
[c]Cumulative amount of water consumed by the crop, in acre-feet.
[d]Loss in basin income per consumptive acre-foot for the crop.
[e]Total loss if each crop phased out, in thousands of dollars.
[f]Cumulative loss if all crops phased out, in thousands of dollars.

irrigated output equals $1.61. For each dollar reduction in irrigated agriculture output under the foregoing assumptions, payments to insurance, real estate, rent, interest, and profits are reduced $1.92. Having no information on the distribution of these financial payments between in-state and out-of-state parties, the authors chose one-third to represent an income loss to the state, the rest of interest, dividends, etc., going to out-of-state parties. The total state income loss in Colorado per dollar reduction in irrigated agricultural output in Colorado would then be $1.61 + $0.64 = $2.25.

Similar calculations for Utah resulted in a Utah multiplier of 1.34, that is, for each dollar of reduction in irrigated grain output, state income will fall $1.34.

These multipliers have been derived from state input-output tables, each representing the entire state. Colorado, on average, has a more integrated, more extensive economy, leading to its higher multiplier. However, the Upper Basin areas of the two states are somewhat isolated from the more highly developed parts of each state and are, in fact, closely linked because of physical proximity. Thus, multiplier effects should be similar throughout the basin, rather than differing across the (arbitrarily designated) state lines. We judge that Colorado's multiplier is too high for the region and Utah's is somewhat low (partly because of lack of information on some of the financial payments). We have chosen the average of the two multipliers, 1.80, to estimate the state income impacts throughout the Upper basin.

It is now possible to evaluate the effects on the Upper Basin economy if water is transferred out of agriculture. The estimated income losses are expressed per acre-foot of reduced consumptive use in Table 5.23. The first line for each subbasin gives the cumulative amount of water used by the crops, and the second line shows the loss in basin income per consumptive acre-foot for that particular crop. (Note: the order of crops by value is that shown in Tables 5.6–5.13.) The third line represents the cumulative loss in basin income as the various crops are progressively phased out. The average loss of basin income per acre-foot of consumptive use for each subbasin is given in the last column, the range being $75 to $160. This range represents what the local area might expect to lose in income as water is moved out of agriculture. It seems unlikely that offsetting investments in new industries will take place in the same area. From the state point of view, the water withdrawn might support new industries in the state, but whether it will is quite uncertain. If the water is transferred out of state, the above state income losses are likely.

The private values of water consumed, ranging from zero to $72 per acre-foot (Table 5.13), contrast sharply with possible state income loss of \$57–\$490 (Table 5.14). We can anticipate sharply differing views among many who will want to sell water and state officials concerned with state effects.

References

Anderson, Terry L. 1983a. *Water Crisis: Ending the Policy Drought.* Baltimore: The Johns Hopkins University Press.

Anderson, Terry L. 1983b. *Water Rights: Scarce Resource Allocation, Bureaucracy, and the Environment.* San Francisco: Pacific Institute for Policy Research.

Colorado River Basin Salinity Control Forum. 1984. *Water Quality Standards for Salinity: Colorado River System.* Las Vegas, Nevada.

Daubert, John T., and Young, Robert A. 1980. "Recreational Demands for Maintaining Instream Flows: A Contingent Valuation Approach." Working Paper 80-1. Colorado State University, Department of Economics, Fort Collins.

Dracup, J.A. 1977. "Impact on the Colorado River Basin and Southwest Water Supply." In National Research Council, Panel on Water and Climate, *Climate, Climate Change, and Water Supply.* Washington, D.C.

Fradkin, Philip L. 1968. *A River No More: The Colorado River and the West.* Tucson, Arizona: University of Arizona Press.

Gardner, Richard L. 1983. "Economics and Cost Sharing of Salinity Control in the Colorado River Basin." Ph.D. dissertation. Colorado State University, Fort Collins, Colorado.

Gardner, Richard L. 1985. "Water Marketing – Idaho Style." Presented at the Annual Conference of the American Water Resources Association, Tucson, Arizona, Aug. 13.

Gisser, Micha, et al. 1979. "Water Trade-Off between Energy and Agriculture in the Four Corners Area." *Water Resources Research, 15*(3), 529–538.

Gray, S. Lee, and McKean, John R. 1975. An Economic Analysis of Water Use in Colorado's Economy. Completion Report Series 70. Colorado State University, Environmental Resources Center. Fort Collins, Colorado.

Greenley, Douglas A., Walsh, Richard G., and Young, Robert A. 1982. *Economic Benefits of Improved Water Quality: Public Perceptions of Option and Preservation Values.* Boulder, Colorado: Westview Press.

Gross, Sharon P. 1985. "The Galloway Project and the Colorado River Compacts: Will the Compacts Bar Transbasin Water Diversions?" *Natural Resources Journal, 25*(4), 935–960.

Hamilton, Joel R., and Lyman, R. Ashley. 1983. "An Investigation into the Economic Impacts of Subordinating the Swan Falls Hydroelectric Water Right to Upstream Irrigation." Moscow, Idaho: University of Idaho, Idaho Water and Energy Resources Institute.

Hartman, L.M., and Seastone, Don. 1970. *Water Transfers: Economic Efficiency and Alternative Institutions.* Baltimore: Johns Hopkins Press for Resources for the Future, Inc.

Howe, Charles W. 1977. "A Coordinated Set of Economic, Hydro-Salinity, and Air Quality Models of the Upper Colorado River Basin with Applications to Current Problems." In Blair T. Bower, ed., *Regional Residuals Environmental Quality Management Modeling.* Washington, D.C.: Resources for the Future, Inc., 125–156.

Howe, Charles W., Alexander, Paul K., and Moses, Raphael J. 1982. "The Performance of Appropriations Water Rights Systems in the Western United States during Drought." *Natural Resources Journal, 22.*

Howe, Charles W., and Easter, K. William. 1971. *Interbasin Transfers of Water: Economic Issues and Impacts.* Baltimore: Johns Hopkins Press for Resources for the Future, Inc.

Howe, Charles W., and Lee, Dwight R. 1983. "Organizing the Receptor Side of Pollution Rights Markets." *Australian Economic Papers, 22*(41), 280–289.

Howe, Charles W., and Murphy, Allan H. 1981. "The Utilization and Impacts of Climate Information on the Development and Operations of the Colorado River System." In National Research Council, Climate Board, *Managing Climatic Resources and Risks,* App. C. Washington, D.C.: National Academy Press.

Howe, Charles W., and Orr, Douglas V. 1974. "Effects of Agricultural Acreage Reduction on Water Availability and Salinity in the Upper Colorado River Basin." *Water Resources Research, 10*(5), 893–897.

Howe, Charles W., Schurmeier, Dennis R., and Shaw, William D., Jr. 1972. "Innovations in Water Management: An Ex-Post Analysis of the Colorado–Big Thompson Project and the Northern Colorado Water Conservancy District." University of Colorado, Department of Economics, mimeograph.

Howe, Charles W., Shurmeier, Dennis R., and Shaw, W. Douglas, Jr. 1986. "Innovative Approaches to Water Allocation: The Potential for Water Markets." *Water Resources Research, 22*(4), 439–445.

Howe, Charles W., and Young, Jeffrey T. 1978. "Indirect Economic Impacts from Salinity Damages in the Colorado River Basin." In Jay C. Andersen and Alan P. Kleinman, eds., *Salinity Management Options for the Colorado River.* Report P-78-003. Logan, Utah: Utah Water Research Laboratory.

Hundley, Norris, Jr. 1986. "The West Against Itself: The Colorado River – An Institutional History." In Gary D. Weatherford and F. Lee, eds., *New Courses for the Colorado River: Major Issues for the Next Century.* Albuquerque, New Mexico: University of New Mexico Press.

Jacoby, Gordon C., Jr. 1975. "An Overview of the Effect of Lake Powell on Colorado River Basin Water Supply and Environment." Lake Powell Research Project Bull. 14. University of California at Los Angeles, Institute of Geophysics and Planetary Physics.

Kleinman, Alan P., and Brown, F. Bruce. 1980. *Colorado River Salinity: Economic Impacts on Agricultural, Municipal, and Industrial Users.* Denver: Water and Power Resources Service.

Maass, Arthur, and Anderson, Raymond L. 1978. *And the Desert Shall Rejoice: Conflict, Growth, and Justice in Arid Environments.* Cambridge, Massachusetts: The MIT Press.

Meyers, Charles J. 1966. "The Colorado River." *Stanford Law Review,* Nov., 1–75.

Nadeau, Remi A. 1950. *The Water Seekers.* Garden City, New Jersey: Doubleday and Company.

Narayanan, Rangesan, Padungchai, Sumol, and Bishop, A. Bruce. 1979. "An Economic Evaluation of the Salinity Impacts from Energy Development: The Cast of the Upper Colorado River Basin." Water Resources Planning Series UWRL/p-79/07. Utah State University, Utah Water Research Laboratory, Logan, Utah.

Oyarzabal-Tamargo, Franciso, and Young, Robert A. 1978. "International External Diseconomies: The Colorado River Salinity Problem in Mexico." *Natural Resources Journal, 18*, 77–89.

Spofford, Walter O., Jr., Parker, Alfred L., and Kneese, Allen V., eds. 1980. *Energy Development in the Southwest: Problems of Water, Fish, and Wildlife in the Upper Colorado River Basin.* Washington, D.C.: Resources for the Future.

Stroup, Richard L., and Baden, John A. 1983. *Natural Resources: Bureaucratic Myths and Environmental Management.* San Francisco: Pacific Institute for Policy Research.

Terrell, John Upton. 1965. *War for the Colorado River.* Glendale, California: The Arthur Clark Co.

U.S. Department of Agriculture, Soil Conservation Service. 1977. *On-Farm Program for Salinity Control: Final Report of the Grand Valley Salinity Study.* Washington, D.C.: U.S. Government Printing Office.

U.S. Department of Commerce, Bureau of the Census. 1982. *Census of Agriculture.* Prepared for the U.S. Department of Agriculture.

U.S. Department of the Interior. 1983. *Quality of Water: Colorado River Basin.* Progress Report 11.

U.S. Department of the Interior. 1985. *Quality of Water: Colorado River Basin.* Progress Report 12.

U.S. Department of the Interior, Bureau of Land Management. 1977. *The Effects of Surface Disturbance on the Salinity of Public Lands in the Upper Colorado River Basin: 1977 Status Report.* Washington, D.C.

U.S. Department of the Interior, Bureau of Reclamation. 1981. *Annual Report.* Appendix III, *Project Data.*

U.S. Department of the Interior, Bureau of Reclamation. 1982. *Salinity Update – A Quarterly Report on the Colorado River Water Quality Improvement Program.*

U.S. Department of the Interior, Bureau of Reclamation. 1983a. *Salinity Update – A Quarterly Report on the Colorado River Water Quality Improvement Program.*

U.S. Department of the Interior, Bureau of Reclamation. 1983b. *Colorado River Water Quality Improvement Program, Status Report.*

U.S. Department of the Interior, Bureau of Reclamation. 1984. *Salinity Update – A Quarterly Report on the Colorado River Water Quality Improvement Program.*

U.S. Department of the Interior, Bureau of Reclamation. 1985a. *Study of Saline Water Use at the Jim Bridger Power Plant: Final Report.*

U.S. Department of the Interior, Bureau of Reclamation. 1985b. *Salinity Update – A Quarterly Report on the Colorado River Water Quality Improvement Program.*

U.S. Department of the Interior, Bureau of Reclamation, Water and Power Resources Service. 1980. Plan of Study: Appraisal Investigation, Saline Water Collector System, Colorado River Water Quality Improvement Program, Lower Colorado Region, Boulder City, Nevada.

U.S. Department of the Interior, Bureau of Reclamation, Water and Power Resources Service. 1981. *Project Data.* Denver, Colorado.

U.S. Environmental Protection Agency. 1971. *The Mineral Water Quality Problem in the Colorado River Basin.* San Francisco.

Upper Colorado Region Group, for the Water Resources Council Pacific Southwest Interagency Committee. 1971. *Upper Colorado Region Comprehensive Framework Study.* Appendix VIII, *Watershed Management,* Denver, Colorado.

Vaux, H.J., Jr., and Howitt, Richard E. 1984. "Managing Water Scarcity: An Evaluation of Interregional Transfers." *Water Resources Research, 20*(7), 785–792.

Weatherford, Gary D., and Jacoby, Gordon C. 1975. "Impact of Energy Development on the Law of the Colorado River." *Natural Resources Journal, 15.*

Young, Robert A., and Leathers, K.L. 1978. "Economic Impacts of Selected Salinity Control Measures in the Upper Colorado: A Case Study of the Grand Valley, Colorado." In Jay C. Andersen, and Alan P. Kleinman, eds. *Salinity Management Options for the Colorado River.* Report P-78-003. Logan, Utah: Utah Water Research Laboratory.

6 Growth and water in the South Coast Basin of California

HENRY J. VAUX, JR.

The South Coast Basin of southern California includes the second largest urban area in the United States as well as the two largest cities in California. In addition to the major centers of Los Angeles and San Diego, there are numerous other urban and suburban communities. The 1980 population of the entire region was 12.01 million, compared with a prewar (1940) total of only 2.9 million. Over the past 40 years, the dramatic growth in population, which has averaged 10 percent annually, has been fueled by a variety of factors, including a favorable climate and the rise of defense and aerospace-related industry. This growth was achieved despite the severe limitations of local water supplies.

Mean annual precipitation in the region averages only 14 inches. Over the period of record, annual precipitation has been quite variable, ranging between 5 and 38 inches annually. In addition, the area has a typically Mediterranean climate in which rainfall occurs predominantly between November and April. As a consequence, there exists not only a dearth of locally generated water supplies but an incongruity between the winter period, when those supplies are more readily available, and the summer period, when water demands are at a peak.

The modern history of the region has been characterized by the development of supplemental water supplies and the storage facilities necessary to regulate water flows so as to redress the natural imbalance between periods of peak supply and peak demands. The physical manifestations of this development include three major aqueducts that, with their associated storage facilities, permit the South Coast Basin to import water from the Colorado River, the Central Valley of California, and the Owens Valley to the northeast. A major justification for the development of all these facilities rested on the proposition that water is necessary to support continued population growth and related economic development. This view was probably most succinctly stated by the legendary William Mulholland, who observed at one point during the controversial development of the Owens Valley supply: "If we don't get the water, we won't need it" (Kahrl, 1982).

Despite the availability of substantial quantities of imported water, the South Coast Basin is faced with the prospect of an intensifying water scarcity. The basin's population is expected to continue to grow, although at rates substantially below those prevailing in recent decades (Metropolitan Water District, 1982). At the same time, recent legal developments have diminished or threaten to diminish the amount of water currently available for importation. Inasmuch as the current prospects for developing additional supplies are dim, the region is threatened with a situation in which demands may continue to increase steadily while existing dependable supplies diminish.

The apparent intensification of water scarcity in the South Coast Basin may at first appear puzzling. The state possesses a rich water endowment, with an average annual runoff of almost 71 million acre-feet (maf). The statewide net use is only 34 maf annually. These figures are deceptive, however, because the regions in which population and irrigated agriculture are concentrated are not those where water supplies are plentiful. The growth of irrigated agriculture and the large urban centers has been supported, in part, by public investment in a system of water impoundment and conveyance facilities that have served to rectify the imbalance between areas where water is plentiful and areas where water demands are concentrated. Today, new water supplies cannot be developed and transported easily or inexpensively. The problem, then, is a scarcity of relatively inexpensive water rather than an absolute scarcity.

Prospective water supply deficits in urban areas may also be hard to understand in view of the fact that municipal and industrial water use amounts to but a small fraction of total water use. California's agricultural sector accounts for 85 percent of the total annual net use, and municipalities and industry use only 15 percent. This large disparity suggests that urban water demands could be easily met by reallocating a small fraction of the water used by agriculture to urban uses. Yet, the institutions that have governed water development and use in California tend to discourage such reallocation. Water institutions were initially created to provide security of tenure in water use as part of a larger strategy to attract settlers to the West. Although the West is now well settled, institutional and public biases favoring security of tenure persist. These biases create a policy thrust toward resolving water scarcities solely through the continual augmentation of new supplies. Thus, opportunities for profitable water reallocation between uses and regions are largely overlooked, and competition for scarce new water supplies increases. This competition, coupled with the escalating costs of new water development and intensifying pressures on the public purse at all levels, suggests that the traditional means of augmenting the basin's water supplies may be less viable than they once were.

The perceived need for additional water supplies to service the population of the southern California region is at the heart of the latest round of California's water wars. At issue in this current round is whether water supplies will continue to be augmented in response to anticipated population and economic growth or whether some other means will have to be found to ameliorate the pervasive scarcity of water. The implications of this issue for a major urban area dependent on imported water supplies have not always been clear. The major water supply agencies of the South Coast Basin have typically construed their mission as predominantly a service obligation. That is, the obligation requires water to be provided to an expanding population so that historically prevailing levels of per capita use are preserved, the impacts of drought are minimized, and prices remain as low as possible (see, for example, Metropolitan Water District, 1983). These agencies usually assume that population and economic growth is determined by exogenous factors and that failure to support that growth with abundant supplies of water will result in economic stagnation of the region as well as in personal hardship of its residents.

In this chapter, the implications of a constant or shrinking water supply for the South Coast Basin are explored, several alternative means for adapting to water scarcity are identified and analyzed, and a number of conclusions are drawn about the viability of major urban settlements in semiarid areas. The following section contains an abbreviated history of the growth and development of water resources for the basin. The current and prospective water supply and demand situations are then characterized. Alternative adaptations to water scarcity, including changes in pricing rules and the development of quasi-markets, are identified and analyzed. The last section contains some concluding comments.

The historical setting

Imports from the Owens Valley

For the city of Los Angeles, explosive population growth was not exclusively a postwar phenomenon. For example, the 1880 population of 11,183 increased nearly fivefold to a little over 50,000 by 1890. Dur-

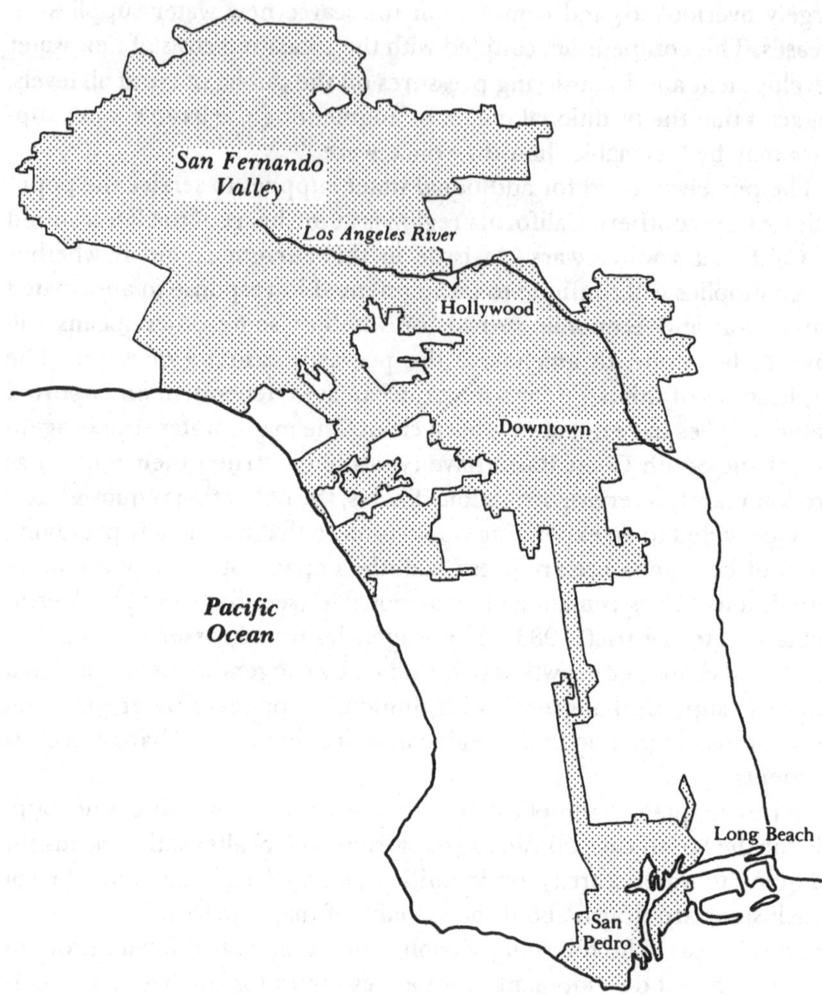

Figure 6.1. City of Los Angeles.

ing the latter years of the nineteenth century, water was supplied by a private firm, the Los Angeles City Water Company. The predominant source of domestic water was the Los Angeles River, which drained a watershed of approximately 500 square miles, including the San Fernando Valley and the coastal plain on which the city is located. (See Figure 6.1.) The source of the river was a great aquifer underlying the San Fernando Valley that served to regulate seasonal runoff from the surrounding mountains so that the flows of the Los Angeles River remained relatively constant throughout the year (Ostrom, 1953).

The inadequacy of the supply from the Los Angeles River first became apparent during the early years of the twentieth century. Between 1900 and 1905, the population grew from 102,000 to a little over 250,000. This rapid growth occurred in the final years of what was publicized as an 11-year drought cycle during which annual precipitation fell to as little as 40 percent of its long-term average (Lynch, 1931). In one period, storage declined so precipitously that only a ten-day supply remained available. By this time, the city of Los Angeles had assumed responsibility for providing the water supply. In this setting, city water officials recognized a compelling need to develop additional supplies in order to avert an extreme water shortage toward which the population growth was leading. William Mulholland, superintendent of the Los Angeles City Water Department recognized the potential for developing the waters of the Owens Valley, some 235 miles to the north, and conveying them to Los Angeles in an aqueduct to be constructed by the city.

The controversies surrounding Los Angeles's development of water resources in the Owens Valley have been widely chronicled by the entertainment industry, apologists, and serious scholars (see, for example, Kahrl, 1982). The acquisition of lands with appurtenant but unquantified rights to surface water was accomplished almost completely out of public view. Charges of fraud, chicanery, conflict of interest, and deception of the public were frequent and, in many instances, of substance. However, owing largely to the perseverance of Mulholland, the aqueduct was completed and water deliveries began in 1913. The cost was $4 million less than Mulholland's original estimate of $25 million. With a capacity of nearly 450 cubic feet per second (cfs), the aqueduct was capable of delivering approximately 320,000 acre-feet of water per year, far more than could be used domestically at the time.

The decision to annex the San Fernando Valley and other areas immediately adjacent to the city permitted the surplus waters to be used for irrigation until such time as the demand for domestic use in those areas developed. Domestic demand grew more rapidly than anticipated, however, and surface water deliveries through the aqueduct were substantially less than Mulholland had predicted because drought years followed its opening. These two factors prompted the city to begin acquiring water-bearing lands in the Owens Valley in order to gain access to the valley's substantial groundwater reserves. Pumped groundwater could then be used as a substitute for shortfalls in runoff during drought periods to ensure that the aqueduct, together with newly developed storage facilities, could be operated to supply the quantities of water originally envisioned by Mulholland. Despite the development of these groundwater supplies, rapid growth and the drought caused the city to continue searching for additional water sources.

By the early 1920s, attention had focused on the possibility of obtaining supplementary supplies from the Colorado River. Potential canal routings were identified and surveyed. In 1924 the city filed an application to appropriate 1,500 cfs of surplus waters from the Colorado River. City officials recognized, however, that many years might elapse before Colorado supplies could be made available locally. In addition, there were concerns over the plans of Owens Valley residents to develop some surplus supply for their own use. Although Los Angeles owned more than 80 percent of the valley lands, city officials were concerned that local development of surplus waters could adversely affect the quantities of water ultimately available for export to the south.

In an effort to bolster its own supplies while forestalling local residents from developing firm claims to water that might otherwise be used by Los Angeles, the city sought to solidify its claims to the waters of the Mono Basin immediately north of the Owens Valley. The juxtaposition of the Mono Basin and the Owens Valley is illustrated in Figure 6.2. The Mono Basin, like the Owens Valley, is a closed basin with no ultimate outlet to the sea. Mono Lake, the central feature of the basin, is a remnant of an ancient inland sea, and it acts as an evaporating pan with salt content increasing inexorably over time. As a consequence of the uncertain water quality within the lake, the city of Los Angeles obtained rights to much of the water in the basin's streams as well as the rights-of-way that would ultimately be necessary to convey the water into the Owens Valley system.

Capturing the Colorado

In the meantime, efforts to bring water to the South Coast Basin from the Colorado River intensified. These efforts differed fundamentally from those in the Owens Valley in that they involved a host of municipalities in the basin all confronted with the same problem of water scarcity. With the exception of Los Angeles, none of these communities had either existing water rights outside the basin or access to sufficient capital to permit development of remote supplies. Although the city of Los Angeles recognized explicitly the economic desirability of allowing development to continue in surrounding areas, it also saw that some portion of the water that could ultimately be developed from the Colorado might usefully supplement its Owens Valley supplies (Ostrom, 1953).

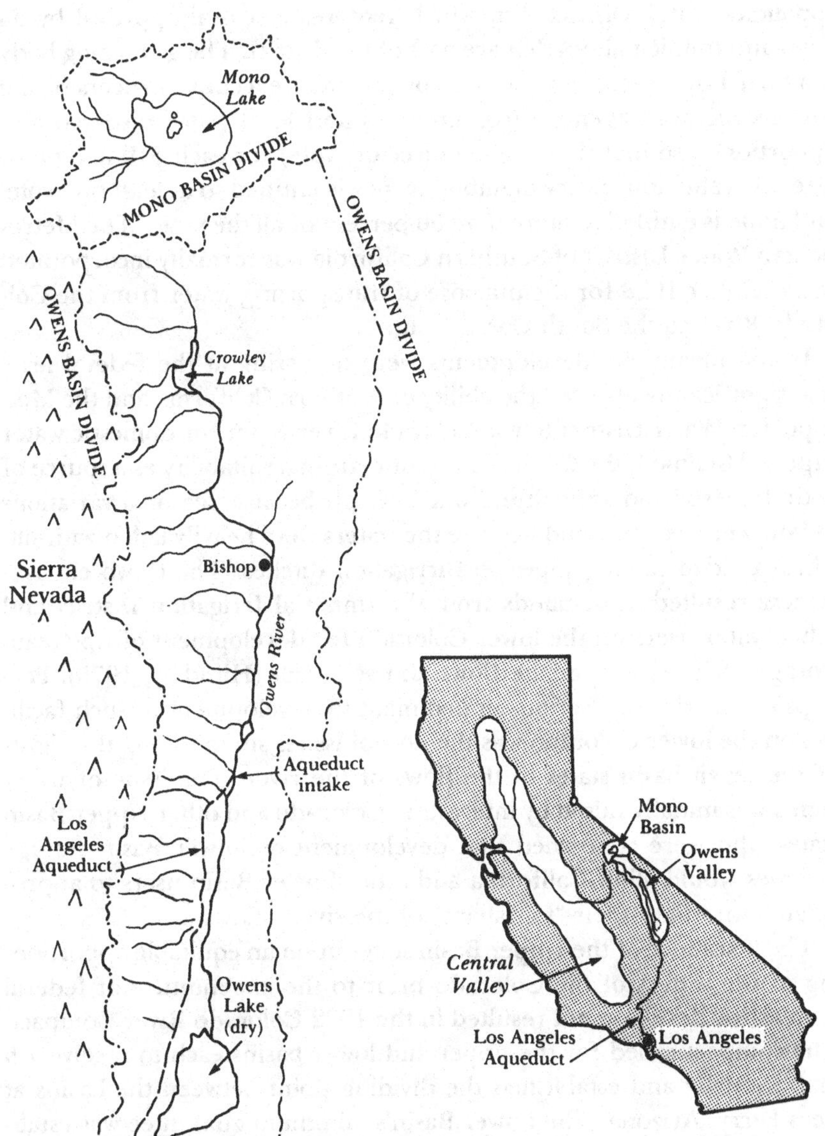

Figure 6.2. Mono Basin and Owens Valley.

Despite other communities' lingering suspicions of Los Angeles's motives, the Metropolitan Water District of Southern California was formed for the purpose of developing and distributing the waters of the Colorado River. The district, which has the status of a local unit of government, was incorporated under the provisions of California's Metropolitan Water District Act, which requires electoral approval by all cities and municipalities that are part of the district. The governing body is a board of directors, appointed by the chief executive officers of the various member agencies (communities and local water districts) and apportioned so that there is one director (vote) for each $10 million of assessed valuation. Every member agency is entitled to at least one vote, and none is entitled to more than 50 percent of all the votes. The Metropolitan Water District of Southern California was formally incorporated in December 1928 for the purpose of transporting water from the Colorado River to the South Coast Basin.

In the meantime, developments were occurring at the federal level that significantly affected the ability of southern California and the Metropolitan Water District to use Colorado River water for domestic water supply. Untamed, the Colorado was of marginal suitability as a source of both domestic and agricultural water supply because seasonal variations in flow were extreme and because the waters were heavily laden with silt, which tended to clog pipes and irrigation ditches. These two circumstances resulted in demands from the Imperial Irrigation District and other water users on the lower Colorado for development of upstream storage facilities to regulate flows and settle silt (Hundley, 1975). Perhaps the single most serious impediment to development of such facilities on the lower Colorado was the host of issues surrounding the rights of the seven basin states to the flows of the river. The issue of water rights was initially raised by interests in Colorado and other Upper Basin states who were concerned that development of Lower Basin storage facilities would allow California and other Lower Basin users to appropriate more than their "fair share" of the river's flows.

The insistence of the Upper Basin states upon an equitable apportioning of the waters of the Colorado prior to the expenditure of federal monies for development resulted in the 1922 Colorado River Compact. The compact called for the upper and lower basins each to receive 7.5 maf annually and established the dividing point between the basins at Lees Ferry, Arizona. The Lower Basin's minimum guarantee was established at 75 maf over any ten-year period and not in terms of a minimum annual entitlement. In this way, the Upper Basin was assigned the burden of any deficiency in flows caused by a dry cycle. The Lower Basin was given the right to make beneficial use of an additional 1 maf, a provision whose ambiguity ultimately took 35 years to resolve. In addition, the Lower Basin was also given the right to any surplus flows (Hundley, 1975).

The apportionments of river flows among the states occupying the two subbasins were accomplished separately. Because there was little development pressure in the Upper Basin, apportionment to those states did not occur for nearly three decades. Ultimately the Upper Colorado River Compact, ratified in 1948, allocated Upper Basin flows among Colorado, Wyoming, Utah, New Mexico, and the small area of Arizona lying in the Upper Basin. In the Lower Basin, the pressure for development in the 1920s was intense. The Lower Basin states, Arizona, California, and Nevada, were unable to negotiate a mutually satisfactory allocation. Under pressure from California, Congress included in the Boulder Canyon Project Act a suggested apportionment that was derived from the terms of ratification of the Colorado River Compact. This apportionment assigned 4.4 maf to California, 2.8 maf to Arizona, and 300,000 acre-feet to Nevada. Arizona and California were to divide any surplus.

From the outset, Arizona was dissatisfied with both the Colorado River Compact and the apportionment spelled out in the Boulder Canyon Project Act. Twice during the 1930s it petitioned the Supreme Court to overturn the Boulder Canyon Project as an invasion of its right as a state to appropriate waters from the Colorado River. In both instances, the court denied Arizona's claims, citing the legitimacy of the federal involvement in dividing the river and arguing that Arizona's rights were inherently undefined because it was not a signatory to the compact. Almost a decade and a half later, in 1944, Arizona did agree to ratify the Colorado River Compact, but only when it became apparent that continued opposition could result in the loss of water to which it would otherwise be entitled under the terms of the compact. At the same time, Arizona signed contracts with the Secretary of the Interior providing for the development of 1.2 maf of its share of the Colorado. Arizona continued to object, however, to the Lower Basin allocation contained in the Boulder Canyon Project Act. By the 1950s, this continued opposition and the associated legal and political maneuvering had created significant uncertainty over whether California's entitlement would be sustained in the long run (Hundley, 1975).

The Boulder Canyon Project Act authorized the development of a mainstream storage facility at either Boulder Canyon or Black Canyon in northwestern Arizona and construction of the All-American Canal, through which Colorado River water could be transported to the Imperial and Coachella valleys. The act required the execution of contracts for water and power between the Secretary of the Interior and the user agencies as a prerequisite to beginning construction of the facilities. The water contracts presented something of a problem because of Arizona's

Table 6.1. *Colorado River priorities for California*

Priority	Agency	Annual quantity (maf)
1	Palo Verde Irrigation District	3.85
2	Yuma Project (California Division)	
3	Imperial Irrigation District Palo Verde Irrigation District	
4	Metropolitan Water District	0.55
5	Metropolitan Water District City and/or County of San Diego	0.55 0.112
6	Imperial Irrigation District Palo Verde Irrigation District	0.30
	Total	5.362

Source: Ostrom, 1953.

continuing opposition to both the compact and the Lower Basin allocation suggested by Congress. To permit construction to begin, the Secretary of the Interior entered into preliminary agreements with California agencies based on three assumptions: (1) the allocation of Lower Basin waters among California, Arizona, and Nevada in the Boulder Canyon Project Act was appropriate; (2) 10.5 maf of water would be available annually at Hoover Dam; and (3) the additional 1 maf of appropriation permitted in the Lower Basin and any other surplus waters were to be divided equally between Arizona and California (Ostrom, 1953).

Subsequently, this action led to the signing of contracts for the delivery of water from Hoover Dam to water agencies in California in accordance with the so-called Seven Party Agreement. The agreement, which is summarized in Table 6.1, stipulates the priorities of delivery and the quantities to which California users were entitled. The Metropolitan Water District became a party to the agreement as the inheritor of the city of Los Angeles's original 1,500 cfs claim. However, Metropolitan's priorities were preceded by those of agricultural water users along the lower mainstream of the river. This point is especially significant in view of the fact that the Seven Party Agreement allots 5.362 maf among California users, 962,000 acre-feet more than that suggested in the Boulder Canyon Act. Moreover, this latter allocation includes Metropolitan's fourth priority of 550,000 acre-feet but not the fifth. Thus, under the terms of the Seven-Party Agreement, the Metropolitan Water District was dependent upon surplus flows in order to obtain its full allotment (Ostrom, 1953).

To acquire its allocation, the Metropolitan Water District initiated construction of the Colorado River Aqueduct in 1931 after obtaining electoral approval for a bond issue of $220 million to defray the costs. Although markets for these bonds were nearly nonexistent during the Great Depression, federal loans and aid in marketing the bonds were successful, and water from the Colorado was first delivered for use in Metropolitan's service area in June 1941. While the aqueduct was being built, Metropolitan's board of directors encouraged nonmember communities in the South Coast Basin to join the district before the aqueduct was completed. Largely because of a cycle of wet years there were no takers, but the stage was set for future annexations with the advent of a dry cycle in 1944. The most important annexation occurred in 1946, when the San Diego County Water Authority merged the water rights of the city of San Diego with those of the Metropolitan Water District. The effect was to include almost the entire western half of San Diego County within Metropolitan's service area (Ostrom, 1953). Most annexations were completed by 1954, and the last occurred in 1963 (Metropolitan Water District, 1984).

The service area of the Metropolitan Water District is pictured in Figure 6.3. Its role is primarily as a supplier of supplementary water to its 27 member agencies, most of whom have local sources of groundwater and some surface water. It is important to recognize that the city of Los Angeles is one of the member agencies entitled to purchase supplementary water from Metropolitan. Although the city has made disproportionately large financial contributions toward the capital costs of the aqueduct, it has used little Colorado River water. Since 1941, it has received only 5.7 percent of all the water delivered, and in recent years its use has averaged only 2.0 percent of total deliveries (Metropolitan Water District, 1984). The reason is that the city's local and Owens Valley supplies are both less expensive and of significantly higher quality than Colorado River supplies.

The population of Metropolitan's service area grew rapidly in the post-World War II era, but developments in the Colorado Basin cast doubt on the long-term reliability of Metropolitan's supply. Despite its approval of the Colorado River Compact in 1944, Arizona still had no assurance of water supplies from the Colorado River. In 1947, the Central Arizona Project (CAP) plan to bring Colorado River water to the growing Phoenix and Tucson areas was first unveiled. Early efforts to authorize the CAP for federal construction failed because of the possibility that water would be unavailable for the project until Arizona's claims to river flows were clarified. This fact, coupled with the signing of

Figure 6.3. Service area of the Metropolitan Water District.

the Upper Colorado Basin Compact, created fears that competing states might successfully appropriate water that would otherwise be allocated to Arizona. As a result, the state petitioned the Supreme Court once again in 1952. Arizona argued that it would be irrevocably harmed by any further delay in settling its dispute with California over the allocation of Lower Basin water. This case, *Arizona* v. *California et al.* (373 U.S. 546, 1963; 376 U.S. 340, 1964), took 11 years to resolve. At issue was interpretation of the Colorado River Compact and the allocation suggested by Congress in the Boulder Canyon Project Act.

Specifically, Arizona had long argued that the intent of the framers of the compact was to award it 2.8 maf, plus the waters of all streams tributary to the Colorado in Arizona. On the other hand, California took the position that part or all of Arizona's use of tributary waters counted against the 1 million additional acre-feet declared available for appropriation in the Lower Basin. This issue was at the heart of myriad formal and informal negotiations between the states that occurred over nearly three decades. An added complication was introduced in the suit, when the federal government intervened to protect its own reserved rights as well as those of Indians living on 25 reservations in the Lower Basin. Indian water rights had not been previously considered in the division of the river's waters, but it was clear that a case might be made that those rights were substantial and could have first or high priorities (Hundley, 1975; Veeder, 1969).

This suit and the long delay in resolving it cast doubt on the certainty of Metropolitan's supply from the Colorado River. It was clear that should the court decide that California's allocation was limited to 4.4 maf, Metropolitan's fifth priority right of 550,000 acre-feet could be jeopardized. Metropolitan was thus faced with the possibility that its firm entitlement could be effectively halved, or worse, if the court decided to throw out all previous allocations and begin anew.

The State Water Project

Into this climate of uncertainty was introduced the California Water Plan, which embodied a host of storage and conveyance facilities designed to ship surplus water from water-rich northern California to the more arid environs of southern California. The plan envisioned the sequential development of surplus waters to support both urban and agricultural growth throughout the state over the last third of the twentieth century and the first two decades of the twenty-first.

In 1959 bond issues in the amount of $1.75 billion to finance the first facilities of what was to be known as the State Water Project (SWP) were submitted to the voters for approval. The proposition was highly controversial, with northern Californians opposed on general principles, labor opposed because landholding corporations such as Southern Pacific stood to be enriched, and early environmentalists opposed because of the destructive impact the project would have on free-flowing rivers. Initially, the Metropolitan Water District also opposed the SWP. Although it cited many reasons publicly, there appear to have been two fundamental explanations. First, the city of Los Angeles, whose water supply at the time was more than ample, was reluctant to underwrite the development of new, more expensive water supplies. Acquisition of such supplies by Metropolitan would not benefit the city directly but would serve to increase the rates charged to city water users. With 50 percent of the votes on Metropolitan's board, the city was in an influential position.

Second, some members of the Metropolitan board believed that construction of the SWP would weaken California's case in the *Arizona* v. *California* litigation.

The countervailing view held that the SWP was required to protect Metropolitan from an adverse decision by the Supreme Court. The larger argument, however, was that California could never hope for enough federally developed water to support anticipated growth statewide and that the failure of the state to provide this water would ultimately strangle economic growth. Governor Pat Brown, in perhaps his most impressive political triumph, was ultimately able to develop broad support for the project, and several days before the election, the Metropolitan Water District joined by signing the necessary water supply contracts. Board members from the city of Los Angeles were ultimately persuaded to support the proposal because the governor had effectively isolated them as the only major water-using group in the southern part of the state to oppose it. The city did not want to be left at the station should the proposal pass, which it did, narrowly (Bean, 1968; Kahrl, 1982).

Construction of the state water project began in 1961. The primary facilities, illustrated in Figure 6.4, were the Oroville Dam on the Feather River and the California Aqueduct, which ran from the delta of the Sacramento and San Joaquin rivers to the South Coast Basin. Oroville Dam provided further control of the flows of the Sacramento River, thereby making additional water available for export from the delta. The California Aqueduct was designed to serve agricultural users on the west side of the San Joaquin Valley as well as domestic users in the South Coast Basin. As usual with such projects, it was overscaled to meet anticipated growth, although some of the basic impoundment facilities necessary to meet that growth were deferred.

Contracts for SWP water are tied to a schedule of entitlements that increase to a maximum level by 1990. Metropolitan's entitlement in 1980, for example, was 1.057 maf; it increases annually to the 2.011 maf maximum by 1990. In addition, Metropolitan could theoretically obtain part of any surplus waters in the system. As a consequence, it is difficult to identify precisely the quantities of water that could be delivered to Metropolitan at any given time. The question is further complicated because the total amount of water that SWP could deliver in any one year is uncertain. At the time the project was constructed, the Metropolitan Water District expected that deliveries would ultimately reach 2.01 maf, enough to support growth well into the twenty-first century.

The last event in the historical development of water supplies for the

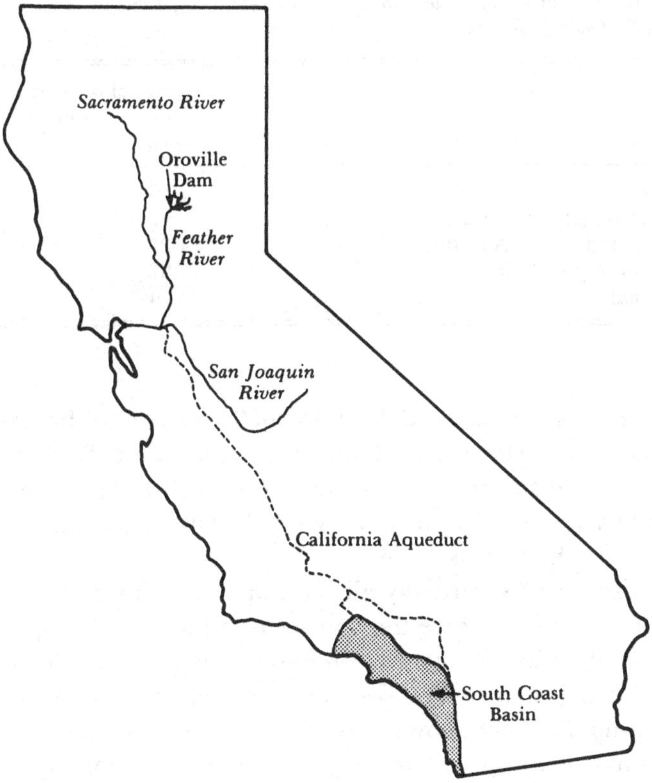

Figure 6.4. Primary facilities of the California State Water Project.

South Coast Basin was the expansion of the city of Los Angeles's aqueduct leading from the Owens Valley. By the middle 1950s, the city had succeeded in acquiring substantial water rights in the Mono Basin together with the rights-of-way needed to convey that water into the Owens Valley Aqueduct. No effort had been made to develop the collection and conveyance facilities because the city had no immediate need for additional water supplies at the time. In 1956 the California Department of Water Resources noted that Los Angeles was using only 320,000 acre-feet of the 590,000 potentially available to it. This announcement and follow-up actions by the department suggested that the state was preparing to intervene and to aid in developing surplus water for use by Owens Valley residents.

Largely out of fear of losing title to water that it had already acquired, the city began construction on a second barrel for its aqueduct in 1963. This barrel, with a capacity of 210 cfs, was completed in 1971. Com-

Table 6.2. *Maximum water entitlements and delivery capacities for sources of supply to the South Coast Basin*

Source	Original or maximum entitlement or capacity (maf)
Local	1.10
Los Angeles Aqueduct	0.47
Colorado River Aqueduct	1.21
State Water Project	2.01
Total	4.79

bined, the two barrels could deliver 480,000 acre-feet to the city of Los Angeles annually. However, exports from the Mono Basin were expected to account for only 50 percent of the water to be transported in the second barrel, so the remainder had to be obtained from groundwater pumped in the Owens Valley.

By the early 1970s, virtually all the impoundment and conveyance facilities that currently serve the South Coast Basin were in place. The nearly seven decades of water development activities were characterized by two recurring themes. The first was the notion that water scarcity in the region should not be allowed to constrain economic growth. This was the fundamental and predominant justification for water supply development. The second underlying theme can be characterized as preemptive development. That is, construction of the second barrel of Los Angeles's aqueduct to the Owens Valley as well as some of the negotiations with Arizona over Colorado River allocations were motivated by the desire to forestall water claims by others so that the water in question would be available for future use in the South Coast Basin.

During this 70-year period, the water agencies of the South Coast Basin developed or contracted for water supplies that totaled 4.79 maf, as shown in Table 6.2. Had these supplies remained unimpaired, they could have supported the projected population of the South Coast Basin well beyond the year 2020 at per capita use levels somewhat above the historical trend (California Department of Finance, 1980). However, despite the efforts of both the city of Los Angeles and the Metropolitan Water District to acquire water rights and entitlements with a minimum of legal encumbrances, the dependable supply has been significantly reduced by the political and legal conflict that followed virtually all their water-development activities.

The current situation

Available water supplies

The Supreme Court's decision in *Arizona* v. *California* was the first of a number of events that constrain the water supply of the South Coast Basin today. Broadly speaking, the Supreme Court ruled in Arizona's favor by awarding it 2.8 maf from the Colorado River plus the rights to all Arizona waters that are tributary to the Colorado. This decision effectively awarded Arizona the additional 1 maf Lower Basin allocation because, it had previously been argued, Arizona's use of its tributary waters should be counted as an offset against the additional 1 maf made available for appropriation in the Lower Basin by the Colorado River Compact (Hundley, 1975). The court also sustained federal claims to water for five Indian reservations along the lower Colorado and, although it did not explicitly quantify and allocate those claims, ruled that these Indian rights were superior to non-Indian rights irrespective of whether non-Indian appropriations had occurred at an earlier date.

The immediate effect of this decision on California's diversions was minimal, because at the time (1964) Arizona had no physical means of using its share. Thus, the Metropolitan Water District could continue to take 1.212 maf (including the rights acquired from the city of San Diego) and its share of surpluses until such time as Arizona developed the facilities necessary to divert waters from the Colorado River. With its court-mandated allocation in hand, Arizona sought, and in 1968 received, authorization for federal construction of the Central Arizona Project. Construction of the project, which was designed to divert 1.2 maf of Colorado River water to the rapidly growing population centers of Arizona, began in 1973. The first segment, serving the Phoenix area, was placed in operation in 1985. The remaining segments are scheduled for completion prior to 1990 and will serve Tucson and intervening agricultural regions.

It was understood that the advent of an operational CAP would reduce Metropolitan's firm allocation of Colorado River water from 1.212 maf to 0.55 maf. Uncertainty over further reductions was created by the unallocated Indian water rights because, by the Supreme Court's decision, rights with higher priorities than those in the Seven Party Agreement of California users would be charged against Metropolitan's fourth priority right, as stipulated in that agreement (Metropolitan Water District, 1983). Although some uncertainties still remain, a 1982 Supreme

Court decision awarded 52,200 acre-feet to the Indians. Thus, Metropolitan's firm allocation was further reduced to 495,000 acre-feet. This allocation supersedes Arizona's and is therefore unlikely to be diminished further in the absence of additional Indian awards.

The magnitude of the city of Los Angeles's supplies from the Mono Basin and Owens Valley has also become somewhat uncertain. Although the city had indicated that only modest amounts of water would be pumped from Owens Valley aquifers, by 1972 it was pumping nearly 150,000 acre-feet per year. Groundwater pumping thus accounted for nearly one-third the water exported from the area. Owens Valley residents believed that the city had been less than forthright, if not deceptive, in announcing its plans for operation of the second aqueduct. Moreover, the situation was further aggravated by the city's unwillingness to agree to restrictions on groundwater pumping that would diminish the quantity of water delivered through the aqueducts. The city was also unwilling to deny flatly the prospect that a third aqueduct might in time be built, an event that would almost assuredly place further pressure on groundwater and other sources supplying local users in the Owens Valley (Kahrl, 1982).

The city's attitude led to an outbreak of a second water war in the Owens Valley; its primary manifestation was a lawsuit filed by Inyo County (where the Owens Valley is located) seeking to require the city to assess the environmental effects of groundwater pumping before permitting further extractions. What has emerged is a legal stalemate in which the risks of losing a definitive legal decision dominate the thinking of both sides. In the meantime, the city continues to extract groundwater with the approval of a lower court at rates sufficient to meet the total flow requirements of the two aqueducts. Uncertainties remain over whether the city can continue to pump at these rates indefinitely and whether finite legal limits may ultimately be imposed on its pumping rates (Kahrl, 1982).

Almost coincidentally, operation of the second aqueduct was threatened when environmental groups challenged the city's diversion from streams feeding Mono Lake. This diversion has two impacts on the lake, both potentially disastrous for wildlife. Salinization of Mono Lake is inevitable because, in a closed basin, there is no pathway for exporting natural salts. However, the streams tributary to Mono Lake have very low salt concentrations, and their inflows retard the rate at which the lake becomes saline. City diversions from these streams thus accelerate the rate of salinization of the lake. The lake provides habitat for a species of brine shrimp that is the basic food supply for gulls, grebes, and other migratory shore birds. The brine shrimp can survive only in a relatively narrow range of salt concentrations; any increase therefore advances the time of extinction of the brine shrimp in the lake. And, without brine shrimp, the bird populations will disappear (California Department of Water Resources, 1979).

In addition to accelerated salinization, the lake's lower level caused by diversions from the tributary streams has other environmental effects. Mono Lake's Negit Island supports the second largest rookery of California gulls in the world. The declining lake level has created a land bridge to Negit Island across which the rookery can be attacked by coyotes and other terrestrial predators. A fence has been erected as a short-term protection measure. For the longer run, an intergovernmental task force has recommended that Los Angeles reduce its diversions in the Mono Basin from 100,000 acre-feet to 15,000 acre-feet annually. This reduction of 85,000 acre-feet represents 18 percent of the Los Angeles aqueduct supply and 6 percent of the city's total demand.

A suit filed by environmental groups seeking a legal ruling on the quantity of water that the city can divert from the Mono Basin has been settled in favor of the environmental groups. The California Supreme Court has held that the city's Mono Basin rights may be restricted under the public trust doctrine. This doctrine requires the state to protect the public trust when valuing the environmental attributes of Mono Lake and when comparing those values with the value of water exports to Los Angeles. The implications of this ruling for the Mono Basin diversions are not clear, pending further study and legal rulings.

The uncertainty over the long-term supplies in the Mono and Owens basins may not be quickly resolved because of the time-consuming legal maneuvers available to both sides. If court decisions ultimately resolve the Mono Lake case and the Owens Valley groundwater pumping controversy in a fashion completely unfavorable to the city of Los Angeles, the total supply from the two aqueducts could be reduced as much as 200,000 acre-feet. This reduction would pose no particular threat to the adequacy of water available to the city because of its entitlements to water from both the Colorado River and the State Water Project through the Metropolitan Water District. However, it would create broader problems because of the imminent reduction in supplies from the Colorado and because of major difficulties associated with completion of the State Water Project. In effect, shortfalls in deliveries to the city through the two Owens Valley aqueducts could be visited, in part, upon other Metropolitan users currently relying on Colorado River water or SWP water.

The plan for the SWP initially called for construction of only those storage facilities necessary to meet reasonable prospective demands for water. Additional facilities were to be built when and if the growing demand for water provided sufficient justification. The current debate over water policy in California focuses on the need, or lack thereof, for these additional facilities. The SWP can now deliver to the Metropolitan Water District a dependable supply of 1.13 maf. Deliveries in excess of this amount can be made except during extended drought periods. Thus, for example, 1.51 maf can be delivered in normal years. Metropolitan's contractual entitlement to SWP water is currently 1.56 maf, and it will reach a maximum of 2.011 maf in 1990. However, there is no guarantee that entitlement quantities will actually be delivered because availability depends upon construction of additional storage and conveyance facilities (Metropolitan Water District, California, 1983).

The California Department of Water Resources projects that if no additional SWP facilities are built, deliveries from the project cannot meet either entitlements or contractors' requests for water (which may be less than the entitlement) during a dry year, beginning in 1985. In average years, deficiencies in the project water supply will begin in 1990, and by 1995 the supply will be deficient even in wet years (California Department of Water Resources, 1983b). The Metropolitan Water District estimates that these shortfalls will reduce the water supplies available to them 400,000–500,000 acre-feet in a dry year. In an average year, the reduction would be perhaps one-tenth of that amount. Moreover, these reductions would occur at a time when Metropolitan would be relying to an increasing extent on supplies from the SWP because of curtailed Colorado River supplies (Metropolitan Water District, 1983).

Proposals for construction of the additional facilities needed to avert this shortfall have failed to garner the political support necessary for passage. The next phase of the SWP requires both additional reservoir capacity and a water transport facility across the delta area at the mouths of the Sacramento and San Joaquin rivers that will improve the efficiency of water use within the delta region. Water quality in the delta is subject to a rigorous standard prescribed by the State Water Resources Control Board in order to protect domestic and agricultural water users in the area from salt water that might otherwise intrude from San Francisco Bay. Water for the California Aqueduct is drawn from the southern delta (which is not subject to saltwater intrusion). Large diversions into the aqueduct reduce freshwater flows into San Francisco Bay, thereby allowing salt water to advance beyond the intakes of water users in the lower portion of the delta. In order to meet the water quality standard, diversions to the California Aqueduct must be restricted in time of low flows so that outflows from the delta are sufficient to repel the salt water. A channel through or around the delta would allow water flows to be managed in a way that would reduce the quantities of water needed to repel saltwater intrusion, thereby making additional water available for diversion into the aqueduct.

Two recent proposals for a cross-delta facility and additional reservoir storage have been defeated in the political arena. A development package whose central feature was the peripheral canal around the delta was overwhelmingly defeated in a statewide referendum in 1982. A 1984 proposal by the governor calling for a through-delta canal and some additional offstream storage south of the delta was stymied in the legislature. The reasons for these defeats are varied and complex. However, it seems apparent that the public and its representatives are not convinced that a legitimate need for the water exists in southern California. Calls for evidence that southern Californians are economizing on water use are frequent. Environmentalists are concerned about the effect of such projects on free-flowing rivers and on the amenity values of water. Northern Californians want strong guarantees that their own water supplies, including delta water quality, will not be jeopardized as a result of further exports to southern California. It also seems clear that taxpayers are uneasy about the ultimate financial impacts of future water projects (Mann, 1984).

In the absence of new facilities, the firm yield from the State Water Project will decline somewhat in the future because of the County of Origin Law. Largely in response to Los Angeles's appropriation of water in the Owens Valley, the California legislature enacted this law in 1928 to guarantee regions of origin sufficient water for their own development. The law applies to the SWP but not to the Owens and Mono basins or the Colorado River supply. The effect of the County of Origin Law is to limit exports from a basin to waters that are surplus to the needs of that basin. The definition of surplus is allowed to vary with time so that, theoretically, surplus water could completely disappear (Cooper, 1968). Actual declines in surplus water are likely to be quite modest in the foreseeable future because economic growth in areas of origin is expected to be limited. The Department of Water Resources estimates that declines in surplus waters in areas of origin will be minimal during the next four or five decades. Similarly, Metropolitan estimates that supplies available to it from the SWP would shrink only 30,000 acre-feet in a normal year by 1995 owing to increased use by counties of origin.

The possibilities for expanding local supplies are not especially prom-

Table 6.3. *Water supplies available to the South Coast Basin in 1985 and 2000 for normal years and dry years, by source (million acre-feet per year)*

	Normal year		Dry year	
Source	1985	2000	1985	2000
Local supplies	1.11	1.12	1.11	1.12
Los Angeles Aqueduct[a]	0.47	0.47	0.47	0.47
Colorado River	0.46[b]	0.40[c]	0.46[b]	0.40[c]
State Project	1.51	1.48	1.13	0.99
Total	3.55	3.47	3.17	2.98

[a]Supply may be reduced 100,000–200,000, depending upon the outcome of litigation currently pending against the city of Los Angeles.

[b]This figure differs from Metropolitan's firm allocation of 0.495 maf by the amount of estimated conveyance losses in the Colorado River aqueduct.

[c]Differs from Metropolitan's firm allocation of 0.495 maf by the amount of estimated conveyance losses plus an allocation of cooling water for two new power plants that will serve the South Coast Basin.

Source: Metropolitan Water District of Southern California, 1983.

ising. Existing surface water sources have been developed virtually to capacity, and most groundwater basins are managed under court supervision at steady-state levels. Any departure from present groundwater management policies would produce supplementary water supplies on only a temporary basis. Exotic sources such as seawater desalination hold little promise because they remain prohibitively expensive. Recycled sewage effluent may ultimately supplement local supplies, but it is currently unattractive for two reasons. Recycled wastewater will not be used for domestic purposes in the foreseeable future because of the potential hazards to health, hazards that are not completely understood. Although wastewater could be used for industrial purposes and for urban irrigation, the costs of the treatment necessary to render it suitable for these uses make it unattractive. The Metropolitan Water District estimates that the use of recycled wastewater could grow 5,000 acre-feet per year during and beyond 1990. However, these estimates are not especially reliable given the uncertainties surrounding potential health effects and the costs of treatment (Metropolitan Water District, 1983).

The current water supply picture for the South Coast Basin is characterized first by the fact that the quantities of water available for import now appear to be significantly smaller than what was envisioned at the time these supplies were developed. A second important characteristic is the uncertainty about the ultimate long-term availability of water from both the Owens-Mono Basin and the State Water Project. The outcome of further litigation related to diversions from both the Mono Basin and groundwater pumping in the Owens Valley could reduce the city of Los Angeles's supply as much as 200,000 acre-feet. Uncertainty over supplies from the State Water Project focuses on resolution of the political stalemate that currently forestalls development of additional supplies.

The firm supplies available to the South Coast Basin for 1985 and 2000 are presented in Table 6.3. Figures are included for both normal years and a dry year in which conditions correspond to those in the drought of 1928–1934. The figures show that for normal years supplies will decline 80,000 acre-feet between 1985 and 2000. For dry years the decline from current average year availability amounts to 380,000 acre-feet in 1985 and 490,000 in 2000. Both the Metropolitan Water District and the California Department of Water Resources estimate that these supplies will be inadequate to serve the growth in demand expected to occur in the South Coast Basin between now and 2000. Demands for water in the South Coast Basin are considered next.

Current and prospective demands for water

The Metropolitan Water District estimates that the South Coast Basin population will grow from 12.3 million in 1980 to 16.5 million in 2000, a 34 percent increase. From this population growth, Metropolitan projects an associated increase in water demands from 3.06 maf in 1980 to 3.61 maf by 2000 (Metropolitan Water District, 1982). A comparison of the projected demand figures with the totals shown in Table 6.3 indicates that supplies will be insufficient to meet demand even in a normal year by 2000. The basis for the projected growth in demand can be understood by examining Metropolitan's methods for deriving its figures.

Historically, Metropolitan has projected water demands by combining a population growth projection with per capita use rates. For the most recent projection, the difference in use rates from area to area was emphasized. Within Metropolitan's service area, per capita daily use averages 227 gallons, but it varies from a high of 246 gallons in warm inland regions to 182 gallons in the cooler coastal regions (Metropolitan Water District, 1982). This variation is significant inasmuch as most growth will occur in the warmer inland areas. Four additional factors, all of which may reduce daily per capita use, are major variables on which such use depends: (1) water conservation programs mandating the use of water-saving plumbing fixtures, (2) urban redevelopment wherein family homes in older areas are replaced by multiple-dwelling units that have reduced requirements for landscape irrigation, (3) socioeconomic changes related to the large in-migration of Latin Americans and other low-income groups that tend to use water more sparingly than higher-income groups, and (4) cost incentives provided by sewer charges imposed by the sanitation agencies in the early 1970s. To date, sewer charges have induced economizing on water use and the employment of recycling facilities as industries attempt to minimize their liability (Metropolitan Water District, 1982).

Significantly, the Metropolitan Water District does not acknowledge that the price of water affects the quantities demanded. The neglect of potential price effects would not necessarily be important if the real price of water remained constant until 2000. However, water prices are likely to increase for at least two reasons. First, increases in energy prices over the last decade are only now beginning to appear in water prices. The California Department of Water Resources anticipates an 80 percent increase in the price of SWP water delivered to the South Coast Basin by 1986 owing to increases in the price of energy needed to pump the water over the Tehachapi Mountains (California Department of Water Resources, 1983b). Second, new water impoundment and conveyance facilities are likely to be far more costly than existing ones both because the relatively inexpensive facilities have already been built and because the costs of constructing civil works have risen at a disproportionately higher rate than costs in general. Thus, any new development will tend to raise the price of water.

The Metropolitan Water District explicitly recognizes the uncertainty inherent in forecasting water demands. At the same time, the neglect of potential price effects may have resulted in an overestimate of the levels of water demand likely to prevail in 2000. The price elasticity of demand is the conventional measure of the responsiveness (in terms of quantity adjustments) of consumers to changes in price. Several studies have established the fact that municipal water use in general is sensitive to price. For example, Howe and Linaweaver (1967) estimated price elasticities from cross-sectional data obtained from municipalities throughout the United States. They found that water consumption is more price elastic (sensitive) in the humid eastern portions of the country than in more arid western regions; elasticities vary with the type of use, with demands for outdoor uses (sprinkling) more price sensitive than those for indoor uses; and winter demands are more elastic than summer demands. These findings have been generally confirmed by other studies focusing on individual communities (Danielson, 1977; Grima, 1972; Young, 1973). Nearly all this work suggests that the demand for municipal water is relatively inelastic, with elasticities generally smaller than -1.0.

Although there are a number of studies of the price elasticity of demand for urban water in South Coast Basin and other urban areas in the semiarid West, no single work is definitive. One review of the studies concludes that the likely range of elasticities of demand for urban water is between −0.23 and −0.79 (Wahl and Davis, 1986). California Department of Water Resources economists estimated that, in the relevant price range, elasticity falls between −0.23 and −0.47, and they recommended the use of a provisional elasticity of −0.41 for its urban service areas, which include the South Coast Basin (California Department of Water Resources, 1982). These estimates of elasticity are aggregated in the sense that they include both winter and summer demands as well as indoor and outdoor water uses. There appears to be general agreement, however, that consumers would curtail outdoor uses far more easily than they would curtail indoor uses in response to an increase in the price of water in the South Coast Basin.

Vaux and Howitt (1984) used cross-sectional data on water consumption and the provisional elasticity of −0.41 to estimate water demand functions for the South Coast Basin for 1980, 1995, and 2020. A demand function for 2000, interpolated from the 1995 and 2020 functions, can be used to assess the potential effects of changes in the price of water in that year. An analysis of the demand functions for 1980 and 2000 shows that if real water prices remain constant at 1980 levels, demand for water will grow 19.9 percent. This figure is higher than Metropolitan's normal projection of a 17.9 percent demand growth. The difference is explained by Metropolitan's assumption that significant water savings would be realized through its water conservation program and the demand analysis assumption that in the absence of price changes there would be little or no incentive to conserve.

Potential price effects can be illustrated by analyzing the impact of the 80 percent increase in the 1986 price of water predicted by the Department of Water Resources. If it is assumed that energy-related price increases are limited to those occurring by 1986 and that those increases affect only the price of SWP water, then the 80 percent increase in the price of SWP water will result in an overall water price increase of 34.12 percent for a normal year and 26.58 percent for a dry year in the year 2000. The elasticity estimate of −0.41 can be utilized to determine the impact of such an increase on the quantities of water demand. The resulting demand would be 3.105 maf for a normal year and 3.22 maf for a dry year. The difference arises because SWP water constitutes a smaller proportion of the supply in a dry year, and thus the 80 percent price increase means a smaller overall price increase. A comparison of these demand figures with the available supply figures reported in Table 6.3 shows that when the SWP price increase is accounted for, normal year supplies will be adequate in 2000 but dry year supplies will not.

The real price effects related to increasing energy costs that are likely to occur by 2000 will undoubtedly be larger than those predicted by the Department of Water Resource for 1986, and they will have some impact on the costs of all supplies. As a consequence, existing supplies may be adequate well beyond 2000, even for dry years. Howitt, Vaux, and Gossard (1985) examined the impact on water consumption of a rise in the real price of energy from $0.025 per kilowatt hour (in 1980) to $0.08 per kilowatt hour (in 1995), as forecast by Christensen, Harrison, and Kimbell (1982). The study showed that water consumption in the South Coast Basin would be reduced from 3.25 maf in 1980 to 3.00 maf in 1995 despite the continuing population growth, and existing supplies would be more than adequate for both normal and dry years well beyond 2000. Price effects, then, may be an important determinant of water consumption. By neglecting the potential role of prices, Metropolitan could be led to support investment in new water supply facilities earlier than would otherwise be warranted.

The Metropolitan Water District's demand projections also reveal the implicit assumption that population growth is largely independent of the available water supply. Metropolitan's view is that although there are uncertainties inherent in projecting population growth precisely, substantial growth will occur and water must be available to service it. There is little difference between this view and the one expressed by William Mulholland during his early efforts to develop supplemental supplies of water. So long as this philosophy guides the activities of Metropolitan, pressure to develop additional sources of supply will continue. Water supplies in the South Coast Basin are currently adequate, but they will not support continued population growth at historically prevailing rates of use if real prices remain constant. Should population growth continue at projected rates, the region might respond in a number of ways. Some currently enjoy strong support and others have been neglected.

Alternative policies for adapting to water scarcity

The alternative means for adapting to water scarcity include construction of new impoundment and conveyance facilities to supplement existing water supplies, changes in water-pricing policies, and development of quasi-markets through which water could be voluntarily transferred. Although these alternatives could be used singly or in combination to ameliorate water scarcity, they will not be costless. The issue, then, is not whether it is possible for the South Coast Basin to adapt to increasing water scarcity but, rather, what costs will have to be incurred in the adaptation. Moreover, the history of water supply development in the region suggests that the costs should be reckoned broadly to include more than just the financial impacts on water users in the South Coast Basin. The uncertain magnitude of existing supplies is due, in part, to the historical neglect of broader categories of cost, including third-party effects and environmental impacts.

The construction alternative

The construction alternative involves a continuation of historical policies of mitigating water scarcity by constructing new impoundment and conveyance facilities. It is based on the proposition that California has ample water statewide that only awaits development. The most promising construction alternative involves the development of facilities needed to augment SWP's dependable supply. The specific proposals most recently supported by the Metropolitan Water District include a cross-delta facility with additional offstream storage.

The California Department of Water Resources (1983a) states that "the selection and building of an improved Delta water transfer system is the single most important decision to be made to advance the State Water Project." The problem, as outlined earlier, is that the existing capacity of the delta channels constrains the amount of water that can be pumped without saltwater intrusion, adverse effects on fish, and channel scour. The department has proposed four alternative plans involving channel widening and deepening, each of which has several capacity options. From a strictly cost-effectiveness standpoint, the most attractive options could deliver 450,000–500,000 acre-feet annually.

The Metropolitan Water District would be entitled to take up to 47 percent of this incremental yield, or 211,000–235,000 acre-feet annually, with the remainder going to other state contractors. These quantities could cover the year 2000 shortfall projected by Metropolitan for normal years but not for dry years. The estimated costs of these supplies to users in Metropolitan's service area would range between $370 and $390 per acre-foot. They include the annualized capital cost as well as transportation and treatment costs, and they are approximately $150 higher, in constant dollar terms, than the average prices currently paid by urban users in the South Coast Basin (California Department of Water Resources, 1983a, 1983b).

Although most attention has been focused on the need for a cross-

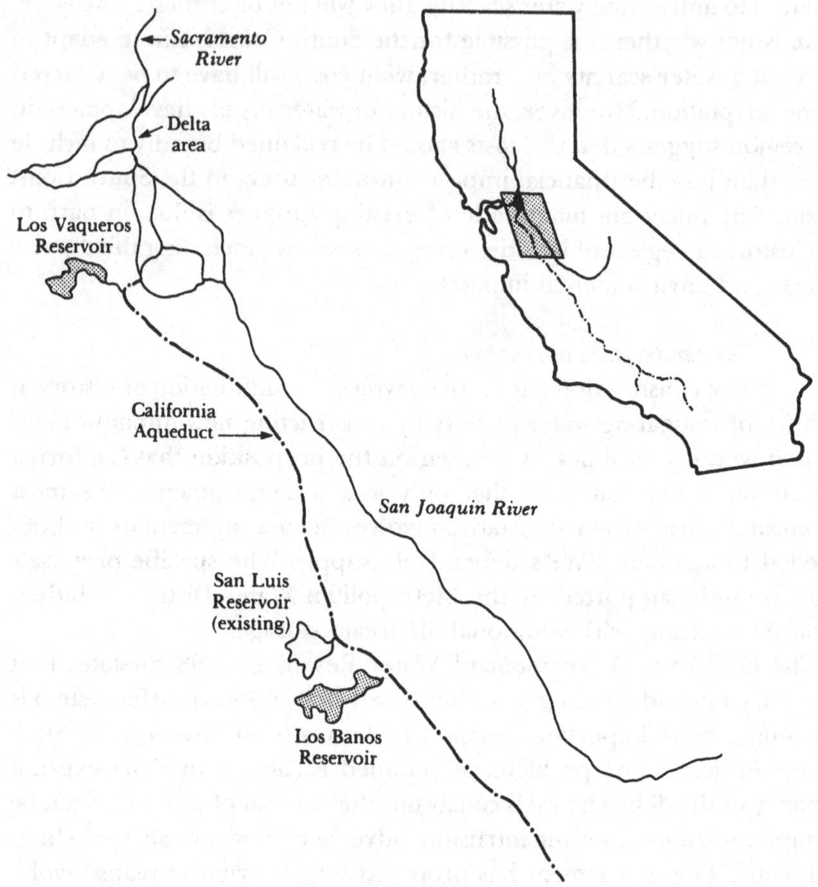

Figure 6.5. Sites of proposed Los Vaqueros and Los Banos Grande facilities.

delta facility, Metropolitan and other state water contractors point out that offstream storage facilities will be needed south of the delta if dry year shortfalls are to be avoided by 2000. Offstream storage is defined as storage constructed on a stream with little runoff for the purpose of storing imported water. Such a facility could be filled with excess delta flows during the wet season for use during the dry season when constraints on delta pumping are most binding. A preliminary investigation by the Department of Water Resources identifies 29 alternatives at nine sites for offstream storage facilities (California Department of Water Resources, 1984).

The preliminary investigations suggest that the two most cost-effective projects would be the Los Banos Grande and Los Vaqueros reservoirs. These two sites are illustrated in Figure 6.5. Los Vaqueros would have an average annual yield of 90,000 acre-feet, and Los Banos Grande could yield 100,000–345,000 acre-feet. The Metropolitan Water District estimates that these two facilities together could provide an additional 300,000 acre-feet of firm yield and that, in a severe drought period, Los Banos Grande could be drawn down to yield 700,000 acre-feet annually over a two-year period. Such a drawdown would reduce subsequent firm yield if there were additional dry years before the reservoir was refilled. Metropolitan indicates that these two projects, with a cross-delta facility and some water-salvaging and banking actions on the Colorado, should render the supply adequate for even a dry year by 2000 (Metropolitan Water District, 1983). Although Metropolitan does not mention costs, estimates of annualized capital costs and transportation costs by the Department of Water Resources plus treatment costs suggest that water from these reservoirs could not be delivered to users in the South Coast Basin for less than $500 per acre-foot.

The demand functions estimated by Vaux and Howitt (1984) show that water users in the South Coast Basin would not be willing to pay the incremental or marginal costs of the supplies made available by either a cross-delta facility or the offstream storage facilities before the year 2020. However, under Metropolitan's current pricing practices, users are not required to pay marginal or incremental costs. Rather, between one-fourth and one-third of the total is charged as fixed costs through property taxes, and the remainder appears as a commodity charge determined by the melded or average cost of all supplies. Because the costs of these new facilities are considerably higher than the costs paid for existing water supplies, melded pricing practices result in rather modest price increases to all users compared to the sharply higher prices that would have to be paid under marginal cost-pricing rules. Shortly after 2000, users in the South Coast Basin would be willing to pay the prevailing melded price after construction of a cross-delta facility but not a melded price that includes the two offstream storage reservoirs (Howitt, Vaux, and Gossard, 1985).

If users are unwilling to pay the costs of new supplies, there remains the possibility that these costs could be underwritten (subsidized) by California residents. In the prevailing political climate, in which there is intense competition for funds, coupled with a declining willingness to make those funds available through taxes, the outlook for state subsidization of water development is not good. Although a prolonged drought might change the political setting, the era of subsidized water development appears to have closed. Unless users are willing to defray the full costs of new development in some fashion, the prospects for such development are dim (Mann, 1984).

In addition to the potential economic problems with new facilities, several environmental problems have not been resolved. Although the plans for a cross-delta facility include assurances that water quality in the delta will be preserved, these assurances will have to be embodied in the operating criteria to satisfy water users in the delta region. Such a guarantee is necessary to assuage fears that, in the event of a major drought in southern California, diversions from the delta would be continued irrespective of the saltwater intrusion that might result. Beyond the saltwater intrusion, there are unresolved concerns about the effects of further diversions from the delta on anadromous fish, particularly salmon and striped bass.

In times of low flows, the pull of the pumps necessary to move water from the delta into the California Aqueduct is sufficiently strong to divert fish toward the pumps. This phenomenon can interfere with spawning runs that have already been reduced by upstream dams. A cross-delta facility may increase this interference, because most of the pumped water would be taken from a single point that spawning fish would have to traverse. The need to protect spawning runs will probably require extensive new fish screens and a reverse-flow fish return system. The installation of these fish-protection facilities will push the cost of water from a cross-delta project into the high end of the estimated cost range or beyond (California Department of Water Resources, 1983a).

Fish screens and return systems are only physically practical for a few cross-delta options identified by the Department of Water Resources. In addition, the effectiveness of fish screens of the size proposed for a cross-delta facility has not been evaluated thoroughly, creating the possibility that it will prove impossible to protect anadromous fish fully. The Department of Water Resources (1983a) indicates that artificial propagation of several major delta species in hatcheries could partially mitigate losses caused by the operation of a cross-delta facility. The costs of such a program have not been estimated, but they would undoubtedly increase the price of delta water to users in the South Coast Basin still further.

The potential environmental impacts of the two offstream storage reservoirs have been assessed only in a preliminary fashion (Department of Water Resources, 1984). Rare or threatened species could be affected, and there could be some impact on archeological sites. Although the preliminary environmental assessment does not identify the possibility of any major environmental impacts, further studies in connection with more detailed planning would be needed to rule them out.

The construction alternative, then, appears to suffer from several disadvantages. The economic feasibility in terms of user willingness to pay is dubious for the offstream storage facilities and may become quite marginal for a cross-delta facility if the costs escalate much beyond the level of current estimates. Environmental impacts of the cross-delta facility may be significant, but those for offstream storage are largely unknown. The effects on third-party users are manifested primarily in the need for guarantees of delta water quality and assurances that northern California's water supplies would not be further jeopardized by construction of these projects. The construction alternative thus appears burdened by the same problems that have reduced existing sources of supply or have made them uncertain. The major distinction is that new sources of supply are economically suspect, and subsidies, the typical means of financing water for users unwilling or unable to pay, are in all likelihood not available.

Demand management through pricing

The singular focus of western water policies on the development of new supplies has led to an almost complete neglect of the possibilities for regulating demand. Although water rationing has been used elsewhere in the West as a means of adapting to extreme drought for short periods, efforts to manage demand in the South Coast Basin have been restricted to public appeals to economize on water in times of drought and to promotion of water-saving plumbing for home use. Rationing to manage demand is probably effective only for short periods when there is a commonly perceived water-supply deficit. The difficulties and costs of extensive enforcement activities required for effective rationing over long periods preclude their use as other than an interim measure. Demand management through pricing has considerable promise both because the enforcement problem is eliminated and because users can adapt to the price of water voluntarily according to their own circumstances.

Water in the South Coast Basin is currently priced through a system of taxes and commodity charges. In recent years, 75 percent of the water revenues of both the Metropolitan Water District and the city of Los Angeles were obtained from commodity charges; the remainder came from property taxes. Over the long run, total annual revenues from these sources about equal the total annual costs incurred to deliver water. Although the pricing structure of these supplying agencies is complex and differentiates among classes of service, pricing policies generally call for prices to be established at levels that just cover cost. As a rule, then, these pricing policies conform to the principle of average cost pricing wherein the unit price of a commodity is set equal to the average cost of producing it (Metropolitan Water District, 1984; City of Los Angeles, Water and Power Commission, 1984).

In the long run, average cost-pricing practices ensure that costs are always covered and there is no surplus or profit, in contrast with a fundamental pricing prescription of economics that prices should be equated with marginal cost. That is, the price should be equal to the cost of producing the last unit of the commodity in question. Marginal cost pricing ensures both that resources are allocated efficiently and that profits or net returns are maximized. This prescription is conditioned by the requirement that average costs must be covered. An enterprise that does not cover average costs over the long run will inevitably fail unless it is subsidized from external sources.

Production of certain types of commodities such as energy and water has sometimes been characterized by the fact that average costs decline as output increases. When it does, marginal costs are always less than average costs, and marginal cost pricing cannot be employed if production is to be financially self-sustaining. Under these circumstances, prices are normally equated with average cost to ensure that all costs are recovered, although profits or returns in excess of costs do not then accrue to the enterprise in question. Although the practice of pricing urban water according to average costs may have begun in response to the decreasing average cost character of early water-purveying activities, it appears to have persisted because of legal provisions to constrain water districts from making profits and because of user pressure to keep the price of water as low as possible (Phelps, Moore, and Graubard, 1978).

Hirshleifer, DeHaven, and Milliman (1960) first argued in a theoretical vein that the average costs of water supply rise as new supplies are developed. Recent empirical evidence from California lends strong support to their argument. The capital costs per acre-foot of annual dependable yield for several existing and proposed California facilities are presented in Table 6.4. The figures show that the real capital costs of the proposed new storage facilities, Los Banos Grande and Los Vaqueros, would be four to six times as high as the cost of Shasta Dam and two or three times as high as Oroville Dam, the main storage facility for the SWP. This evidence suggests, then, that average cost-pricing policies can no longer be justified on the grounds that average costs are decreasing.

The legal provisions that militate against profit making by public agencies are one explanation for the persistence of average cost pricing. In addition, water-supply agencies and water users have a strong incentive to retain average cost pricing in the face of increasing average costs

Table 6.4. *Costs per acre-foot of yield for existing and proposed California water projects (1980 dollars)*

Project	Capital cost per acre-foot of annual dependable yield	Date completed
Shasta Dam	$ 415	1949
Oroville Dam	835	1968
Cross-Delta Facility	850 est.[a]	NA[b]
Los Banos Grande	1,734 est.	NA
Los Vaqueros	2,477 est.	NA

[a]May not include entire cost of fish screens and other facilities to mitigate fishery damage.

[b]NA = not applicable.

Sources: For Shasta and Oroville Dams, Meral, 1982; for the Cross-Delta Facility and Los Banos Grande, California Department of Water Resources, 1983a; for Los Vaqueros, California Department of Water Resources, 1984.

because it serves to keep the price of water artificially low. A simple example is shown in Figure 6.6. The line labeled "Demand" represents a hypothetical water demand function; the figure also shows the long-run marginal cost function and the long-run average cost function. If marginal cost-pricing rules prevail, equilibrium will be established at the point where the demand function intersects the marginal cost function. The resulting price and quantity can be called P_m and Q_m, respectively. By contrast, if average cost-pricing rules are followed, equilibrium will be established where the demand function intersects the average cost function, and the price and quantity are P_a and Q_a, respectively. As the figure shows, when average costs are increasing, the use of average cost-pricing rules results in an equilibrium characterized by lower prices and higher levels of consumption (quantities) compared with the equilibrium that prevails under marginal cost-pricing rules.

Additionally, when the use of average cost-pricing rules results in prices that are lower than the marginal cost of supply, users respond to a false price that understates the real scarcity of water. Users are thereby induced to use water inefficiently in activities whose value is less than the true costs of the water. Thus, when average costs are increasing, the use of average cost-pricing practices results in overdevelopment of water supplies and in inefficient use of a significant portion of those supplies.

The empirical implications of different pricing rules for the South Coast Basin can be estimated by using the demand, supply (marginal cost), and average cost functions reported by Vaux and Howitt (1984) and Howitt, Vaux, and Gossard (1985). The absolute magnitudes of the

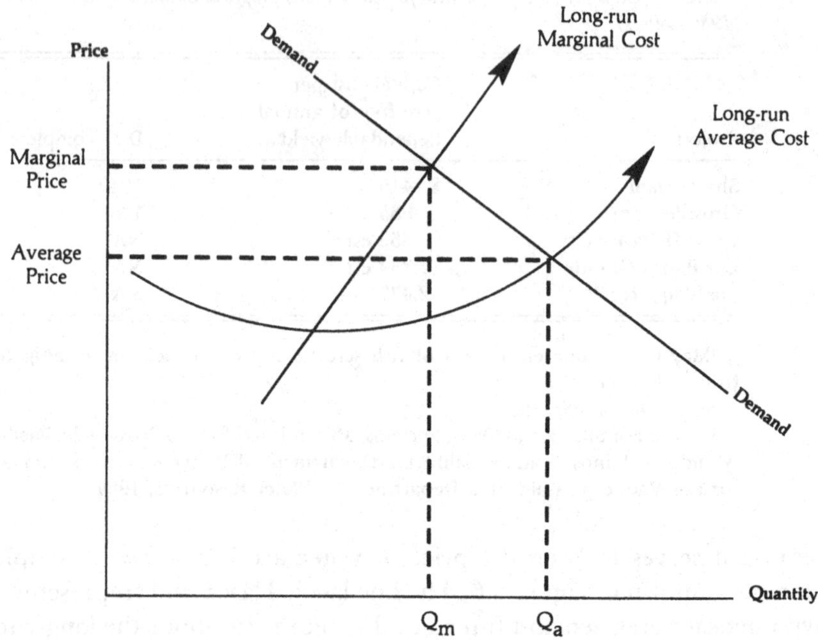

Figure 6.6. Equilibrium prices and quantities with average cost and marginal cost pricing rules (Vaux, 1986, p. 63a).

final results are influenced by the fact that both the marginal and average cost functions are based on the assumption, alluded to earlier, that the price of power needed to convey water will rise from $0.025 per kilowatt hour in 1980 to $0.08 per kilowatt hour in 2000 and will stabilize thereafter. However, it is the relative magnitudes of prices and quantities attributable to pricing policy rather than the absolute magnitudes that are of interest here. Table 6.5 shows the quantities and prices that are obtained from marginal and average cost-pricing policies for 2000 and 2020. It can be seen that average cost pricing causes consumption to be 260,000 acre-feet (8.4 percent) higher in 2000 and 340,000 acre-feet (10 percent) higher in 2020. Marginal cost prices are 13.1 percent higher in 2000 and 12.6 percent higher in 2020.

These figures can be placed in perspective by examining the welfare effects of the higher prices on consumers. The higher price that obtains with marginal cost pricing has the effect of reducing consumption 260,000 acre-feet over what it would be under average cost pricing. The 260,000 acre-foot "savings" cost consumers a total of $35,391,200 because of the higher price paid for the water they do use. Thus, it costs

Table 6.5. *Equilibrium quantities and prices for water in the South Coast Basin under marginal and average cost-pricing rules, 2000 and 2020 (1980 dollars)*

	Year 2000		Year 2020	
Pricing policy	Price ($/acre-foot)	Quantity (maf)	Price ($/acre-foot)	Quantity (maf)
Marginal	$350.72	2.86	$370.33	3.09
Average	310.15	3.12	328.87	3.43
Difference	+$ 40.57	−0.26	+$ 41.46	−0.34

Table 6.6. *Daily per capita use in 2000 and 2020 under marginal and average cost-pricing rules (gallons per capita per day)*

Year	Marginal cost pricing	Average cost pricing	Difference
2000	180.18	196.57	16.39
2020	167.19	185.58	18.39

consumers $136.12 per acre-foot to save the 260,000 acre-feet that could then be used to support additional development. This cost is less, by over half, than the cost of additional water that could be made available from a cross-delta facility. For 2020, the consumer cost per acre-foot saved is $47.93, which compares even more favorably with the costs of new supplies.

The impact of marginal cost pricing on daily consumption is quite modest. Daily per capita use under marginal and average cost pricing rules are displayed in Table 6.6. The difference in use rates is less than 20 gallons per day and amounts to a 9 percent reduction in 2000 and a 10 percent reduction in 2020. These figures illustrate how relatively small changes in daily use, 45 percent of which is for outdoor purposes, can make a substantial impact on total water consumption within the basin.

Demand management through marginal cost pricing of water has several advantages. The reduction in water use that occurs as a consequence of higher prices delays the time when expensive new supplies will have to be developed by stretching existing supplies farther. Moreover, the amount of reduced water use is not simply arbitrary. Marginal cost pricing rules ensure that water is used efficiently. That is, the uses to which water is put yield values that are equal to or greater than the cost of developing and delivering the water. Water uses that are abandoned when the price is raised to the level of marginal costs are inefficient uses, uses that return less in value than the cost of the water. Marginal cost pricing induces economizing on water by providing water users with the incentive to avoid using the resource inefficiently. Users are then allowed to respond voluntarily to the higher prices and adjust water consumption levels according to their own individual circumstances. By contrast, for centralized rationing to be effective, the rules must be simple and cannot account for the diversity of water-consuming habits and circumstances among millions of people. In addition to these advantages, internal price reform does not have adverse environmental impacts, and only direct users of water within the South Coast Basin would be affected.

The major institutional barrier to the adoption of marginal cost-pricing practices stems from the fact that marginal cost pricing results inevitably in the accumulation of profits. Water districts are generally constrained from profit making over the long run, and no mechanisms have yet been developed for distributing profits (Phelps, Moore, and Graubard, 1978). These constraints could be overcome by several means. Strictures against profit making could be removed and profits rebated to users either as direct dividend payments or as contributions to general funds to help defray other public expenses. In a case such as the Metropolitan Water District, where the provision of water is the sole activity, profits might be rebated downward to the 27 member agencies or lower levels of government where general funds do exist.

An alternative proposed by Phelps, Moore, and Graubard (1978) for urban water districts departs somewhat from pure marginal cost pricing but achieves substantially the same results. Under this alternative, water would be priced according to an increasing block rate. The first block would include some "lifeline" quantity of water, the minimum amount needed for basic sanitation and cooking purposes. The price of this block would be extremely low or zero. Succeeding blocks of water would be priced at the marginal cost of producing them so that all consumers using more than the minimum lifeline quantity would be charged the marginal cost of their use. Under such a proposal, the structure of water rates would be similar to the inverted or ascending block tariff used by many utilities and power companies. Profits would be dissipated by underpricing the lifeline block or even giving it away. The costs of the lifeline block would be recouped from the marginal cost charges for succeeding blocks. In this way, strictures against profit making would be observed, prices would conform approximately to marginal costs, and the oft-heard public cry that higher prices discriminate against the poor would be stilled because of the relatively low rates charged for the lifeline block.

It is interesting to note that the Department of Water and Power of the city of Los Angeles has developed a simple variant of this pricing policy. In 1986, the department adopted a peak use pricing structure in which the price of water during the dry summer months is 13 percent higher than during wetter and cooler portions of the year (Georgeson, 1985). This proposal represents the first of a number of changes in pricing policies directed at managing the demand for water more intensively.

Although the use of marginal cost-pricing practices as outlined above would result in more efficient water use within the South Coast Basin, there would be no assurance that use would be efficient within a larger statewide context. That is, an internal policy of marginal cost pricing would result in efficient use of water within the basin so long as the only supplies available are those developed by basin agencies and new supplies for which users might be willing to pay. It is possible, however, that there are developed supplies used elsewhere in the state that might be voluntarily exchanged. Market-like institutions would be required for this arrangement.

The potential of markets and interregional transfers

The apparent absence of water markets in California stems from a system of water laws, institutions, and attitudes that emphasize the development of new supplies and deemphasize the possibilities for reallocating water from low-value uses to higher-value uses (Howitt, Mann, and Vaux, 1982). This system has tended to lock water into specific uses over time. As a consequence, new supplies must be developed in order to service new uses as they arise. In a dynamic economy, the demand for water in various regions and for various activities does not grow evenly over time. Rates of population growth and economic development differ from region to region. When new supplies are not available and water transfers are impeded, it is inevitable that new uses that are highly valued may lack water while some existing low-value uses continue to be served.

This sort of inefficiency can be overcome by creating market-like institutions in which water can be voluntarily exchanged between buyers and sellers. The workability of such institutions would require that the price offered by potential buyers be sufficiently high to compensate the seller fully for withdrawing the water from current or potential uses. That is, after transportation costs are deducted, the seller must receive

Table 6.7. *Water prices[a] and quantities[b] for three regions of California, without trade*

	South Coast Basin		Imperial Valley		Southern agriculture	
Year	Price	Quantity	Price	Quantity	Price	Quantity
1980	$187.32	3.25	$ 6.47	2.91	$29.90	11.65
2000	350.72	2.86	17.80	2.91	39.75	10.46
2020	370.33	3.09	27.67	2.91	44.51	10.65

[a]In constant 1980 dollars per acre-foot.
[b]In million acre-feet.

more by selling or leasing the water than could be realized from devoting it to a productive activity. Such voluntary exchange arrangements would help to ensure that scarce water supplies are allocated to the uses that generate the highest values.

Vaux and Howitt (1984) and Howitt, Vaux, and Gossard (1985) analyzed the economic potential of interregional water trading in California. The state was divided into five water use regions; three of them use water predominantly for agricultural purposes, and the other two are urban regions. The three agricultural regions include the northern and southern halves of the state and the Imperial Valley. The urban regions include the San Francisco–Oakland metropolitan area and the South Coast Basin. Demand and supply functions were estimated for each region, as were the costs of transporting the water between regions. The demand functions shift over time in response to projected population growth, and the supply functions and transport costs reflect an anticipated increase in the price of energy between 1980 and 1995 from $0.025 per kilowatt hour to $0.08 per kilowatt hour. Energy prices remain stable after 1995. The functions and transport costs were then used in a spatial equilibrium model to identify the potential of interregional water trading.

The analysis showed that in 2000 the South Coast Basin would trade profitably with both the southern agricultural sector and the Imperial Valley. The quantities of water traded would increase still further between 2000 and 2020. The impacts of trade can be analyzed by comparing water prices and quantities in each of these regions both with and without trade. Table 6.7 shows the quantities of water used and the prices that would prevail in the absence of trade. The sharp price increases and quantity reductions between 1980 and 2000 are largely attributable to the effect of increasing energy prices on water costs. The price increases between 2000 and 2020 are due to increased competition

Table 6.8. *Water prices[a] and quantities[b] for three regions of California, with trade*

	South Coast Basin		Imperial Valley		Southern agriculture	
Year	Price	Quantity	Price	Quantity	Price	Quantity
2000	$253.15	3.48	$44.15	1.65	$44.15	9.95
2020	255.73	4.03	46.73	1.63	46.73	9.74

[a]In constant 1980 dollars per acre-foot.
[b]In million acre-feet.

for available water supplies, because none of the regions in question is willing to pay the price of developing new supplies.

Table 6.8 displays the results when interregional trade is permitted. The implications for the South Coast Basin are significant because the quantities are significantly larger and the prices are significantly lower. Specifically, the quantities rise 620,000 acre-feet in 2000 and 940,000 acre-feet in 2020. Prices are 28 percent lower in 2000 and 31 percent lower in 2020. The 3.48 maf used in 2000 is somewhat less than the 3.61 maf forecast by the Metropolitan Water District. The difference is attributable to the fact that Metropolitan does not account for any price effects, whereas this analysis includes the price effects related to population growth and the predicted increase in the price of energy. The post-trade price differential between the trading agricultural regions and the southern urban region is exactly equal to the costs of transporting and treating agricultural water for urban consumption.

The effects of trade on per capita daily consumption appear in Table 6.9. The figures illustrate that, with trade, relatively high rates of consumption can be maintained despite the changes in energy prices and population. This is because the price of agricultural water (including transportation costs) is relatively low compared with the willingness of urban users to pay for additional water.

The impact of trade on the exporting regions is also important. When water is traded, these regions use less water and the price increases. In

Table 6.9. *Daily per capita use with and without trade, 2000 and 2020 (gallons per capita per day)*

Year	Without trade	With trade	Difference
2000	180.18	219.24	39.05
2020	167.19	218.04	50.85

the case of the southern agricultural region, price increases are only slightly greater than the increases that would have occurred in the absence of trade. For the Imperial Valley, the price increase is quite sharp in relative terms. It is important to understand that these price increases are accepted voluntarily by sellers or leasers who find that returns from the sale of water are greater than the returns that could be obtained by using the water for irrigation. Thus, for example, in 2000, water users in the Imperial Valley who elect to sell or lease will receive $46.73 per acre-foot for water that costs them only $17.80 per acre-foot. As a result, both importers and exporters gain from trade.

The impact of trade on the economies of exporting regions will probably not be significant. Higher water prices will induce growers to economize on water use. The means of economizing are not limited to taking land out of production. Improvements in the management of irrigation water and in the physical facilities for delivering and applying it can result in significant water savings in many areas. Some growers may be induced to switch away from water-intensive crops of relatively low value to less water-intensive, higher-value crops. These water-economizing activities are likely to increase rather than stifle regional economic activity.

Where land may be taken out of production, regional diseconomies may occur to some extent. Banks and farm implement concerns are examples of secondary, or linked, industries that service direct water-using activities such as irrigated agriculture. If land is removed from production as a consequence of trade, some of these industries may suffer. To the extent that the resources supporting linked industries are mobile and can be employed elsewhere, it is inefficient to lock them into existing uses that would no longer be competitive. Indeed, the decline of industries and communities that are no longer competitive is one of the ways in which the economic system ensures growth and prosperity. If the linked industries prove to be completely immobile, real losses will occur. It is not at all clear, however, that such losses would be realized as a consequence of water trading in California (Vaux and Howitt, 1984).

The direction of trade, quantities traded, and the annual benefits from trade are summarized in Table 6.10. The benefits from trade are substantial, totaling a little over $100 million in 2000 and rising to $143 million by 2020. Benefits increase over time because demands in the South Coast Basin continue to grow while internal supplies remain static. As relative scarcity thus increases, so do gains from trade. The magnitude of the gains from trade can be placed in perspective by noting that an annual return comparable to the total benefits accruing to the South Coast Basin alone in 2000 would require an investment of $1.044 billion

Table 6.10. *Annual benefits from trade among three California regions*

Year and direction of trade	Quantity traded (maf)	Benefits (1980 dollars)	
Year 2000			
Imperial to South Coast	1.26	Imperial	$ 16,600,000
		South Coast	59,843,000
Southern agriculture to South Coast	0.50	Southern agriculture	1,100,000
		South Coast	23,747,000
Total			$101,290,000
Year 2020			
Imperial to South Coast	1.28	Imperial	$ 12,198,000
		South Coast	73,344,000
Southern agriculture to South Coast	1.00	Southern agriculture	1,110,000
		South Coast	57,300,000
Total			$143,952,000

at a real interest rate of 8 percent. The capitalized value to the South Coast Basin rises to $1.63 billion by 2000. These benefits also represent an opportunity cost to water users in the trading regions in the form of unrealized potential trade gains.

Market-like institutions for water trading would also permit the South Coast Basin to adapt relatively easily to dry years or drought periods. Options to take additional water could be sold whereby the seller could continue to use water except in dry years. This arrangement could obviate the need to build expensive new water storage facilities for the sole purpose of protecting against intermittent dry years. It would also ensure that in dry years scarce water supplies would be committed to high-value uses, with low-value uses temporarily abandoned.

Although there are a number of potentially significant barriers to trade, close examination suggests that none is insurmountable. The trades envisioned here could be easily accomplished with existing conveyance facilities. Specifically, the Colorado River Aqueduct provides a means to ship Imperial Valley water to the South Coast Basin, and the California Aqueduct and Metropolitan's own conveyance facilities permit a transfer between southern agriculture and the South Coast Basin. Thus, no additional conveyance facilities would need to be constructed.

Although the many potential legal problems have not been firmly resolved, existing work indicates that most could be overcome. The protection of third-party users who rely on return flows from some previous upstream use is critical. One clear-cut means of protecting third parties involves restricting quantities that can be leased or sold to water that is consumptively used. The determination of consumptive use of thousands of water users could be both costly and time consuming. The problem could be overcome by making water districts the trading parties rather than individual water users. Districts would then be required to compensate internally for third-party effects.

Many western states, including California, require that water be put to beneficial use. Failure to use water beneficially can constitute grounds for forfeiture. Recent changes in California water law make clear that water rights cannot be lost as a consequence of long- or short-term sales or of activities that result in water conservation. These same changes also protect unregulated groundwater resources against premature drawdown owing to trade by prohibiting expansion of groundwater pumping for purposes of transfer (California Water Code, secs. 22250–22257, 22259, and 22109, as amended). One comprehensive analysis of California water law affecting transfers concluded that there are currently no legal provisions that bar water trading so long as third-party interests are protected (California State Legislature, 1985).

The Imperial Valley could present a special case inasmuch as its water was originally allocated under federal law. Wahl and Davis (1985) note the special circumstances surrounding the creation of the Imperial Irrigation District and point to the decision in *Arizona* v. *California*, which confirms that Imperial's rights are self-owned rather than held in trust by the Secretary of the Interior. Although the secretary's permission would have to be obtained before any transfer could occur, the secretary has consistently upheld the primacy of state law once Colorado River water has been diverted for use in the state (Wahl and Davis, 1985).

The potential workability of water trading as a means of augmenting water supplies for the South Coast Basin is illustrated by a proposed agreement between the Metropolitan Water District and the Imperial Irrigation District (Boronway, 1985). Under the terms of this agreement, Metropolitan would provide $10 million annually to the Imperial Irrigation District to fund studies and projects that will result in saving irrigation water. Beginning two years after the first payment, Metropolitan would receive 100,000 acre-feet annually in return, the estimated quantity of water to be saved under the program. The cost of the water to Metropolitan would thus amount to $100 per acre-foot plus conveyance charges. The agreement would remain in force for 35 years; provision is made for further agreements that could augment both the financial contribution and the quantities of water made available to Metropolitan.

The roots of this arrangement lie with a 1983 proposal by the Environmental Defense Fund. This proposal noted that substantial quantities of water delivered to the Imperial Irrigation District were being lost to seepage through the All-American Canal and Imperial's main and lateral canals. Further, it was estimated that lining the canals at a cost of approximately $300 million could save 260,000 acre-feet of water. The Environmental Defense Fund (1983) proposed that the Metropolitan Water District line the canals in return for the resulting water savings. The proposed agreement differs only in that the water salvage projects are not specifically identified. Such projects could include both modification to Imperial's conveyance facilities as well as programs designed to economize on water use at the farm level.

The exchange program enumerated in the proposed agreement differs from the transfer program described in this paper only in that exchanges appear to be limited to water that can be salvaged from the delivery and drainage systems. Trades need not be limited to these sources, however. Additional water could be made available as a result of changes in cropping patterns or even from retirement of some irrigated lands. The proposed agreement, if consummated, represents a substantial beginning, and there is no reason to believe that the water ultimately available for trade would be limited to salvageable quantities from delivery and drainage systems.

Interregional trade, then, appears to offer substantial opportunity to the South Coast Basin for augmenting water supplies both during dry years and in the long run. The substantial economic gains that would result from trade are symptomatic of the inefficiencies with which water is currently used statewide as well as in southern California. Correcting these inefficiencies through establishment of trading institutions would allow the South Coast Basin to maintain current water use patterns at a relatively modest cost despite the projected population growth and predicted increases in the price of energy.

Conclusions

Development of the South Coast Basin was predicated on the capability of water agencies to develop remote supplies to supplement local sources that are inadequate to support the explosive population growth of the twentieth century. William Mulholland and other early water developers foresaw the possibility that competing interests might successfully reduce the quantities of water ultimately available for export to the South Coast Basin. Despite their best efforts to insulate imported supplies from the threats to competing interests, those interests have been able to reduce the level of exports or mount substantial threats to water claimed by the South Coast Basin. The experience with both the Owens-Mono Basin and the Colorado River illustrates how a region dependent on imported supplies subjects itself to the economic and environmental conditions prevailing in exporting areas. The concerns of northern California water interests over further development of delta supplies for delivery to the south through the SWP is only the most recent manifestation of this lesson.

The water supply agencies of the basin have predicted that, without additional supplies, historically prevailing rates of use cannot be sustained much beyond 1990. The reasons are the imminent decline in supplies available from the Colorado River and the anticipated continuation of population growth. The forecasted shortfall of water supplies is premised on the need to preserve historically prevailing patterns and rates of use even during dry years. This premise is consistent with the traditional philosophy that water supply should not be allowed to constrain economic growth. To prevent this shortfall, the Metropolitan Water District and others continue to emphasize the construction of additional water supply facilities. Although additional supplies can be developed, the costs of development are much higher than they were historically, and the competition for remaining supplies to support growth and preserve environmental quality has intensified. As a result, continuation of the historical tradition of responding to scarcity by augmenting supplies appears to be less tenable than it once was for both economic and political reasons.

The inability to develop new supplies need not forestall continued growth or doom the residents of the South Coast Basin to progressively worsening water deprivation, however. The evidence presented in this chapter suggests that the basin can adapt to intensifying water scarcity through changes in water pricing policies, development of institutions to facilitate market-like trade in water, or some combination of the two. Pricing policies that emphasize the marginal costs of water rather than average costs would improve the efficiency of water use within the South Coast Basin and ration an increasingly scarce water supply among growth demands. The resulting reduction in per capita daily use would probably affect outdoor uses but would maintain ample quantities for basic cooking and sanitation needs. Changes in pricing policies, then, provide one means of living with an essentially fixed supply for perhaps three or four decades.

Market-like water trading would provide the South Coast Basin with the opportunity to preserve water consumption rates at a relatively low cost. In the face of continuing population growth, the ability of domestic users in the basin to bid water away from less valuable uses in other parts of the state would also grow. The gains from trade would be substantial to all trading parties and would continue to grow over time. A well-functioning water market would reduce the scale of agriculture in the Imperial Valley; the effects on other agricultural regions would be relatively minor.

Although the use of imported water by the South Coast Basin does not affect the sustainability of the water supply, it does affect other regions that also depend on that supply. The development of new supplies in remote locations promises to have substantial impacts on both environmental quality and economic activity in exporting regions. The pricing and trade options would permit the basin to live within its existing supply or contract for other established supplies without imposing economic losses or degrading the quality of the environment in areas of origin. The issue of how much additional growth is desirable in the South Coast Basin cannot be examined from the water perspective alone. The evidence reviewed in this chapter suggests that the absence of new water supply facilities need not constrain growth but that appropriate water allocation policies could be used as a tool to help manage it.

References

Bean, Walton. 1968. *California: An Interpretive History*. New York: McGraw-Hill Book Company.

Boronkay, Carl, Metropolitan Water District of Southern California. 1985. Memorandum to Board of Directors, "Memorandum for Understanding-Imperial Irrigation District." July 5.

California Department of Finance. 1980. "Population Estimates for California Counties." Report 80 E-2. Sacramento.

California Department of Water Resources. 1979. "Report to Interagency Task Force on Mono Lake." Sacramento.

California Department of Water Resources. 1982. "The Price Elasticity of Demand for Urban Water in the SWP Service Area." Memorandum Report, Division of Planning.

California Department of Water Resources. 1983a. "Alternatives for Delta Water Transfer." Sacramento.

California Department of Water Resources. 1983b. "Management of the California State Water Project." Bulletin 132–83. Sacramento.

California Department of Water Resources. 1984. "Alternative Plans for Off-stream Storage South of the Delta." Progress Report. Sacramento.

California State Legislature. 1985. "Water Trading: Free Market Benefits for Exporters and Importers." Report 058A. Sacramento: Assembly Office of Research.

Christensen, M.N., Harrison, G.W., and Kimbell, L.J. 1982. "Energy." In Ernest Engelbert, ed., *Competition for California Water*. Berkeley: University of California Press.

City of Los Angeles, Water and Power Commission. 1984. "Water and Power, Eighty-Third Annual Report, 1983–84. Los Angeles: Department of Water and Power.

Cooper, Erwin. 1968. *Aqueduct Empire: A Guide to Water in California, Its Turbulent History, and Its Management Today*. Glendale, California: A.H. Clark, Co.

Danielson, Leon E. 1977. "Estimation of Residential Water Demand." Economics Research Report 33. Raleigh: University of North Carolina.

Environmental Defense Fund. 1983. *Trading Conservation Investments for Water*. Berkeley: Environmental Defense Fund, Inc.

Georgeson, Duane. 1985. Assistant General Manager, Los Angeles Department of Water and Power. Personal communication.

Grima, Angelo P. 1972. *Residential Water Demand: Alternative Choices for Management*. Toronto, Ontario: University of Toronto Press.

Hirshleifer, Jack, DeHaven, James, and Milliman, Jerome. 1960. *Water Supply: Economic, Technology, and Policy*. Chicago: University of Chicago Press.

Howe, Charles W., and Linaweaver, F.P. 1967. "The Impact of Price on Residential Water Demand and Its Relation to System Design and Price Structure." *Water Resources Research*, *3*(1), 13–32.

Howitt, Richard E., Vaux, H.J., Jr., and Gossard, Thomas. 1985. "The Costs of Average Cost Pricing: A Lesson from the Water Industry." University of California, Davis, Department of Agricultural Economics. Unpublished manuscript.

Hundley, Norris, Jr. 1975. *Water and the West*. Berkeley: University of California Press.

Kahrl, William L. 1982. *Water and Power*. Berkeley: University of California Press.

Lynch, H.B. 1931. *Rainfall and Stream Run-Off in Southern California since 1769*. Los Angeles: Metropolitan Water District of Southern California.

Mann, Dean E. 1984. "California Water Politics: Future Options." In John J. Kirlin and Donald Winkler, eds., *California Policy Choices, 1984*. Sacramento: University of Southern California, School of Public Administration.

Meral, Gerald. 1982. "Remarks to the Kern County Water and Energy Conservation Symposium." Sacramento: Department of Water Resources. Unpublished manuscript.

Metropolitan Water District of Southern California. 1982. "1982 Population and Water Demand Study." Report 946. Los Angeles.

Metropolitan Water District of Southern California. 1983. "Water Supply Available to Metropolitan Water District Prior to Year 2000." Report 948. Los Angeles.

Metropolitan Water District of Southern California. 1984. "Annual Report for the Fiscal Year July 1, 1983 to June 30, 1984." Los Angeles.

Ostrom, Vincent. 1953. *Water and Politics*. Los Angeles: The Haynes Foundation.

Phelps, Charles E., Moore, Nancy Y., and Graubard, Morlie. 1978. *Efficient Water Use in California: Water Rights, Water Districts, and Water Transfers*. Santa Monica, California: The Rand Corporation.

Vaux, H.J., Jr. 1986. "Growth and Water in the South Coast Basin of California." Draft paper for World Resources Institute.

Vaux, H.J., Jr., and Howitt, Richard E. 1984. "Managing Water Scarcity: An Evaluation of Interregional Transfers." *Water Resources Research, 20*(7).

Veeder, William M. 1969. "Federal Encroachment on Indian Water Rights and the Impairment of Reservation Development." In U.S. Congress, Joint Economic Committee, *Toward Economic Development for Native American Communities*. Joint Committee Print, 91st Congress, 1st sess., pp. 231–298.

Wahl, Richard W., and Davis, Robert K. 1985. "Satisfying Southern California's Thirst for Water: Efficient Alternatives." In Kenneth Frederick, ed., *Scarce Water and Institutional Change*. Baltimore: Johns Hopkins University Press.

Young, Robert A. 1973. "Price Elasticity of Demand for Municipal Water: A Case Study of Tucson, Arizona." *Water Resources Research, 9*(4), 1068–1072.

7 Toward sustaining a desert metropolis: water and land use in Tucson, Arizona

WILLIAM E. MARTIN,
HELEN M. INGRAM,
DENNIS C. CORY, AND
MARY G. WALLACE

A new era in water resources is being proclaimed by those who study and those who participate in water policy. (See, for example, Anderson, 1983; Weatherford and Brown, 1982). Water policy has moved from a period of water resource development to a new age of water management. Instead of building new sources of water supply to meet new demands, existing supplies will be made to serve through a reallocation from low- to higher-value uses and greater efforts at water conservation. The market system is supposed to play a much larger role in the emerging era. Increasing water prices will cause conservation and encourage the sale of water from less productive to more productive water rights holders.

Arizona in general and Tucson in particular provide an opportunity to test against reality the extent to which a new age in water has already dawned and the degree to which the vestiges of previous patterns of decision making and policy continue. The previous era of water development was driven by an overwhelming commitment to growth and development in the West (Wiley and Gotlieb, 1985; Fradkin, 1984). But has the attraction of expansion subsided as water officials turn from construction to management?

The era of development was criticized for its large environmental externalities that drowned free-flowing streams and buried beautiful and irreplaceable canyons under reservoirs. Whether contemporary water management in Arizona is more environmentally sensitive remains to be seen. In the age of water development, beneficiaries seldom paid the costs of development. Key to this analysis is whether the beneficiaries of water policy in Arizona now pay the costs or whether costs are still consigned to an unaware public and future generations.

The future role of government and administrative agencies in water policy is supposed to decrease, partly because the federal government can no longer afford to build and subsidize water projects. State control of water policy is also in question, and in a historic decision the Supreme Court proclaimed that public ownership of water was a legal fiction and that water was like any other commodity: water could be bought and sold across state lines, and states were not to place an undue burden on interstate commerce (*Sporhase* v. *Nebraska*, 458 U.S. 941, 1982).

To what extent does water in Arizona continue to be managed as a public resource by an administrative agency? The answer has three parts. At the state level, the way that the groundwater overdraft problem developed and the strategies the state has adopted to deal with the problem come into play. In Tucson itself are two dimensions of the problems of population growth and the land use changes that have fueled the demand for water and the politics and decision-making processes in Tucson that have kept at bay conflicts over who is to sacrifice as limited water supplies are spread among growing numbers of users.

Statewide Arizona water problem and policy response

Water supplies

The land use and water resource patterns in Arizona can best be understood in relation to its three water provinces. (See Figure 7.1.) The Plateau Uplands are for the most part high, cold in the winter, and dry. Groundwater is held in large quantities in the underlying sandstones, but these aquifers yield their water slowly to wells, so pumped water is expensive. Because precipitation is light, the rate of evaporation is high, and seepage is rapid, surface runoff from and across the Plateau Uplands is small. Population is sparse, and the major land uses are for grazing and timber.

The Central Highlands Province is the principal source of water originating within Arizona. It receives the heaviest precipitation within the state, ranging from 10 to 35 inches annually at various points. It is Arizona's chief area of perennial water supply and the principal source of streamflow and groundwater recharge to the Desert Lowlands Province. Here too population is sparse, and grazing and timber are the major land uses.

The Desert Lowlands Province, consisting largely of desert and warm steppe climates, is the beneficiary of the outflowing supplies from the

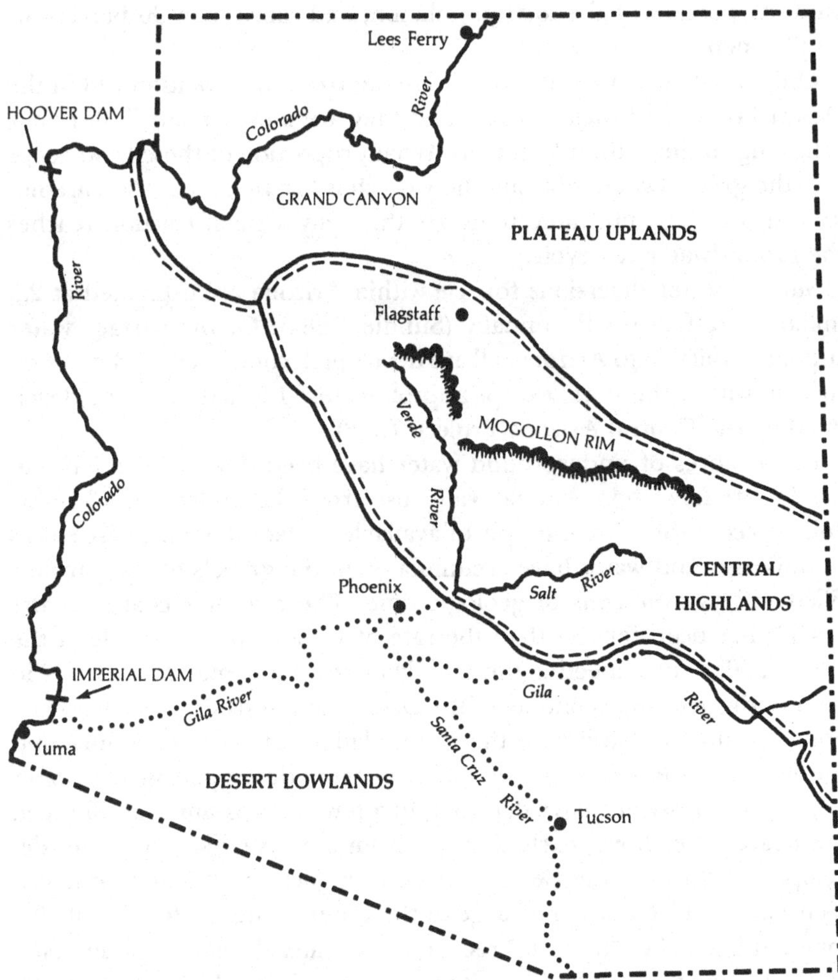

Figure 7.1. Water provinces of Arizona (adapted from Cross, Shaw, and Scheifle, 1960, fig. 1, p. 101).

Central Highlands Province. In addition, the province receives surface supplies entering New Mexico in the Gila River and from the Upper Colorado River Basin in the Colorado River; underground supplies are derived from precipitation on the highlands along its eastern and southeastern peripheries. It is by far the largest user of surface and subsurface water supplies in the state. Of all measured diversions of surface-flowing supplies and of groundwater withdrawals, more than 95 percent occur in this province. Although this province comprises only about 45 percent of the state's total area, it contains more than 95 percent of the cultivated land, 85 percent of the total population, and more than 90 percent of employment.

Only small quantities of water originate from precipitation within the Desert Lowlands Province itself, and they are mostly from flash floods following summer thunderstorms. What proportion of these flows seeps into the groundwater table and how much is lost by evaporation are not known precisely, but indications are that only a small fraction reaches the groundwater reservoir.

Surface water diversions for use within Arizona are estimated at 2.5 million acre-feet (maf) annually (Sundie, 1984). Of the surface water supplies available to Arizona, all are developed, controlled, and diverted for use within the state, except approximately 1.2-maf yet to be developed by the Central Arizona Project (CAP).

Large stores of underground water have been developed for use in the Desert Lowlands. Annual water use from the underground reservoirs exceeds the annual supply of available surface waters. These stores of underground water have accumulated in the gravels of the partially filled valleys over eons of geologic time. The rate of recharge to the gravels has been far less than the rate of withdrawal since at least the early 1940s. Thus, meeting the rapidly increasing demands for water in the rapidly growing economy of the Desert Lowlands Province has been possible only by drawing on the accumulations of many millennia, and the levels of the groundwater tables in most developed groundwater areas of the province have declined, in a few areas as much as 400 feet.

Surface water flows, excluding the Colorado River portion under development through the Central Arizona Project, are 2.5 maf annually. (See Table 7.1.) Natural recharge of the groundwater system is only 0.3 maf, but 2.5 maf of "induced" recharge infiltrates the aquifer after being applied to the surface, and thus it may be reused. In addition, at least 2.2 maf per year are mined from the aquifer (appearing in Table 7.1 as groundwater overdraft).

Economic development, population growth, and water consumption

The economy of Arizona, like the economies of other mountain and desert states of the West, depended initially on direct use of three natural resources – mineral ores, natural forage, and water for irrigation. At first the economy was based largely on mining and grazing. Federal expenditures for administration of the Arizona Territory and for the army's policing and protecting it were the only other important base for the state's economy. Incomes dependent on the use of irrigation

Table 7.1. *Developed water resource supplies in Arizona (acre-feet per year)*

Surface flows		
Colorado River	1,100,000	
Salt River Project	883,100	
Gila River	193,200	
Safford area	96,600	
Duncan area	19,470	
Agua Fria River	31,500	
Other smaller streams	176,130	
Subtotal, surface water flows		2,500,000
Groundwater supplies		
Natural recharge	300,000	
Induced recharge	2,500,000	
Subtotal, groundwater flows		2,800,000
Groundwater overdraft		2,200,000[a]
Subtotal, groundwater supplies		5,000,000

[a]Recent estimates set the groundwater overdraft at 2.5 million acre-feet (Arizona Department of Water Resources, 1984), but they do not include a consistent table of all developed supplies. Obviously, supplies vary from year to year.

Sources: Sundie, 1984; Valley National Bank, 1983.

water began to assume increasing importance with development of the Salt River and Yuma areas.

With further economic development, income came from sources other than grazing, irrigation, mining, and government. In 1929 (the first year for which such statistics are available), only 28 percent of all personal income in the state was received directly from agriculture and grazing (14 percent) and mining (14 percent). An additional 10 percent came from government sources. Income shares remained fairly constant until World War II; population had been increasing slowly until then. But since World War II both population and personal income have been growing at an increasing rate. Between 1945 and 1986 the population of Arizona grew from 590,000 to 3.2 million. The greatest absolute increases were in Maricopa and Pima counties, which contain Phoenix and Tucson, respectively. Maricopa County rose from 230,000 to 1.8 million, and Pima County from 95,000 to more than 0.62 million (Valley National Bank, 1962; University of Arizona, 1986).

The shares of personal income received directly from agriculture and mining are now only slightly over 2 percent each. The share from government has increased slightly, but manufacturing, services, and other sources account for almost 70 percent.

Although the shares of personal income generated by Arizona's econ-

Table 7.2. *Estimated annual water consumption, Arizona, normalized 1970 conditions*[a]

	Acre-feet	Percentage of total
Irrigation	4,294,000	89.2
Municipal and industrial	329,000	6.9
Mining	131,000	2.7
Fish and wildlife	40,000	0.8
Steam-electric power	20,000	0.4
Total depletions	4,814,000	100.0

[a]Normalized means 1970 use adjusted to normal conditions.
Source: Arizona Water Commission, 1977.

omy have changed vastly over time, the share of water used by irrigated agriculture has changed little. (See Table 7.2.) Irrigation use is still nearly 90 percent of the total water consumed. In comparison, municipalities and industry use only about 7 percent, and mining uses less than 3 percent.

Policy strategies responding to groundwater overdraft

As a consequence of continued groundwater mining for irrigated agriculture and of growth in the nonagricultural economy, Arizona's annual water deficit (the difference between the annually renewable supply and the consumptive use) has grown to an officially estimated 2.2 maf. Some unofficial estimates are double that figure. In some areas groundwater is used 30 times faster than it is replaced. This overdraft has lowered groundwater levels as much as 300–400 feet, making it uneconomical for some users to pump the water.

Several strategies might have been pursued to deal with the groundwater overdraft. Some economists endorsed the policy of nonintervention over a decade ago. Without any augmentation of supply and with some water law changes that cleared the way for market transactions in water, they argued, a gradual shift of water from agricultural to nonagricultural uses would occur. As pumping costs rose and buying out agricultural users became common, an equilibrium between pumping and recharge would occur naturally (Kelso, Martin, and Mack, 1973). Alternatively, the state might immediately regulate the amounts of water pumped in critical groundwater areas. However, neither a noninterventionist nor a heavily regulatory strategy was chosen. Rather, a combined supply-augmentation and prospective planning and regulation strategy was chosen. Its two major parts were the Central Arizona Project and the Arizona Groundwater Management Act.

Central Arizona Project. To augment water supplies, Congress authorized the Central Arizona Project (CAP) in 1968. Currently under construction by the Bureau of Reclamation, this project will develop the last remaining surface water supply available to the state. It will transport an allotted 1.2 maf of Colorado River water from Lake Havasu on Arizona's western border to the central agricultural and metropolitan areas. Water reaching the Tucson area, at the terminus of the canal, will have traveled 300 miles and been pumped 2,000 feet uphill through 14 pumping plants. The project began delivering water to the Phoenix area in 1986. Water is scheduled to reach Tucson by 1992. As much as 1.6 maf – more than the amount allocated – might be available for delivery in the project's early years. But in later years, as Colorado River water allotted to the Upper Colorado River Basin is developed, the supply could fall to 0.6 maf, even less in dry years.

The CAP was conceived to supply additional water to expand agricultural acreage. But by the time it was authorized in 1968, the declining water table in the groundwater pumping areas of the state had turned it into an agricultural rescue project. No new land could be irrigated, but agriculture would be rescued by using surface water instead of groundwater, which was perceived as becoming more expensive as the groundwater table fell. An important condition in the authorization act was that for every acre-foot of CAP water delivered to agricultural irrigators, one less acre-foot of groundwater could be pumped.

By the late 1970s, the CAP was increasingly viewed as a rescue project for the growing nonagricultural economy. Agriculture would still be a water recipient, but the quantities received would be reduced as the population grew and the supply of CAP water to agriculture fell. Under the current plan much of the water will be delivered to irrigated agriculture in the early years of the project because there will be more water than nonagricultural users can use. But as municipal and industrial demands grow, much less water will be available for agriculture; in addition, some CAP water will be used to fulfill Indian claims.

The 1980 Arizona Groundwater Management Act. An immediate effect of passage of the 1980 Arizona Groundwater Management Act was formation of the Arizona Department of Water Resources (ADWR). The department is responsible for devising and administering mandatory conservation measures aimed at eliminating groundwater overdraft. Because there are important local differences in water supplies and uses across the state, four "active management areas" (AMAs) were created where groundwater depletion was greatest. In these areas, phased-in reductions in the use of groundwater may be enforced. "Irrigation nonexpansion areas" (INAs) were also created. In these areas, no acreage may be irrigated that was not irrigated between 1975 and 1980, except on Indian reservations. Additional AMAs and INAs may be created.

These management areas are defined relative to hydrologic basins rather than political boundaries. Safe yield, that is, zero long-term overdraft, has been mandated for the Tucson, Phoenix, and Prescott AMAs by 2025. In the Pinal AMA, where only 2 percent of water consumption is nonagricultural, water is to be managed simply to preserve agriculture as long as possible, provided sufficient supplies are protected for future nonagricultural uses. In the other three AMAs, the cities are expected to be the ultimate management beneficiaries.

The mandated goal of zero groundwater overdraft by 2025 will drive Arizona water policy in the coming years. This goal is to be achieved through a series of five- and ten-year plans to be devised by the Department of Water Resources and its AMA directors. In the first phase, all irrigation wells within the four AMAs were registered and farmers obtained "grandfathered" irrigation rights based on the land they irrigated between 1975 and 1980. Water-metering devices are required for each well, and water duties, which may be reduced periodically, have been assigned to each eligible acre. A water duty is the amount of water allowed per acre to grow a particular crop. Under the Arizona Groundwater Management Act, a water duty is the maximum amount of water that may be applied per acre, based on historical cropping patterns for each farm, the consumptive water use for the crops, and assumptions about certain conservation practices yielding a given level of irrigation efficiency. Prescribed efficiency will be increased periodically, and the duty (including both surface water and groundwater) will be reduced.

Pattern of politics underlying strategies

Arguably, the strategies chosen by Arizona to deal with its water problems are both expensive and inefficient. As interest rates and construction costs have escalated, the Central Arizona Project has become enormously expensive, both to the federal taxpayer and to the state, which has agreed to share increasing costs to assure the project's early completion. In addition to this financial burden, the state has undertaken a big regulatory job in administering the Groundwater Management Act. Large planning staffs are required, and inspectors and enforcers must be hired. The decision was to phase agriculture out gradually through increasingly stringent water duties rather than relying on a similar phase-out through market forces.

Why has Arizona chosen these strategies? The answer is in the strength and resilience of the pattern of distributive politics that has dominated water policy throughout the development years. The existence of distributive politics both nationally and in Arizona has been marked by a number of observers (Maass, 1951; Freeman, 1955; Redford, 1969). The overall power of the alliances in water and "how those linkages have become institutionalized over the years" have been noted and analyzed (Pierce and Doerksen, 1976, p. 955). Numerous other studies have confirmed and clarified the nature of subsystem involvement in (and dominance over) national water policy, including the symbiotic nature of the alliances (Wandesford-Smith, 1974), the importance of local support (Ingram, 1971), "the porkbarrel basis of distributive politics" (Mann, 1978), and the ability of the mutually cooperative relationships to withstand challenge (Miller, 1985). The foundations of distributive politics can be found in the coalitions of local water users who once banded together to support supply solutions to their water needs that could be funded at state and federal levels. Differences among users' interests were minimized in deference to the achievement of a larger goal that would serve all. The key to success in distributive politics is to reach a consensus, obtained through mutual accommodation and avoidance of costs imposed on participating groups. In Arizona, cities have joined with mining and agricultural interests to shape water policy to their advantage, particularly regarding issues surrounding the CAP and the Arizona Groundwater Management Act.

Water politics has usually involved subsystems outside larger political institutions, parties, and ideologies. Only a narrow range of interests and experts has been involved in specialized structures insulated from direct public participation. Many key decisions are made well before issues ever reach formal institutions. In fact, major water users will meet behind closed doors, away from public scrutiny, on most important issues.

A classic example of closed decision making occurred during the negotiations preceding the Arizona Groundwater Management Act. Under pressure from the federal government, representatives of mining interests, agricultural interests, and the major cities held private meetings, operating on the assumption that the state legislature would pass the resulting management act without amendments (Connall, 1982, p. 319). At these meetings the cities joined with the mines and "flexed some muscles nobody thought they had" to force agricultural interests to bargain (Connall, 1982, p. 329). The net result of hours of negotiations was a management act that, while setting a framework for future conservation efforts, also essentially protected the interests of the dominant water users in the state. Many would argue that without these meetings a groundwater management act could never have been passed at all. In any case, the management act and the negotiations preceding it are good examples of the closed water policy process. Outside interests penetrate this arena only with difficulty.

The Central Arizona Project has been largely responsible for perpetuating the politics of mutual accommodation. Secretary of Interior Andrus's threat to withhold support for the CAP unless a groundwater law was passed spurred action among water users. As one observer noted, all parties involved "realized the importance of the CAP for the state" and no one "wanted to be responsible for sabotaging it" (Connall, 1982, p. 332).

Throughout Arizona, the CAP has provided a focal point uniting disparate interests. Since its authorization, various "crises" associated with the project have served to galvanize water users and strengthen support. For example, 1985 was the sixteenth consecutive year that major water users traveled to Washington, D.C., to support continued funding of the project (Hoy, 1985). This annual trek to Washington has become something of a ritual. Water lobbyists representing farmers, mining interests, and developers all willingly travel to Washington to "have lunch with [the] U.S. Interior Secretary" and to visit key Congressmen inside and outside the Arizona delegation to make their case (Tony Davis, 1985).

There is presently little opposition to the CAP in Arizona. Although the general public will be affected by higher taxes and lower-quality water as a result of the project, few see an alternative. The CAP has repeatedly been sold as the solution to the water problem and an "absolute lifeblood" for arid Arizona (Hoy, 1985). In fact, opposition at the national level to funding water development in the West has strengthened the coalition supporting the project. The governor and the congressional delegation have repeatedly stressed the need to unite the major water users, especially in support of Plan 6, a flood control and water shortage plan for the Phoenix area. As Governor Bruce Babbitt admonished:

> We are in the midst of a very elaborate discussion process involving analysis among the cities, the farms, the power interests, and agriculture. Not only will we have a consensus by September, but we will have a unanimous consensus at

great personal risk to anyone who is not a member of that consensus (Hoy, 1985).

The message is clear: anyone not participating under these consensus rules will not participate at all.

Excluding interests that cannot be folded into a consensus has a profound impact upon the sorts of issues that are put on the agenda in distributive politics. Only those issues that do not introduce costs that would generate sharp opposition by any participant can be considered. Controversy is tolerable concerning where CAP reservoirs will be located, how the aqueducts will be routed, whether wildlife crossings of canals should be constructed, and whether waterways should be covered. But more profound issues – whether the CAP should be built and population growth stimulated – cannot be addressed. Similarly, the bargaining among interests that resulted in the Groundwater Management Act could not seriously address more stringent regulation, through which some interests might sharply suffer or alternatively make way for water markets where "needy" potential buyers might be held up by "unscrupulous" owners.

The city of Tucson and the Tucson AMA

The city of Tucson and the Tucson AMA are the most water-short area in Arizona. Yet, all within the municipal area have been able to obtain all the water for which they wished to pay because all water used is currently groundwater and the quantity demanded is simply pumped from the ground. Because all supplies come from groundwater, drought is not a factor. But the area is water-short in the sense that there is a finite quantity of water stored in the underlying aquifers, and annual recharge is much smaller than annual consumption. Further, unlike use patterns in other populated and irrigated areas of the state, where agriculture uses almost 90 percent of the water, only 58 percent of the consumptive use is for irrigated agriculture. The immediate potential for land use conflicts related to water is greatest in the Tucson area because the city has fewer agricultural water rights to absorb if population growth continues.

Residents of Tucson have been conscious of a potential water scarcity since the mid-1970s. The first shock was from sharp rate increases levied to meet the financial crisis of the municipal water utility, which had to sink new wells and enlarge the distribution system to meet peak needs of a burgeoning population. The city later embarked upon a publicity cam-

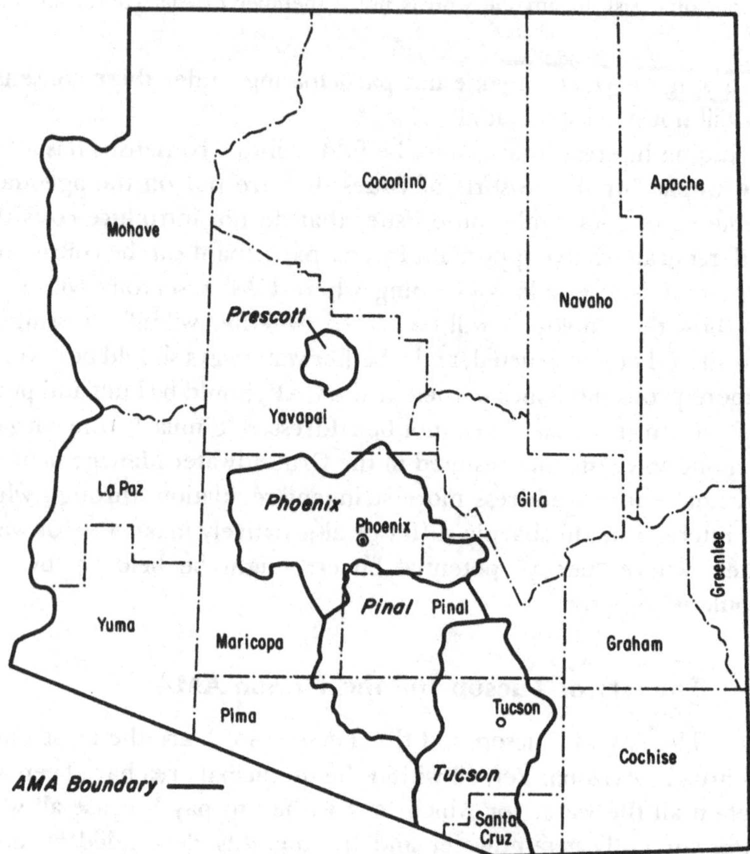

Figure 7.2. Active management areas (AMAs). (Adapted from *Planning for Growth in the Southwest,* by William E. Martin and Helen M. Ingram [Washington, D.C.: National Planning Association, 1985], fig. 2, p. 13.)

paign to reduce water use during the peak hours. Eventually it campaigned to reduce overall per capita water use. Tucson civic leaders pride themselves on being ahead of the rest of the state in facing up to water problems.

The Tucson AMA basically covers the populated areas of eastern Pima County and Santa Cruz County. (See Figure 7.2.) It includes the Tucson Basin on the eastern side, running north into Pinal County and south to the Mexican border. This basin contains metropolitan Tucson, four copper mines, two small communities, several thousand acres of farming north of the city in Pinal County, and portions of the sparsely populated Papago Indian reservations. Much of the irrigated agriculture in the AMA is west of Tucson, across the Tucson Mountains in the Avra Valley. South of the Avra Valley is the Altar Valley, with little population or agriculture; this valley could serve as a future source of groundwater.

The total population of the AMA in 1986 was over 0.62 million (University of Arizona, 1986). The Tucson municipal water system serves about 80 percent of the AMA population and pumps about 80 percent of its water from well fields in and south of the city along the Santa Cruz River (dry since the mid-1940s) in the Tucson Basin. The balance of its supply is pumped from across the Tucson Mountains from under 12,000 acres of formerly irrigated farmlands that were purchased and retired from agriculture to gain their water rights. Although the farmers had had unlimited pumping rights under this land, Tucson is restricted to the farmers' estimated consumptive use. Twenty-eight small water systems and private wells serve the balance of the population.

There is some question as to what water uses and supplies really are. Differences in these perceptions affect individuals' perceptions of the urgency of the water problem, although all agree that the overdraft is substantial.

The Department of Water Resources estimates average annual total pumpage at 179,000 acre-feet for municipal, industrial, and mining use and 230,000 acre-feet for agricultural use in the Tucson AMA from 1975–1980. (See Table 7.3.) Because these uses create incidental recharge and reusable effluent, consumptive use totals only 317,000 acre-feet. Current annually renewable water supplies are officially estimated at 9,000 acre-feet of reused wastewater, 83,000 acre-feet of incidental recharge, and 68,000 acre-feet of net natural recharge for a total of 160,000 acre-feet. Thus, as of 1984, the annual groundwater overdraft for 1975–1980 was estimated at 249,000 acre-feet, or 79 percent of consumptive use (Arizona Department of Water Resources, 1984).

Almost none of these quantities is known with certainty. Except for the municipal utility companies, pumpage has not been measured. Some estimates have been based on electricity use; for agriculture they are based on reported crop acreage and educated guesses about water applied per acre. As mentioned earlier, much of the recharge may take years to reach the groundwater table and therefore be available for pumping. Groundwater budgets implicitly assume instantaneous usable recharge; where groundwater tables are very low, recharge may be available for use only in geologic time. Thus, the realistic overdraft could be much higher than 79 percent of total consumptive use. Much of the

Table 7.3. *Tucson Active Management Area water use and potential supplies, annual average, 1975–1980*[a] *(acre-feet)*

	ADWR estimate	%
Total water use[b]		
Municipal	92,000	23
Industrial	30,000	7
Mining	57,000	14
Agricultural	230,000	56
Total	409,000	100
Consumptive use		
Municipal	60,000	19
Industrial	30,000	9
Mining	43,000	14
Agricultural	184,000	58
Total	317,000	100
Incidental recharge		
Municipal	23,000	28
Industrial	0	0
Mining	14,000	17
Agricultural	46,000	55
Total	83,000	100
Water supplies		
Central Arizona Project	0	0
Reused effluent	9,000	2
Net natural recharge[c]	68,000	17
Incidental recharge	83,000	20
Overdraft	249,000	61[d]
Total	409,000	100

[a]Municipal, industrial, and mining categories are disaggregated from a single category called Municipal and Industrial, based on unpublished Arizona Department of Water Resources (ADWR) data (Snow, 1985).

[b]Includes all conveyance losses.

[c]Excludes water lost owing to phreatrophytes and basin outflows of 62,000 acre-feet.

[d]Overdraft is 79 percent of consumptive use.

Source: Arizona Department of Water Resources, 1984.

estimated incidental recharge from irrigation, occurring in heavily pumped areas away from streambeds and washes, surely will not be available soon.

However, accepting the ADWR estimates of recharge and consumptive use, the current annual overdraft on the groundwater table is 249,000 acre-feet, about 79 percent of total consumptive use and 135 percent of agricultural consumptive use alone.

To put this estimate of the overdraft in perspective, one must compare it to the quantity of recoverable water stored in the ground in the area. Griffin (1980) summarized the available data. Productivity of the Tucson Basin's aquifer falls off rapidly in some parts of the basin at depths below 600–800 feet. In other parts good-quality water can be drawn at 2,000 feet (Johnson, 1979). Babcock (1979) and Anderson (1979) suggest low productivity of the aquifer below 600–1,000 feet. Griffin concluded that it may be possible to draw at least 10 maf from Tucson Basin sediments.

As to the Avra and Altar valleys, Griffin (1980, p. 262) writes:

> [A]bout seventeen million acre-feet are in storage in Avra Valley to a depth of 1,200 feet and a further ten million acre-feet are stored in the sediments of Altar Valley. The amount of this "recoverable water" that would, in fact, be practicable (economically) to recover is a matter of conjecture, but perhaps half of this water could be recovered by the development of extensive wellfields in Avra and Altar Valleys.

Thus, approximately 10 million recoverable acre-feet are stored in the Tucson Basin, and another 13 maf could be recovered from the adjoining valleys. At the *current* rate of depletion as estimated by the Department of Water Resources, the supply would last about 90 years. Something must eventually be done about this overdraft if the area is to remain habitable, especially if the population continues to grow at the rapid rate currently expected by state and city planners.

Population growth in a desert environment

Tucson is a spread-out city growing rapidly outward toward the edges of the bowl formed by the low mountain ranges that surround it. The foothills of the Santa Catalina range on the east and the Tucson mountains on the west are now dotted with resorts and housing developments where a decade ago only cactus and hiking trails existed. The surrounding environs of Tucson are inviting places to visit and to live. Tucson is within the northern reach of the great Sonoran Desert, which is rich with exotic animals and plants, including the giant saguaro cactus. A small part of Tucson's desert environs is protected by the Saguaro National Monument and by Catalina and Tucson Mountain state parks. By far the largest share of the outlying desert is privately owned and is rapidly urbanizing.

Pima County, which contains most of the population of the Tucson AMA, is expected to grow to more than 750,000 people by 1990, an increase of 42 percent since 1980. Seventy-five percent of that growth will be from immigration (University of Arizona, 1984). The expected population of the Tucson AMA in 2025, the year that the Arizona

Table 7.4. *Percentage of the Tucson Metropolitan Area in high- and low-intensity uses and changes since the preceding study year*

Year	High intensity (%)	Low intensity (%)	Change: % of region converted	Change: Average annual conversion (acres)
1967	27.1	72.9	NA	NA
1971	31.0	69.0	3.9	2,147
1973	33.8	66.2	2.8	3,084
1976	39.6	60.4	5.8	4,259

NA = not applicable.
Source: Willis, Cory, and Cao, 1982.

Department of Water Resources has mandated zero groundwater overdraft, is 1.6 million, an increase of 323 percent for the base population used to generate the water use data shown in Table 7.3 (Arizona Department of Water Resources, 1984). Most of this growth will be in the Tucson Metropolitan Area (TMA), the metropolitan area of the AMA.

Recent land use changes in the Tucson Metropolitan Area

A recent study of land use changes in the Tucson Metropolitan Area documents the dramatic impact of rapid population growth on land use patterns and trends (Willis, Cory, and Cao, 1982). As population growth has accelerated in recent years, low-intensity land uses in the TMA have been converted to high-intensity uses. This conversion nearly always involves some irreversible change in land use. For example, once high-rise apartments are built on a piece of land, such low-intensity uses as grazing and farming are out of the question, and the options for future conversion are considerably reduced. However, the same parcel held in a low-intensity land use could be converted in a myriad of ways.

Changes in low-intensity land uses. As defined here, the TMA comprises 220,300 acres (344.2 square miles). Table 7.4 shows that the percentage of the area in low-intensity use is substantially higher than that in high-intensity use. The TMA extends well beyond the Tucson city limits, and it is in this fringe area that most low-intensity land is located. However, few areas of Tucson have been completely converted to high-intensity uses, even those that are the most densely developed.

Annual conversion rates grew at an increasing rate from 1967 to 1976. From 1967 to 1971, roughly 2,147 acres of land area (3.4 square miles) were developed annually. The annual area converted increased to 3,084 acres (4.8 square miles) between 1971 and 1973. After 1973, the conversion rate continued to climb, averaging 4,259 acres (6.7 square miles) per year until 1976. By 1976, the annual rate of conversion of land to high-intensity use was nearly twice that between 1967 and 1971.

Conversion of low-intensity land uses and population growth. Development is influenced by external factors that encourage or discourage buying, selling, and alteration of property. One of these influences is population density and growth. An influx of people into an area increases demand for land allocated to residential, commercial, industrial, and institutional uses. From 1967 to 1971, population increased an average of nearly 8,000 residents per year. From 1973 to 1976, annual migration into the area was more than double this level. The greatest influx, however, occurred between 1971 and 1973, when population grew an average of 21,000 residents annually.

Development of high-intensity land uses is a reflection of demographic changes. As population increases, land conversion increases along with the demand for housing, goods, and services. The effect of an influx of new residents on the housing market could be delayed, as it seemed to be between 1971 and 1973. Some newcomers live temporarily in existing structures until new houses become available. The full effects of an increasing population on development may not be felt for several years.

Preliminary evidence from a recently completed study indicates that annual population growth averaged 4 percent for 1976–1980 and conversion rates continued to be dramatic. An estimated 6,550 acres were taken out of such low-intensity uses as grazing, agriculture, desert, and scenic uses (Porterfield, 1984). Moreover, additional low-intensity acreage has been retired in nearby Avra Valley; the city of Tucson has purchased 16,000 acres of agricultural land and plans to buy an additional 8,000 acres in order to obtain water rights (Burchell, 1983).

Clearly, the last 20 years of the Tucson land use experience were characterized by increasing strain on low-intensity land uses as high-intensity uses expanded to accommodate urban needs. If population in the TMA continues to increase as projected, accommodating urban expansion while maintaining an intertemporally efficient allocation of water supplies to a changing land use will force other policy issues to the fore.

Ecological implications of urban expansion. The rapid conversion of low-intensity land uses in the Tucson AMA has generated a variety of ecological stresses. The Saguaro National Monument is unique in the Southwest; its preserved desert environment is rich in plant and animal life and has an unparalleled stock of saguaro cactus. Recent evidence suggests that vegetation in this area is being damaged by rising ozone concentrations. The problem could be exacerbated further if the county acts on plans for dense development of the area neighboring the monument, plans that affected neighborhood associations are strongly resisting. A more general ecological concern is deteriorating water quality in the region. Water quality problems have intensified with the rising urban population in recent years owing to the mishandling of wastes and the increased flooding and erosion problems associated with removal of water-absorbing low-intensity land uses.

Municipal water use

The Tucson Water Department classifies its customers as single-family residences, duplex and triplex residences, multiple-family residences, commercial users, and industrial users. Virtually all nonresidential accounts are classified as commercial users: colleges, hospitals, government offices, parks and golf courses, and other commercial enterprises. The city delivers virtually no unmetered water except for firefighting; even water used for irrigating street medians is metered and is included in the totals for commercial use. (The industrial rate is given to few municipal users. Industrial use, as used in Table 7.3, refers to self-supplied industries.) Total pumpage by the water department, including firefighting and losses, is shown in Table 7.5.

As of 1978, consumption by single-family residences accounted for 64 percent of the water delivered by the city's water utility. Their consumption rises strongly during the spring, peaks in June, and then declines during the fall. From May to September, single-family residences use 80 percent more water than they do in the six winter months. Garden watering, particularly lawn sprinkling, is the principal cause of this difference; the other main components are the use of evaporative coolers and evaporation from swimming pools.

Multiple-family residences use 9 percent of the water delivered by the city. The seasonal variation in use is similar to that of single-family residences. But with no or communal lawns and with refrigerated cooling, they use less water in summer than detached homes; consumption by multiple-family residences in the summer is only 40 percent higher than in the winter. Duplexes and triplexes accounted for only 2 percent of metered deliveries. In these homes, consumption in the summer is 4 percent more than it is in the winter.

Table 7.5. *City of Tucson Water Department's pumpage, fiscal years 1965–1966 to 1983–1984*

Fiscal year	Average daily pumpage (million gal/day)	Number of active services	Service population[a]	Pumpage per capita (gal/day)
1965–1966	41.5	60,934	232,000	179.2
1966–1967	41.8	62,383	237,000	176.3
1967–1968	43.8	63,483	241,000	181.6
1968–1969	48.3	68,063	259,000	186.7
1969–1970	51.2	71,241	264,000	194.2
1970–1971	54.1	78,177	289,000	187.0
1971–1972	57.8	84,177	314,000	184.2
1972–1973	60.1	95,105	352,000	170.8
1973–1974	75.4	99,604	369,000	204.6
1974–1975	67.6	102,813	380,000	177.7
1975–1976	70.0	105,595	370,000	189.4
1976–1977	60.5	109,480	383,000	157.9
1977–1978	59.4	113,105	396,000	150.1
1978–1979	60.8	117,777	412,000	147.5
1979–1980	66.2	122,514	429,000	154.4
1980–1981	69.7	125,367	439,000	158.9
1981–1982	74.0	127,650	447,000	165.6
1982–1983	70.2	132,895	465,133	151.0
1983–1984	70.2	133,410	474,436	148.8

[a]Assumes 3.7 people per service through 1974–1975; 3.5 people per service are assumed thereafter.

Source: Tucson Water Department, 1984.

Commercial customers used 17 percent of the water delivered by the city. Irrigation, particularly of parks and golf courses, and cooling raise commercial use 50 percent in summer over that in winter. Customers allowed water at the reduced "industrial" rate accounted for 7 percent of metered deliveries. Almost all was used by a large public school district. Because playing fields are irrigated, industrial consumption shows more seasonal variation than any other class of consumption – summer use is more than twice winter use.

Water consumption by municipal users may be divided into indoor and outdoor uses. Indoor use is the nonconsumptive use of water for general sanitary purposes, and outdoor use is the consumptive use of water for watering lawns and gardens and for feeding evaporative coolers. The distinction is important for understanding water conservation programs. Savings of indoor water are not necessarily real savings at all because indoor water flows to the sewage treatment plant, where it may be reclaimed and made available for reuse. If net pumpage is to be

Table 7.6. *Division in water consumption between indoor and outdoor uses, Tucson, 1978*

Class of building	Indoor use (%)	Outdoor use (%)	Total use (%)	Ratio of indoor to outdoor use
Single-family	62	69	64	62:38
Duplex-triplex	3	2	2	76:24
Multiple-family	11	5	9	79:21
Commercial	19	14	17	72:28
Industrial	5	10	7	46:54
Total	100	100	99[a]	65:35

[a]Column totals less than 100 percent because of rounding.
Source: Griffin, 1980.

reduced, it is outdoor water use that must decline. Saving indoor water saves money but not water. To the extent that reduced indoor water use lowers the water utilities' costs, and if those cost savings are passed on to the consumer through lower water bills, indoor water conservation can actually encourage overdraft.

Outdoor use has not been measured specifically, but it may be estimated as the difference between the flow of sewage into the city's sewage treatment plant and metered deliveries. The breakdown of indoor versus outdoor use is shown in Table 7.6. Thirty-five percent of all water delivered was used outdoors, and 70 percent of this outdoor consumption was by single-family residences. The city's water department suggests that evaporative coolers account for about one quarter of outdoor water consumption.

Given the current trend toward more multiple-family dwellings, outdoor use per capita relative to indoor use will decline slowly over time without mandatory controls, education activities aimed at changing people's tastes and preferences, price changes, or other positive actions.

Municipal water demand

Martin et al. (1984) estimated the effects of changing Tucson water prices on the quantities of water purchased by single-family households. For a random sample of 2,159 single-family residences, monthly data were obtained for 1974–1979 from records of the Tucson Water Department. The quantity of water delivered to a single-family residence in each month was regressed on the corresponding marginal real price of water, the evapotranspiration less the quantity of rain, and the full cash value of the home in 1979. All variables were highly significant statistically. The quantity used varies inversely with price; it has a positive relationship with evapotranspiration less rain and with the value of the house (which is a surrogate for household income as well as house size).

The most important variable for predicting water use in this desert area was evapotranspiration minus rain (EVT − R). (See Figure 7.3.) "Water use closely follows the rise and fall of EVT − R in all years. . . . In fact, most of the *individual* variance in behavior is explained by EVT − R in the statistical demand equation. . . . Yet, [the figure] clearly illustrates . . . that the pattern of water use is substantially lower in recent years" (Martin et al., 1984, p. 64).

The change in water use between years is at least partially caused by changes in water price. The average price elasticity of demand was estimated at −0.26; that is, a 1 percent change in the real price of water implies a 0.26 percent inverse change in the average quantity demanded. Although the Tucson City Council raised nominal water prices in 1974 and in each year between then and 1981, the public was not fooled by nominal price increases (Martin et al., 1984). Rather, consumer behavior related to real (inflation-adjusted) prices. Real prices either fell or remained constant between 1978–1979 and 1981–1982, and pumpage per capita per day crept back up (see Figure 7.3). In the two years 1982–1983 and 1983–1984, rate increases were no larger than in the immediately preceding years, but they were real rate increases because they exceeded the rate of inflation. Pumpage per capita declined, as would be expected, especially because the real price increase coincided with above-average rainfall.

The price of municipal water

Tucson's water pricing policy is traditional in that it seeks only to cover the average cost of a unit of water delivered. Charges are based on current operating costs and repayment of amortized fixed capital investment. The actual rate-making procedures, also traditional, are based on the American Water Works Association (AWWA) procedures as described in its manual, *Water Rates*. Water rate schedules based on these procedures are intended to charge customers for water according to the cost of serving them. Traditionally, the AWWA procedures have been applied so as to give rate schedules in which the price of a unit of water was constant over all levels and times of use, or even where the price per unit falls as a customer's consumption exceeds certain levels each billing period. Such was the case for rate schedules introduced between 1925 and 1970 (see Figure 7.4).

In recent years, the AWWA procedures have been applied to create

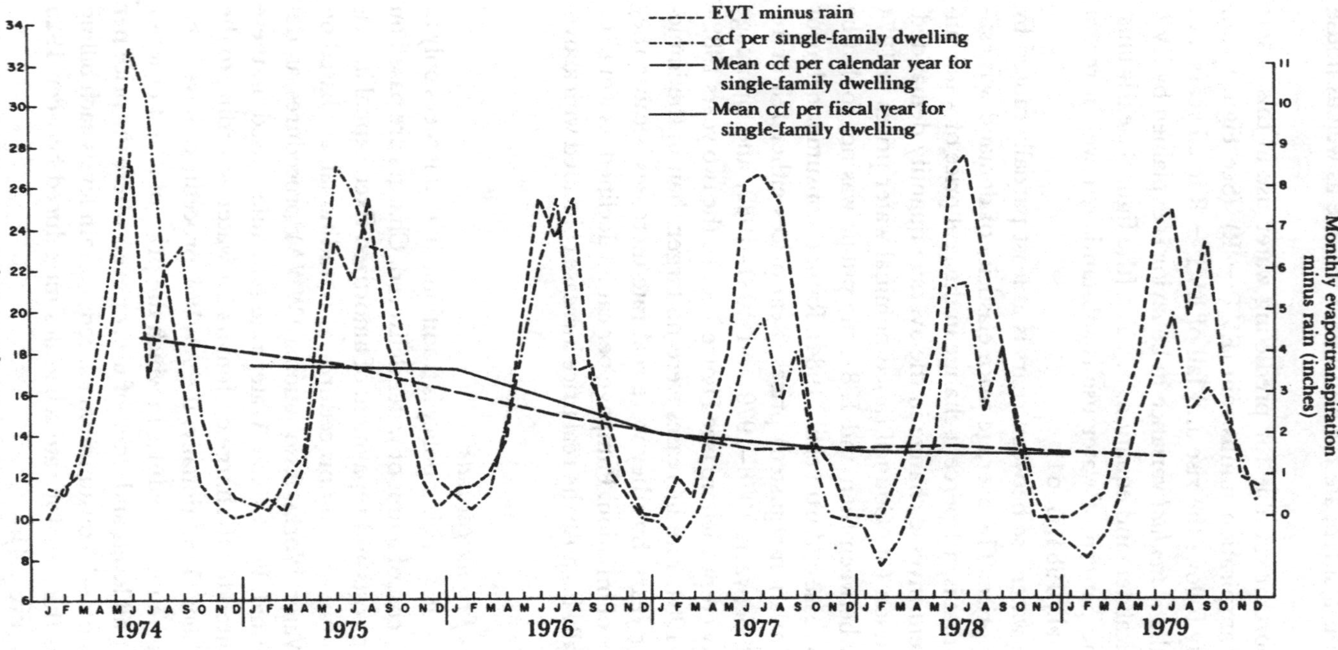

Figure 7.3. Monthly water deliveries per single-family dwelling, in relation to evapotranspiration minus rain, Tucson, 1974–1979 (sample data; ccf = 100 ft³). (From William E. Martin, Helen M. Ingram, Nancy K. Laney, and Adrian H. Griffin, *Saving Water in a Desert City*, fig. 5-2, p. 63. © 1984 Resources for the Future, Washington, D.C.; reprinted with permission.)

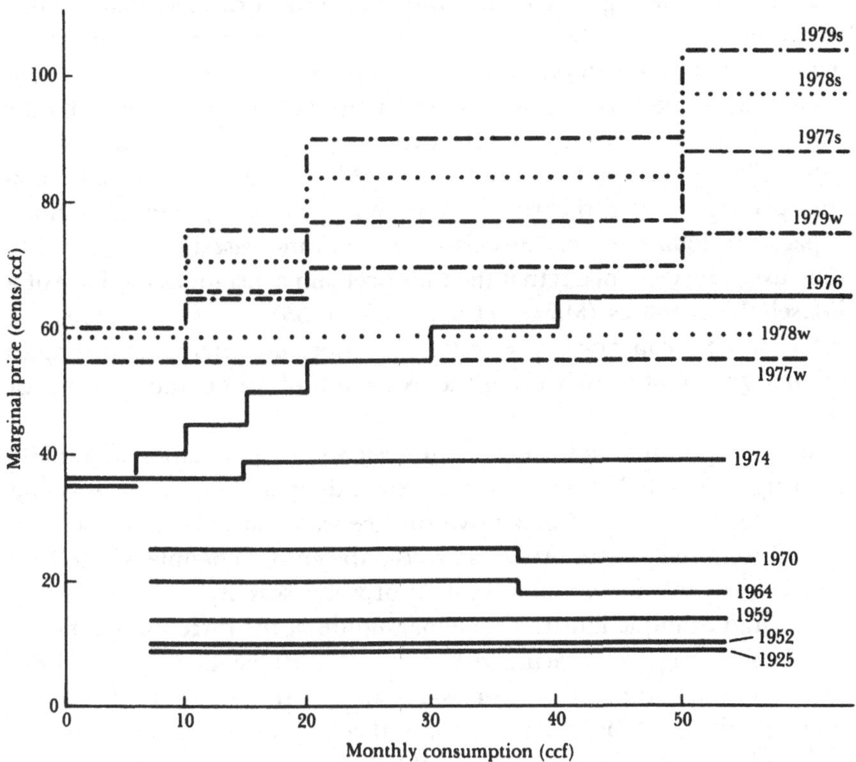

Note: Year of introduction shown alongside rate schedule; *s* and *w* denote summer and winter schedules.

Figure 7.4. Water rates for single-family residences in Tucson, 1925–1979 (ccf = 100 ft³). (From William E. Martin, Helen M. Ingram, Nancy K. Laney, and Adrian H. Griffin, *Saving Water in a Desert City*, fig. 4-2, p. 53. © 1984 Resources for the Future, Washington, D.C.; reprinted with permission.)

new rate schedules in which water prices increase for additional use blocks and rates are higher in summer than in winter. Tucson began this practice in 1974 and continues it today. Novel in form, the schedules are traditional in that the "costs" used in the procedures are not economic costs but are the water utility's revenue requirements. Although these requirements do represent money that must be paid to bondholders and meet current operating costs, yearly revenue requirements do not represent the real cost of providing service as the AWWA procedures claim. Revenue requirements are based on current costs and past capital expenditures. They cover the utility's historical average cost per unit of water delivery, *not* the marginal costs of providing a customer with an additional unit of water (Martin et al., 1984).

In Tucson the higher summer rate does reflect the fact that the distribution system must be larger than necessary to deliver water in winter. To the extent that summer rates are in effect only in the peak distribution months, they are a move toward marginal cost pricing of peak-day capacity. In fact, they are in effect too long and thus subsidize winter rates. The increasing block schedule would be a move toward marginal cost pricing only if the ratios of required peak-day and peak-hour capacity to total use were associated with an increase in a household's total use. The evidence is that the ratios remain constant as total use of a household increases (Martin et al., 1984, p. 55). Thus, the increasing block rate has conservation effects, but it does not charge according to the marginal cost of delivering the extra unit of water: the true cost of service.

More important, however, in an area where population is rapidly growing, municipal water delivery is expanding, and officials are saying that water is scarce and expensive surface water supplies must be purchased from the Central Arizona Project, the entire schedule is much too low. It gives consumers no indication of water scarcity.

A rough example illustrates the magnitude of the difference between historical average cost pricing in Tucson and possible marginal cost pricing. The 1983 summer Tucson water rate in the mean use block for single-family residences (10–20 hundred cubic feet [ccf] per month) was about $1 per ccf. That amount equals $435 per acre-foot. Of the $435, approximately $45 was the variable cost of pumping groundwater and $390 the cost of administration, distribution, and capital repayment. Thus, the water itself cost about $45, the cost of extracting it from the ground. (This high-quality water needs no purification.) The real cost of CAP water to the state as a whole is not known. The project is still evolving, and contracts are being negotiated. It was suggested in 1982, however, that raw CAP water might be delivered to Tucson at a price of about $100 per acre-foot. Other costs such as those for purification (as distinguished from distribution costs) could raise that price to about $250 per acre-foot (Stephen Davis, 1982). Thus, marginal cost pricing for the water itself, in addition to the costs of administration and distribution, implies a charge of $250 per acre-foot to the consumer rather than the current charge of $45.

The suggested price of $100 per acre-foot looked too high to Arizona municipalities when they compared it to their current opportunities, $45 per acre-foot in Tucson and less elsewhere. Although presumably water is scarce and additional quantities are needed, the cities balked at signing contracts. To sell the water, the Central Arizona Water Conservation

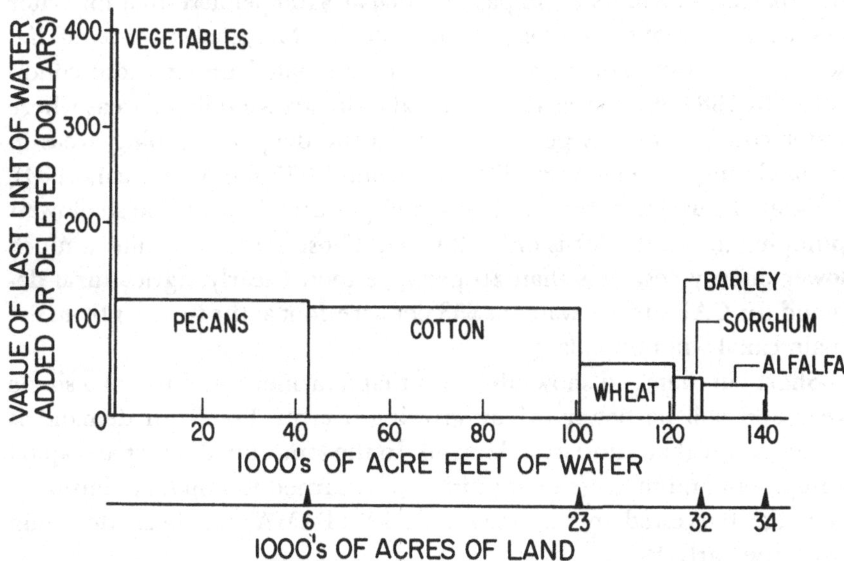

Figure 7.5. Demand for irrigation water in Pima County, 1984 (Bush and Martin, 1986).

District (CAWCD) agreed to deliver CAP water at about $58 per acre-foot in the project's early years. Agricultural users would be charged the same amount.

Tucson agreed to accept water at this price (Meissner, 1984; Burchell, 1985). But continuing negotiations with CAWCD have led to even lower prices (CAWCD, 1986). It is clear that the city will continue to charge a weighted average price for water to keep the price as low as possible.

Agricultural demand for water

Short-run demand for agricultural irrigation water in Pima County, which includes most agriculture in the Tucson AMA, is illustrated in Figure 7.5 for 1984. For each crop a gross revenue per acre was calculated on the basis of average per-acre yields and per-unit crop prices. All short-run (variable) costs of production *except* the cost of water were subtracted from total gross revenue. The remainder constitutes the maximum amount that a farmer could pay for water and still cover his variable costs of production. For the sake of simplicity it is assumed that production costs and revenues for each crop remain constant over every acre planted.

In the short run – that is, as long as commodity prices and input prices remain at current levels and until the time that the well and pump must be replaced – farmers could pay as much as $100 per acre-foot for water and keep most of their acreage in production. Marginal crops of alfalfa, sorghum, barley, and wheat would be eliminated under those conditions. In 1984 alfalfa was the marginal crop, grown only in areas where water cost $30 or less per acre-foot. In the deeper pumping areas in Pima County, where pump lifts are around 375 feet (Hathorn, 1984), associated pumping costs are about $60 per acre-foot. In the shallowest pumping areas, the lift is only 120 feet. Those areas also enjoy a much lower energy cost, less than $6 per acre-foot. Clearly, agricultural demand for CAP surface water at $58 per acre-foot at the turnout from the main canal will not be large.

Short-run demand shows the most that a farmer could pay in a single year and still be better off by growing a crop. Long-run demand is always lower than short-run demand. In the long run, a return to capital equipment and management must also be earned to remain in business. Acreage harvested in the Tucson AMA (TAMA) has been declining since the early 1970s.

Water use in Pima County in 1982 was estimated at only 149,000 acre-feet, which had been applied to about 33,000 acres of land (Arizona Crop and Livestock Reporting Service, 1982). Adding in the smaller acreages in Santa Cruz and Pinal counties, total harvested acreage in the AMA was less than 39,000 acres, and water use totaled about 175,000 acre-feet. Grandfathered rights were based on the largest number of irrigated acres in any year from 1975 through 1979. In that period, acreage was largest in 1975. The base conditions on which water use was calculated for the Tucson AMA as shown in Table 7.3 was 46,000 acres – the average acreage for 1975–1980. Apparently, the decline in agricultural demand for water is already more than 15 years ahead of Tucson AMA projections, simply because of relative crop output and input prices. Agriculture's portion of the overdraft as shown in Table 7.3 may be overstated by about 40,000 acre-feet.

Water and land issues for the coming years

Economic development in Arizona began with agricultural water development, but agriculture today directly generates less than 3 percent of all personal income although it still uses almost 90 percent of all water. Strategies to rescue agriculture, begun because the water table was falling, have evolved into strategies to encourage rapid population and industrial growth. The strategies are to augment the surface water supply through construction of the Central Arizona Project while mandating water conservation by all sectors of the economy through regulatory controls. Obviously, something must be done about water use, especially in the Tucson area, where known available supplies could be exhausted within the coming 90 years. At the same time, however, water has always been available at low cost, and it continues to be priced without regard to its impending physical scarcity relative to the growing population. Water has never been economically scarce.

What of the future? Will CAP and water conservation prove effective strategies for maintaining the Arizona economy and environment? Analysis of the Tucson Active Management Area affords some answers.

Physical and ecological implications of the success of the Groundwater Management Plan

The Department of Water Resources has projected water use and water supplies for the Tucson AMA assuming no effects of the 1980 groundwater law (Arizona Department of Water Resources, 1984). (See Table 7.7.) Population is projected to increase continuously, tripling by 2025 to 1.59 million. Irrigated agriculture would at first increase as the Papago Indians irrigate new farmland with CAP water. The Groundwater Management Act and CAP prohibitions on expanding irrigated acreage do not apply to Indian lands; in fact, the Indian Water Rights Settlement Act seems to *require* expansion. Because existing irrigated acreage is projected to decline, total irrigated acreage in the AMA will begin to fall after 1990. Most of this projected decline occurs as the population expands onto currently irrigated lands. The projection does not include the possibility of additional agricultural land purchases by the city water utility in order to obtain the consumptive use water rights. Currently irrigated agriculture would decrease by about 45 percent by 2025, with a total decline in irrigated agriculture of 33 percent to about 30,000 acres. "The projections assume that no change will occur in municipal per capita consumption [the 1975–1980 rate of 167 gallons per capita per day] . . . and [self-supplied] industrial use increases at the same rate as population growth" (Arizona Department of Water Resources, 1984, p. 21).

The projections of population expansion and the decline of agriculture, both passive occurrences, are based on two other projections. First, delivery of Central Arizona Project surface waters is expected to arrive in the early 1990s. The quantity of water allocated to the Tucson AMA for municipal and industrial use is to expand over time, and the quantity allocated to irrigated agriculture both in the Tucson AMA and throughout the state will decline. Second, almost *all* sewage effluent is to be

Table 7.7. *Tucson Active Management Area water use and potential supplies, 2025 (acre-feet)*

	ADWR estimate	%	Reduce total use 21%[a]
Total water use[b]			
Municipal	297,000	49	235,000
Industrial	88,000	15	69,000
Mining	69,000	11	55,000
Agricultural	148,000	25	117,000
Total	602,000	100	476,000
Consumptive use			
Municipal	130,000	34	103,000
Industrial	88,000	23	69,000
Mining	52,000	13	41,000
Agricultural	118,000	30	93,000
Total	338,000	100	306,000
Incidental recharge			
Municipal	21,000	31	17,000
Industrial	0	0	0
Mining	17,000	25	13,000
Agricultural	30,000	44	24,000
Total	68,000	100	54,000
Potential water supplies			
Central Arizona Project	239,000	36	239,000
Municipal, industrial, and mining	(173,000)	(30)	(173,000)
Agricultural	(28,000)	(5)	(28,000)
Indian	(38,000)	(6)	(38,000)
Reused effluent	146,000	24	115,000
Net natural recharge[c]	68,000	11	68,000
Incidental recharge	68,000	11	54,000
Overdraft	82,000	13[d]	0
Total	602,000	100	476,000

[a]Assumes consumptive use and incidental recharge retain ADWR proportions to total use.

[b]Includes all conveyance losses.

[c]Excludes water lost owing to phreatrophytes and basin outflows of 62,000 acre-feet.

[d]Overdraft is 79 percent of consumptive use.

Source: Arizona Department of Water Resources, 1984. For ADWR estimates, municipal, industrial, and mining categories are disaggregated from a single category called Municipal and Industrial, based on unpublished Arizona Department of Water Resources data (Snow, 1985).

captured and reused. Effluent would be used by agriculture, parks and golf courses, and possibly mines. As the population grows and more water is used, more effluent is available for reuse. This assumption about effluent capture and reuse is reflected in Table 7.7, which shows municipal consumptive use as only 44 percent of total use. In the base year projection (Table 7.3), municipal consumptive use was shown as 65 percent of total use. Unused effluent was assumed lost and was not reflected in the table (Snow, 1985).

On balance, these passive occurrences and positive actions (population growth and agricultural decline; expanded municipal water use and reuse of sewage effluent) would result in an annual groundwater deficit of 81,000 acre-feet per year by 2025, the year for which the 1980 act specifies achieving zero overdraft. (See Table 7.7.) Given that population continues to grow at projected rates and the projected agricultural acreage is not retired other than by normal attrition when urban development occurs on agricultural lands, what must be achieved under a regulatory plan aimed at achieving "safe yield" by 2025?

In specifying a regulatory plan, the alternative of pricing water at its real cost is excluded. Under these assumptions the goal is to eliminate a projected 2025 groundwater deficit of 81,000 acre-feet per year. As Table 7.7 shows, an across-the-board reduction in total use of about 21 percent could bring the Tucson AMA water budget into water balance in 2025, assuming that consumptive use and incidental recharge retain their proportions to total use (see column 1). That assumption must be examined sector by sector, and the implied effects on each sector clarified. How to achieve the decrease must be specified, and the possibility that the overdraft could begin again if population growth continues must be kept in mind.

Municipal use. A 21 percent reduction in municipal use implies a 62,000 acre-foot per year reduction, or 34 gallons per day per capita (gpdc). The AMA plan states that municipal use is to be reduced to 140 gpdc by 2025 – at 21 percent, a reduction of only 27 gpdc – presumably with larger reductions coming from industry, mining, or agriculture. The 21 percent reduction (as shown in Table 7.7) is an across-the-board reduction impinging proportionately on both indoor and outdoor water use. Consumptive use of 96,000 acre-feet remains at 43 percent of total use, as in the ADWR estimates of column 1. This percentage is only slightly higher than the ratio of outdoor use to total use, as shown in Table 7.6.

A 21 percent proportional reduction in both indoor and outside use, down to 132 gpdc, does not seem unreasonable. After all, usage was brought down to 147.5 gpdc in 1978–1979. The crucial variables, however, are the ability to capture and reuse all effluent from the sewage treatment plants and the actual reentry of incidental recharge into the usable water table. Total incidental recharge of 54,000 acre-feet is almost as large as the estimated use for either mining or industry. Sewage effluent of 115,000 acre-feet is almost equal to total agricultural use and to the sum of industry and mining.

The projected water budget is in precarious balance. If effluent and incidental recharge are not created and captured, some uses must be reduced to keep groundwater overdraft at zero. If effluent and municipal incidental recharge are to balance the water budget, outdoor water use must be reduced. Water saving must be concentrated on single-family dwellings' lawns and trees, and commercial and "industrial" use in parks and playgrounds. A 35 gpdc reduction in outdoor use would leave little more than 20 gpdc for greenery, cooling, and firefighting. Cooling alone uses 8–10 gpdc – almost no water would remain for trees and lawns.

Some suggest that the sewage effluent be used for municipal landscaping, as is already required for golf courses. But that action does not provide a solution to the water deficit. If municipal effluent is used for municipal greenery, it is not available to supply the projected industrial, mining, and agricultural uses.

Industrial use. The type of industrial water use anticipated by the AMA plan in 2025 is not clear. Current industrial users supplying their own water include turf-related users (schools, golf courses, and cemeteries), electric power generators, and sand and gravel operations. In the plan, incidental recharge is shown as zero. Without incidental recharge, outdoor water use must be close to total consumptive use. Obviously, reducing water use 21 percent requires greenery to be eliminated and industrial processes using outdoor water to recycle that water. It is clear that much of their supply must be sewage effluent, because CAP water plus natural recharge roughly equals the municipal requirement.

Mining use. Water use in mining will grow from 57,000 to 69,000 acre-feet per year under the Tucson AMA projection. Copper mining is currently a depressed industry, and few expect mining to expand. Thus, reduced water use by the mines may be a natural economic process. Should they prosper, however, use of sewage effluent would seem to be required if the municipal sector is to have fresh water.

Agricultural use. Agricultural acreage in 2025 is estimated by TAMA as 29,500 acres. With water use based on current per-acre applications, total use would be 148,000 acre-feet. Given that the Indian reservation is to receive 38,000 acre-feet of CAP water (possibly to be traded for sewage effluent), Indian agriculture might be as much as 7,600 acres; it is agricultural acreage that does not currently exist. Agriculture off the reservation would have declined to 21,900 acres from the base level of 46,000 acres.

If all agriculture, including Indian agriculture that may not be affected, were to reduce total water use 21 percent, 117,000 acre-feet per year would still be required. That requirement is equal to the total supply of sewage effluent. Should agriculture use the allocations of CAP water shown in Table 7.7, 51,000 acre-feet of sewage effluent would be used by the other sectors to achieve overdraft balance.

Perhaps a more realistic scenario is that agricultural acreage, at least off the reservations, will essentially disappear by 2025 because acreage has been declining more rapidly than the Tucson AMA projected. If all non-Indian agriculture disappeared, even if other sectors did not cut consumptive use 21 percent, the water budget would balance in 2025.

Summary and implications

The projected water supply and water use budget for 2025, the year the 1980 Groundwater Management Act requires achieving a zero groundwater overdraft, is in precarious balance. If all economic sectors are to survive along with rapid population growth, the crucial conditions are receiving the allotment of CAP water, having the ability and desire to capture and make use of all sewage effluent, and correctly estimating both incidental and net natural recharge. If any one of these conditions is lacking, water use in particular sectors of the economy must be reduced, regardless of how effective such conservation measures as the 21 percent across-the-board reduction in consumptive use are. Mining and agriculture, obvious sectors to be eliminated, are now both in economic difficulties because product prices are low, but they could revive.

Although a balanced water budget for the Tucson area might be achieved in 2025 through across-the-board reductions in use even if population continues to grow, the effects on individuals' lifestyles and on the total environment would be severe. Essentially, all greenery within the metropolitan area would disappear. While some agriculture could remain, especially on the Indian reservations, where regulatory AMA water conservation measures probably cannot be enforced, the chance of agriculture's being profitable is small. Required reduction in consumptive use, along with required increases in field efficiency, would raise production costs and decrease flexibility. Any remaining agriculture contributes to the overdraft, and any land retired from production becomes covered with tumbleweed. Further, in 2025 there will be no other potential source of effective water conservation except irrigation agriculture. If population growth continues as projected, the overdraft will continue following its momentary halt.

The short-run effects of the regulatory water use strategy selected by the state of Arizona, as illustrated in the Tucson area, appear favorable to almost every group. The long-run effects may not be. Under the current AMA plans, water remains abundant and inexpensive (priced below its marginal cost for all users). Groundwater is free except for pumping and distribution costs. New surface waters, subsidized by the federal government, will be sold to the public below the local marginal cost because the cost is averaged in with current supplies. Water remains abundant in the early years of the regulatory plan, with municipal and industrial users facing only marginal per capita reductions and agricultural users receiving grandfathered irrigation rights, with total water duties larger than their current use.

In the short run, this choice encourages water use today, hurries new use through population growth and industrial development, and essentially assigns water little future value. Arizona's community leaders are actively encouraging growth and, with limited regulations and low prices, are sending signals of having their water problems well in hand.

The longer-run consequences of this policy are only now being recognized by a minority of the population. Growth in the Tucson area is obviously being accompanied by traffic congestion and air pollution. But although this consequence is to be expected, the disappearance of most greenery from both the public and private municipal areas and the transformation of irrigated agricultural lands to abandoned fields will surprise many. Once abandoned, farmlands do not return to natural desert with trees and shrubs but instead produce only dust and tumbleweeds.

The ultimate necessity of using all sewage effluent has also escaped any public recognition. The projected water budget for 2025 implies that 38 percent of consumptive water use will be sewage effluent out of the wastewater treatment plant. Further, in 2025, Tucson could be a huge city with no future water supplies except those from any remaining agricultural or mining uses or from once again overdrafting the groundwater reserve.

None of these possibilities implies disaster, but they should be recognized. The possibility of a more economically efficient plan for the future has also been ignored. In general, using land and water resources efficiently over time requires converting low-intensity land uses to urban uses at a decreasing rate until a sustainable stable city size consistent with the area's carrying capacity is reached. The evidence documented here suggests that the rate of urban expansion has been too high owing to subsidized water prices. In addition, adherence to zero overdraft by the year 2025 will halt urban expansion abruptly and detrimentally, adding to the undesirable ecological consequences resulting from initial growth. Current policies will have to be adjusted to allow for slower growth over a longer time if efficient use of land and water resources is to become a reality. (See Appendix 7.1.)

The politics of growth and water in Tucson

It is common for natural resources to be exploited inefficiently and for public policies to be less than optimally designed. Policies and their impacts grow out of political processes in which stakes are frequently perceived differently from the way a policy analyst sees them. Some objectively important interests may be poorly represented politically. In Tucson, a pattern of politics has resulted in policies that do not restrain, and do sometimes encourage, the rapid increase in urban water users and the accelerated conversion of desert and farm into residential and commercial developments.

In U.S. politics, interests with direct, and perhaps financial, stakes in an issue commonly overwhelm interests that are more indirect, commonly shared, and concerned with long-term welfare. However, numerous examples of the mobilization of indirect interests in issues that appeal to human concerns about the quality of life and future opportunity exist. Throughout the 1970s on a national level, citizens, politicians, and public policies were activated to protect air and water quality from the more direct financial concerns that benefited from degradation. More recently, strong constituencies have been built to protect communities against toxic substances and groundwater pollution. The obvious question is why the present and future sacrifices involved in higher-than-optimal rates of growth in Tucson have not rallied political activity.

To help answer this question, in the spring of 1985 we interviewed 25

Table 7.8. *Organizational affiliations of respondents to questionnaire*

Environmental groups
Center for Law in the Public Interest
Nature Conservancy
Sierra Club
Southwest Environmental Service
Tucson Audubon Society
Governmental entities
Arizona State Legislature
Central Arizona Water Conservation District
Pima County Board of Supervisors
Tucson Active Management Area
Homeowners associations
Fort Lowell Neighborhood Association
Northwest Homeowners Association
Pima Federation of Homeowners
Tucson Mountains Association
Other organizations
Arizona–Sonora Desert Museum
League of Women Voters
Southern Arizona Water Resources Association
University of Arizona

people from a wide range of organizations and interests active in Tucson water politics. These individuals were selected from organizations that might be expected to understand the diffuse negative costs that current water policies place on the environment and on present and future residents. All those interviewed had a reputation of being active in water policy. (See Table 7.8.)

On average, each interview lasted one hour, and with the exception of four public officials, all respondents were asked the same questions. Discussion was encouraged and those interviewed were able to inject new topics. All were asked general questions about water and growth policy in Tucson, including who is influential in decision making. (See Appendix 7.2.) Their opinions were sought on Tucson's rate of growth, the current water rates, the acceptability of using water policy to control growth, and needed changes in water policy. Because it seemed likely that environmental groups might be especially receptive to changes in water policy, they were asked about environmentalists' positions, their access to decision makers, and their strategies.

Attitudes toward growth and its relationship to water

Tucson is growing much too fast, according to the respondents, an attitude that comes as no surprise considering the individuals' affiliations. In answer to the question, "What do you think of Tucson's current rate of growth?" one person said simply, "It is appalling." Another said thoughtfully,

> I think it is too high mainly because it is not well planned. It's sort of haphazard and all over the place and there's no real evaluation going on of our ability to support the kinds of additional population which growth is entailing. Now, you know, maybe if someone really did study it carefully, they might be able to show that we can support "x" number more people without causing environmental damage down the road, but until I see something that really documents that, I think it is a mistake just to plunge ahead.

Although our respondents were negative about growth, many were not at all sanguine that it could be controlled other than through a natural process. In response to the question, "What factors, if any, do you think will serve to limit growth in the Tucson area?" one person said, "I think when the quality of life starts to decline and word gets around from all the people who've moved here to all their relatives that the traffic is terrible, the air is not good to breathe and, most importantly, that there aren't any jobs, perhaps." Another commented, "[As] Tucson becomes a less desirable place to live, some of the magic leaves because there are too many of us – that can start to brake growth." One environmentalist admitted, "There are very few people in Tucson now talking about limiting growth. Instead, we've taken the other tactic. We've said, 'Growth is really important for Tucson. We want to have a bigger city so we can become richer.' So people are just rubbing their hands gleefully at all the extra new cars they're going to be selling."

Respondents doubted that Tucson's growth would be actively controlled and considered the use of water policy to accomplish such an aim particularly unlikely. In the view of one especially dubious respondent, the use of water policy as a tool for controlling growth was too indirect. She said, "No, I don't think it's a very good tool. It's too backdoor. If you want to control growth, there are ways to do it . . . but it has to be direct about it and using water is an indirect way. . . . The cause is too far removed from the effect." This opinion was by no means unanimous. Many said that water policy certainly could be used, even ought to be used, but admitted that it isn't yet. The major reason mentioned to explain the divorce of water from the growth issue was the recall election of 1977.

The 1977 recall election

In 1975–1976, a majority of city council members, labeled New Democrats, campaigned on issues of controlled growth, planning, and environmental protection. Faced with a financial crisis in the water department because of a high rate of growth in water-delivery facilities, the council members adopted much higher water rates. To finance expansion of the water system, these rates included an assessment of lift charges to houses in the higher areas of the Tucson valley and higher rates for higher water use.

The higher rates proved enormously unpopular with the general public, partly because the council failed to inform the public adequately prior to the rate hike. Further, the new billing began in June, when water use is normally quite high. Some residents received bills that were as much as 400 percent higher than previous bills, and because of the complex billing procedures, they were unable to compute their bills. A drive to recall the New Democrats began, and in January 1977 four recall challengers won on a platform of rolling back water rates. The recall drive was headed by prominent businessmen and supported by the Chamber of Commerce.

The new council members, however, did not reduce the water rates. Once in office, the new members learned that higher rates were essential to facilitate growth of the water system and population growth, not to control population growth as they had suspected the New Democrats of trying to do. Although the new council did eliminate the lift charges and reduce connection charges, water rates remained above their 1975 levels. Since 1976 water rates have increased incrementally each year, often less than the rate of inflation and never in steps steep enough to generate controversy. Instead, the council adopted a voluntary conservation program called Beat the Peak, designed to induce city residents to reduce peak demand and forestall expensive system enlargements. This strictly voluntary strategy allows those who save to feel civic spirited. Beat the Peak is compatible with growth because it creates the image of a water-conscious city coping with its problems. Only later was an educational program begun – Slow the Flow – a program actually aimed at reducing total water use. This program was also voluntary.

A number of lessons can be drawn from the recall election. First, the New Democrats made easy targets of themselves. The timing of the water rate increase and the failure to communicate adequately with the public revealed political naivete. To challenge the prevailing politics successfully would have required a broad and dedicated constituency. Such a constituency might have been built upon a clear understanding of the long-term costs of growth and the immediate and long-term costs of high water use. However, such public support did not exist in 1976, even among environmentalists who "really weren't aware that they could be or should be involved in water issues."

On the other side, pro-growth interests displayed real political skill. The retention of higher water rates and the Beat the Peak campaign showed their ability to adjust and innovate. By taking up the banner of conservation, albeit saving water for growth, the business, financial, and real estate interests robbed their opponents of a potential rallying point. Tucson now has the image of a water-aware city. The Arizona Groundwater Management Act serves much the same role in defusing environmental and slow-growth opposition statewide.

The defeat of the New Democrat city council members – plus the subsequent defeat of a leading controlled-growth-oriented member of the Pima County Board of Supervisors the following year – had a dampening effect on discussions of growth and the use of water policy to limit growth. The supervisor who supported controlled growth was replaced by a decidedly pro-growth candidate, and Tucson residents who were advocating controlled growth and planning simply dropped out of the political process. As one respondent noted, "We certainly lost some leaders. The people who were recalled were leaders and they simply dropped out." Until recently, few political candidates or interest group leaders have raised the growth issue. As one observer noted, the prevailing perception is that anyone talking in terms of stopping growth would "find themselves back on the streets."

Tucson has gone through several shifts of sentiment from pro-growth to anti-growth over the last several decades. The current swing is toward planning for growth. The election of November 1984, when pro-growth Board of Supervisors members were defeated by candidates who might regulate growth, is evidence of this swing in moods. For the first time in roughly eight years, rezonings are being denied and "developers are being told 'no.'" A relatively new political force in the city, homeowners associations, has combined with environmentalists to form a coalition to help elect these supervisors.

But even the political figures who are calling for planned growth are loath to link growth management and water policy. Emphasis in the Tucson area has been on using water policy to encourage growth. The conservation program is intended to conserve water so that a higher rate of growth is possible. For many business interest and civic leaders, the perception that water is well-managed in Tucson is important for keeping investment dollars flowing into Tucson.

Who has power in water policy and who does not?

The public officials and interest group leaders interviewed all agreed that the major water user interests were most active in making Tucson water policy. In the words of one water official, "I would say the larger water users in the basin are the ones with the strongest voices now. The people who use large quantities – Cortaro-Marana Irrigation District, Farmers Investment Company (FICO), the Tucson Water Department, the mines – even though the mines are just in terrible economic shape right now – they're still fairly influential."

The public group most often mentioned as influential was the Southern Arizona Water Resources Association (SAWARA). It sprang up in Tucson in the early 1980s with the dual objectives of bringing the Central Arizona Project to Tucson and promoting municipal water conservation. Its board of directors includes prominent business persons, lawyers, and a few environmentalists. It has a professional paid staff, and it receives an enormous amount of publicity through its Be Water Aware campaign. The organization's influence was traced to the support of powerful business interests and a broad coalition of groups.

When respondents were asked specifically about environmentalists active in water, the names of two individuals in the Southwest Environmental Service and the Center for Law in the Public Interest were repeatedly mentioned. One public official said that these two "were head and shoulders" above everyone else. Their influence was traced to their level of expertise and their knowledge of the political process, not to any mobilization of public support. In the discussion with these two and other individuals, it became clear that the major water issue from an environmentalist point of view was water quality, not water supply and allocation. As one respondent said, "Well, of course we're pushing water quality legislation, regulation, reviewing permits, and working with everybody else we can find as far as the water quality issue goes. . . . I really do not see environmentalists *per se* that active in the supply issue at this point."

The possibility of increasing water rates to reduce water use did not elicit much active commitment. Most interest group representatives agreed that water rates should be increased, but they did not mention water rates as the element they most wanted to see changed in water policy. A clue as to why informants felt this way can be found in the remark of one who said, "In terms of overdraft, you could cut out all of the urban use and still have a problem because of agriculture. I think certainly the urban dweller is paying a heck of a lot more than the industrial and agricultural users of groundwater." Perhaps the most influential environmentalist active in water commented, "I think a steady increase is probably sensible. Actually we're paying more than the water costs. City Water's got money coming out of its ears. Tucson Water is jacking up the water price. . . . They have got all this money. They are building beautiful big buildings, the palace of water."

Mixed feelings were expressed by the interest group leaders when asked whether water conservation would encourage growth in Tucson. Several saw the connection as quite indirect. Others clearly believed that conservation promoted growth. For example, the local leader of a nationally prominent environmental group said, "Present water conservation methods are intended to lead to growth. . . . They are trying to provide more water for an expanding population. That is why they're doing all the things they're doing – buying farms in Avra Valley and getting the CAP – is to encourage continued population growth. Conservation as it is intended at this point is solely for promoting population growth." None of the respondents went so far as to argue against conservation, however. Many contended that lush lawns and swimming pools were not appropriate in the desert.

Environmentalists, these interviews show, lack a consensus about water supply and allocation policy. At this point they do not see themselves as active in these issues; neither do they feel that they lack political influence. Many environmentalists believe that they have good access because "Tucson politicians are sensitive to environmental issues." However, responsiveness to environmental concerns varies greatly from one level of government to another. Further, the influence environmentalists can exert is constrained by the sorts of issues on the agenda and by pressures to go along with other interest groups.

Orientation of decision-making arenas

Of the relevant institutions that govern water use, the Pima County Board of Supervisors is currently most open to public interest groups, particularly environmentalists. This access was gained in 1984 when pro-growth supervisors lost in a heated election to new board members who are less supportive of uncontrolled growth in the area. Because the board controls most open land in the Tucson area and controls such issues as sewer hookups and flood plain development, it may become an important access point in the future.

In contrast, the city of Tucson has been less responsive to environmental concerns. As the largest contractor for CAP water and the body that sets water rates, the council can greatly influence water policy. Only one of the present city council members has had any interest in the environment, and it is peripheral. The council in general is reluctant to take up broad issues of growth and planning, much less in water policy. Water quality could become a greater city issue although it remains linked to economic growth, specifically to the idea that Tucson must demonstrate that it has clean water so that it may grow and attract businesses.

The Arizona state legislature is decidedly not interested in environmental issues. As one legislator noted, environmental issues are "our least motivating concern – we very seldom see anybody stand up and talk about what's good for the environment because you know what's environmentally the best is not the best for certain economic interest." The state legislature historically has been much more responsive to economic interests in the state, including cotton, copper, cattle, and the Phoenix and Tucson chambers of commerce. As for water issues, the state legislature has rubber-stamped decisions made by interest groups.

Environmentalists occupy seats on the board of directors of the Southern Arizona Water Resource Association (SAWARA) and have recently been elected to the Central Arizona Water Conservation District (CAWCD), the organization created to repay the federal government for Central Arizona Project water. Such access does not necessarily mean influence. SAWARA favors water quality legislation but has not considered increasing Tucson's water rates, reducing the quantities of water used, or limiting the influx of new users. Similarly, CAWCD decisions do not reflect environmental influence. One environmentally sensitive official who is a member of both boards stated that she was allowed a position because the prevailing groups "realize that they have to bring these kinds of people, environmentalists, in." It is difficult to be in a tenuous minority given only token representation. These officials are hesitant to raise larger issues that they seem destined to lose, thus jeopardizing the influence they may have on less important matters.

Environmentalists in Tucson have fought hard to achieve some access to the water policymaking system, and in doing so they have had to limit their agenda. "The power structure is giving some sort of lip service and token concession" to environmentalists. Politicians will attend environmental meetings, policymakers address some environmental concerns, and in some instances environmentalists can use this access to influence a water policy decision. On the whole, however, environmentalists exert only marginal influence. As one environmentalist on a prominent board stated, "the price of being let in is the understanding that you will not rock the boat too much."

One of Tucson's most influential environmentalists noted in a recent newspaper interview that "issues such as the direction of growth are yesterday's questions" (Tony Davis, 1984). Often the focus is on such relatively minor issues as placement of wildlife bridges over the CAP and the location of water-storage facilities. Groundwater quality was chosen as a key issue partly because the federal government, which has to be involved, has a stronger bias in favor of the environment. The principal environmentalist active in water issues lobbied hard to achieve a sole source aquifer designation for Tucson from the Environmental Protection Agency. She explained, "I focused on issues where I had federal law behind us, where I could use the federal law as a club and that was water quality and air quality. . . . If you have something that's regulated by federal law, you don't really need the political backing."

Mobilization of the public on issues of growth and water supply is not a strategy in which the water activists put much store. They fear a rejection, such as the 1977 recall election, and they believe that the passage of the Groundwater Management Act of 1980 and the water conservation campaign in Tucson have created a public illusion that water supply problems are being dealt with adequately.

Overview of water politics in Tucson

The pattern of politics existing in Tucson is similar to that prevailing statewide. Emphasis is on mutual accommodation and agreement. The policies pursued seem to let everyone gain.

Water policies are made by a relatively narrow set of interests. When asked for the names of Tucson residents outside major user groups who were influential in water policy, respondents named fewer than 25. They also said that major users dominate decision making. The few environmentalists active in water issues tend to gravitate toward water quality issues, for which federal backing is available. Those who had achieved official positions on committees and boards said that they felt constrained in the kinds of positions they could take or issues they could raise. They felt that their acceptance depended upon not raising divisive issues.

Despite Tucson's reputation for being attuned to the desert, there is little public debate about water supply issues. The public is not conscious of possible adverse present or future consequences of water policy. Growth is considered inevitable. Water conservation is considered "good" and not particularly related to environmental amenities.

Conclusion

Patterns of the past

Appearances suggest that Arizona in general and the city of Tucson in particular exemplify the shift from water development to water management. The Central Arizona Project is regarded as the final supply solution although it is admitted that supply alone cannot solve the groundwater overdraft problem. Arizona has agreed to manage its water resources through a groundwater management law that mandates safe yield of aquifers by 2025. There is some recognition that costs should be borne by beneficiaries, and Arizona has agreed to share some of the costs of the project. Although water rates for urban users are still far below the marginal cost of water, Tucson rates have sometimes been raised enough to alter water use behavior. The safe-yield goal of the groundwater law reflects concern with future generations and with some of the environmental externalities related to overdraft.

Despite the appearance of change, however, many characteristics of traditional water policy have survived and are evident in the new management era. The same impetus toward growth and development, despite the environmental, social, and water allocation problems associated with expansion, drives current decision making. In fact, the Arizona Groundwater Management Act was passed mainly to perpetuate an expanding population and economy. There is no indication that the physical limits of human settlements in the desert are better recognized today than when John Wesley Powell first warned about aridity as a limit to western expansion. Water planners for Tucson, for example, envision accommodating a population nearly three times the current size by 2025, with further expansion expected thereafter.

The negative consequences of such a water policy continue to be slighted. In Tucson, the number of residential water users is increasing rapidly as the desert is converted into residential and industrial developments. Farmlands are being retired to provide water for cities and are abandoned to tumbleweed for the decades it will take for desert flora and fauna to reestablish themselves. Homeowners are sacrificing yards; grass in school grounds and parks, and trees along roadways, are being replaced by gravel to save water.

A characteristic of the previous development era was that problems were not solved but instead passed on to the future. Construction led to more construction – a hydropower dam required a regulatory dam downstream. Water development in California initiated construction in other states eager to put their water to work before it became impossible to wrench it from the Californians. New dams led to salinity problems with Mexico that a desalination plant is supposed to solve.

The current Arizona water policy also is passing problems on to the future. In 2025, many more people will be expected to make do with finite water resources. Reaching a safe yield in Tucson depends upon total reuse of effluent, a technical and social feat not yet accomplished elsewhere in the country. Moreover, should Tucson's population continue to grow after 2025, overdrafting of groundwater will resume.

Implementation of the Arizona Groundwater Management Act depends primarily upon the operation of a regulatory agency, not reallocation through the market. Surface waters are highly subsidized. Groundwater still has no cost except that of delivery. Both agricultural and urban water users are paying much less than the marginal cost of water.

The Arizona Groundwater Management Act, a highly regulatory law, requires a large planning and enforcement staff. The burden on the Arizona Department of Water Resources will grow rather than lessen because increasingly stringent plans must be written and enforced, perhaps over mounting opposition. Whether staged implementation will work remains unclear. As one interviewee remarked, "The real test of the act will be when people are finally asked to sacrifice, because planning is one thing and implementing is another."

The distributive pattern of politics associated with the era of water development continues throughout Arizona. The major interests are accommodated in water decisions and are not required to make large sacrifices. There are great pressures on reaching agreement, and interests that cannot be accommodated or who raise divisive issues are excluded from decision making. The public is not well-informed about or involved in making decisions about water resources, and the language of water discourse tends to be highly technical. Yet, the issues are often simple or could be simplified.

Opportunities for the future

The likelihood of substantial policy and political change depends upon the extent to which the costs of current policy come to be perceived. Quite possibly, divisions will occur among the interests now in the prevailing political coalition. For example, agriculturalists could find the limitations on their groundwater use, the heavy cost of CAP water, and their lack of access to effluent so inimical to their interests that they will break ranks with other water users. It is also possible to imagine that the tourist industry, upon which Arizona and Tucson rely heavily, will come to object to changes that make the state and community less attractive. Should divisions among those most closely associated with present water policy occur, others may be brought into decision making.

It is also possible that change may come about through the mobilization of a broad public concern about the present and future quality of life in the state and in Tucson. As one respondent to the questionnaire noted, "We need people to stand up everywhere and say, 'What good is growth?'" Such mobilization could perhaps be spurred by an event such as the discovery of further groundwater pollution. It could also occur because of a realization that the quality of life of residents was deteriorating at the same time that the cost of living was rising. Mobilizing the public will take leadership that is willing to make direct and emotional public appeals even though such a strategy would be costly in terms of access and influence within the present distributive subsystem.

Appendix 7.1: Optimal intertemporal water allocation: efficiency considerations

As population levels rise in the Tucson AMA, increasing pressure is brought to bear on the region's limited land and water resources. As the demand for residential, commercial, and industrial use of land grows, market forces leading to the conversion of rural low-intensity land uses intensify. In Tucson, low-intensity land uses are predominantly for open space, grazing, and outdoor recreation – all of which require less water than alternative urban high-intensity uses. The allocative questions facing policymakers, then, are, What is an efficient use of land and water resources over time, and when should further conversion of low-intensity land and the associated increase in water requirements be terminated?

An intertemporally efficient conversion policy would maximize the discounted net benefit stream generated from the use of the land over the planning horizon. Symbolically, the allocative problem is one of determining the rate of low-intensity land use conversion over time that will

$$\text{Max} \int_{0e}^{T-rt} [\text{NB}_\text{H}(t, y'(t)) + \text{NB}_\text{L}(t, y'(t))]\, dt$$

$$\text{s.t. } y(0) = 0, \qquad y(T) \leqslant \bar{L}, \qquad y'(t) \geqslant 0$$

where

NB_H = net benefits from high-intensity land use,
NB_L = net benefits from low-intensity land use,
$y(t)$ = cumulative conversion of low-intensity land use acreage,

$y'(t)$ = conversion rate of low-intensity land use acreage,
$\bar{L}$ = the fixed amount of acreage in the region,
r = a constant discount rate, and
t = time.

Irreversibility is incorporated into (1) by the constraint that $y'(t) \geqslant 0$.

At any point in time, the marginal net benefits of high-intensity land use (MNB_H) and the marginal net benefits of low-intensity land use (MNB_L) are decreasing functions of the acreage allocated to each use, reflecting current supply and demand conditions. Thus, as the rate of low-intensity land use conversion ($y'(t)$) increases, MNB_H are positive and decreasing while MNB_L are negative and decreasing. That is, there are positive but diminishing returns to converting land to high-intensity use and corresponding negative and increasing losses associated with foregone rural uses of land.

Over time, increasing population and real incomes should cause the MNB_H schedule to increase, but as the disamenities of urban expansion occur (for example, congestion, noise, air pollution, and the lack of green open space), the rate of increase should fall. Similarly, assuming that low-density land uses are normal goods, the MNB_L schedule should rise over time, only at an increasing rate as the stock of land available for grazing, open space, and outdoor recreation is reduced by conversion. The combined result of these secular net benefit trends is that efficient conversion rates will decline and tend toward zero over time.

Consider Figure 7.6. The necessary conditions for determining efficient conversion and associated water use rates over time that will result in maximizing the discounted net benefit stream are summarized by $Y_0(t)$ in Figure 7.6. Additions to the stock of high-intensity land uses and corresponding increased rates of water use occur at a decreasing rate, terminating at time T_0 leaving $\bar{L}L_0$ acres in low-intensity use.

Had policies been adopted early in the Tucson water management experience to encourage efficient use of land and water resources, a smooth long-run transition to a sustainable steady-state allocation would be occurring. The results of this research, however, suggest that such a transition is not occurring, and in fact, water policy in the Tucson AMA can best be described as schizophrenic. On the one hand, by subsidizing the price of water and ignoring the external costs associated with conversion of low-intensity land uses (for example, increasing noise, congestion, air pollution and lack of green open space), the area has made urban growth artificially attractive. That is, if the full cost of urban expansion were borne by the beneficiaries of growth, conversion and

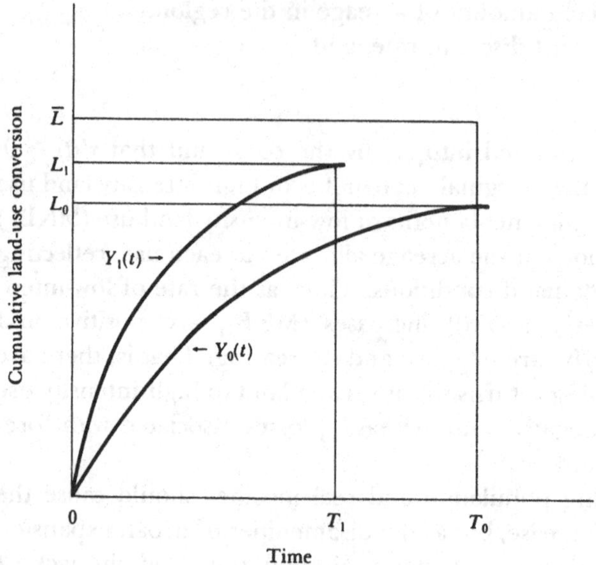

Figure 7.6. Efficient urban expansion over time.

rates of water use would decline. Failure to adopt full-cost policies has tended to place the Tucson area along a path like $Y_1(t)$ compared to an intertemporally efficient growth path ($Y_0(t)$). On the other hand, with the year 2025 the mandatory deadline for realization of zero groundwater overdraft, this research suggests that drastic degradation of the municipal environment will be required to guarantee compliance.

The combined impact of these contradictory policy initiatives is illustrated by conversion path $Y_1(t)$. Comparing $Y_1(t)$ to an efficient rate of urban expansion, three conclusions can be reached. First, excessive water use and conversion of low-intensity land uses have characterized the recent Tucson AMA experience. Second, projected to the year 2025, it appears likely that the stock of rural low-intensity land will be too small ($\bar{L}L_1 < \bar{L}L_0$) and the use of groundwater stocks too large. Third, enforced compliance with the target date of 2025 for zero groundwater overdraft will require harsh measures and will likely result in an austere municipal landscape.

Further, efficiency arguments about land and water use ultimately merge with intergenerational equity concerns for the Tucson area. The conclusions cited above are predicated on discounting future net benefits with some "appropriate" discount rate. These conclusions are further strengthened when the discount rate is reduced or eliminated, a procedure advocated by those who argue that one generation's net benefits should not be discounted by another generation when questions concerning the intertemporal use of society's natural resources are being addressed.

Appendix 7.2: Questionnaire on water management and population growth

Introduction

We are preparing a report about water management and population growth in Arizona. I am going to ask you some very broad questions about growth in the Tucson Basin and strategies that environmentalists are pursuing in relationship to growth.

I want to assure you that your responses are confidential and that you will not be identified individually in our reports unless you would like to be.

Question 1

What is your involvement in environmental issues in the Tucson area?

Question 2

What resources do you think are available to environmental activists in the Tucson Basin?

What about volunteers? Public support? Media support? Money?

Is money a major constraint? How and why?

Question 3

How would you evaluate the local environmental movement, especially in regard to water in the following two areas – leadership and organization?

Question 4

How important is water as an environmental issue in the Tucson Basin? What aspects of the water issue are most important? How would you compare the importance of water issues in the Tucson Basin with other environmental issues like land use and air quality?

Question 5

What would you identify as the goals of the environmental movement in regard to water?

Question 6

What would you say have been the accomplishments of the environmental movement in water-related issues?

Question 7

Among environmentalists, is there a focus on water as an important issue? If no, why not?

Question 8

Do you think environmentalists are influential in water-related issues in the Tucson Basin? Why or why not?

Are they more or less influential in water-related issues than in other environmental issues?

In what ways can environmentalists influence water policy?

Do you think environmentalists have low expectations about their influence on water-related issues?

Question 9

Is water a difficult issue in which to participate? Are water decisions made by small closed groups?

Do environmentalists have a lot of access to decision makers on water-related issues? Why or why not?

Question 10

Generally, how influential are environmental groups compared with other interest groups? Is environmental group influence changing?

Are environmental groups more or less influential than the following groups:

a. Mothers Against Drunk Driving (MADD)
b. Health cost containment
c. Labor
d. Real estate
e. Banking
f. Homeowner/neighborhood associations
g. Victim's rights

Question 11

Who do you think are the most influential environmentalists in this area? Why?

Question 12

What strategies are environmental groups pursuing in relation to water? Do you think these are the most preferable strategies? If not, what would be preferable?

What strategies would you say environmentalists have pursued in the past in these issues: water supply–CAP; flood control; groundwater management act–TAMA plans; sewer hookups?

Would you say there has been a lack of workable strategies in the past?

Question 13

Do you support the Central Arizona Project?

Question 14

How effective is the Arizona Groundwater Management Act?

Question 15

What would you like to see changed about current water policy?

Question 16

What are the biggest barriers to changes in water policy? Are there political impediments?

Question 17

What do you think will be issues in the future in regard to water?

Question 18

Do you think the price of water in Tucson is about right, too high, or too low? Comments?

Question 19

Will water conservation lead to more growth in the Tucson Basin? Why or why not?

Question 20

What do you think of Tucson's current rate of growth?

What can environmentalists do about the high rate of growth? What policies or strategies would work?

Question 21

How do you think a high rate of growth affects the environment?

Question 22

Is water policy a good tool for controlling growth in the Tucson Basin? What are some of the positive and negative implications of using water to control growth? How does it compare with other strategies?

Question 23

What factors, if any, do you think will serve to limit growth in the Tucson area?

Question 24

What was the impact of the 1977 recall of the Tucson City Council on the local environmental movement? Did environmental strategy change as a result? Why or why not?

References

Anderson, T.W. 1979. Project Chief, Southwest Alluvial Basins Regional Aquifer System Assessment, with the U.S. Geological Survey at Tucson. Personal communication.

Arizona Crop and Livestock Reporting Service. 1982. *Arizona Agricultural Statistics*. Phoenix.

Arizona Department of Water Resources. 1984. "Proposed Management Plan, First Management Period 1980–1990." Tucson Active Management Area.

Arizona Water Commission. 1977. *The Arizona State Plan: Phase II*. Phoenix.

Babcock, H.M. 1979. Chief (Retired), Water Resources Division, U.S. Geological Survey at Tucson. Personal communication.

Burchell, Joe. 1983. "Tucson to Resume Buying Avra Farms for Water Rights." *Arizona Daily Star*, Sept. 12, p. B1.

Burchell, Joe. 1985. "Council OKs Changes to Correct Problems in City Court." *Arizona Daily Star*, Jan. 29, p. B1.

Bush, D.B. and Martin, W.E. 1986. "Potential Costs and Benefits to Arizona Agriculture of the Central Arizona Project." Technical bulletin. Tucson: Arizona Agricultural Experiment Station, Jan.

Central Arizona Water Conservation District. 1986. "Management of Surplus Colorado River Water." Memorandum, Oct. 21.

Connall, Desmond D., Jr. 1982. "A History of the Arizona Groundwater Management Act." *Arizona State Law Journal*, 2, 313–344.

Cross, Jack L., Shaw, Elizabeth H., and Scheifle, Kathleen, eds. 1960. *Arizona: Its People and Resources*. Tucson: University of Arizona Press.

Davis, Stephen E. 1982. Tucson Water Department. Personal interview.

Davis, Tony. 1984. "Planned Study Will Offer Compromise on Growth, Local Civic Group Hopes." *Tucson Citizen*, May 13, p. 1B.

Davis, Tony. 1985. "CAP Budget Proposal Brings Usual Response." *Tucson Citizen*, May 25, p. 1C.

Fradkin, Philip L. 1984. *A River No More*. Tucson: University of Arizona Press.

Freeman, J. Leiper. 1955. *The Political Process: Executive-Bureau-Legislative Committee Relationships*. Garden City, New Jersey: Doubleday and Co.

Griffin, Adrian H. 1980. "An Economic and Institutional Assessment of the Water Problem Facing the Tucson Basin." Unpublished Ph.D. dissertation. University of Arizona, Tucson.

Hathorn, Scott, Jr. 1984. *Arizona Field Crop Budgets*. University of Arizona, Tucson, Cooperative Extension Service.

Hoy, Ann. 1985. "U.S. Is Urged to Give CAP Extra $20 Million in Budget." *Arizona Republic*, May 29, p. 4B.

Ingram, Helen. 1971. "Patterns of Politics in Water Resource Development." *Natural Resources Journal*, *11*(1), 102–118.

Johnson, R.B. 1979. Chief Hydrologist, Tucson Water Department. Personal communication.

Kelso, Maurice M., Martin, William E., and Mack, Lawrence E. 1973. *Water Supplies and Economic Growth in an Arid Environment: An Arizona Case Study*. Tucson: University of Arizona Press.

Maass, Arthur. 1951. *Muddy Waters*. Cambridge, Massachusetts: Harvard University Press.

Mann, Dean E. 1978. "Water Planning in the States of the Upper Basin of the Colorado River." *American Behavioral Scientist*, *22*, 237–276.

Martin, William E, and Ingram, Helen M. 1985. *Planning for Growth in the Southwest*. Washington, D.C.: National Planning Association.

Martin, William E., et al. 1984. *Saving Water in a Desert City*. Baltimore: The Johns Hopkins University Press for Resources for the Future.

Meissner, Steve. 1984. "New Rates to Encourage Earlier Use of CAP Water." *Arizona Daily Star*, Jan. 6, p. B1.

Miller, Tim R. 1985. "Recent Trends in Federal Water Resource Management: Are the 'Iron Triangles' in Retreat?" Special Symposium Issue of Water Resources Policy, R. Kenneth Godwin and Helen M. Ingram, eds., *Policy Studies Review*, *5*(2).

Pierce, John, and Doerksen, Harvey R. 1976. *Water Politics and Public Involvement*. Ann Arbor, Michigan: Ann Arbor Science.

Porterfield, Shirley L. 1984. "Factors Affecting Rural-Urban Land Conversion: An Empirical Analysis of the Tucson Metropolitan Area, 1975/76–80." Master's Thesis. University of Arizona, Department of Agricultural Economics.

Redford, Emmette. 1969. *Democracy in the Administrative State*. Fair Lawn, New Jersey: Oxford University Press.

Snow, Lester. 1985. Area Director, Tucson Active Management Area. Personal communication.

Sundie, Dennis. 1984. Arizona Department of Water Resources. Personal communication.

Tucson Water Department. 1984. Unpublished statistical data.

University of Arizona, College of Business and Public Administration, Division of Economic and Public Research. May 1984, January 1986. *Arizona's Economy*.

Valley National Bank, Economic Research Department. 1962, 1978, 1983. *Arizona Statistical Review*.

Wandesford-Smith, Geoffrey. 1974. "On Doing the Devil's Work in God's Country: Legislators and Environmental Policy." In Stuart Nagel, ed., *Environmental Politics*. New York: Praeger.

Weatherford. Gary, and Brown, Lee, eds. 1982. *Water and Agriculture in the Western United States*. Boulder, Colorado: Westview.

Wiley, Peter, and Gotlieb, Robert, 1985. *Empires in the Sun: The Rise of the New American West*. Tucson: University of Arizona Press.

Willis, Mary B., Cory, Dennis, and Cao, Than Van. 1982. "Conversion of Low-Intensity Land Uses, Tucson Metropolitan Area, 1967–76." Research Report 16. University of Arizona, Department of Agricultural Economics.

8 Water management issues in the Denver, Colorado, urban area

J. GORDON MILLIKEN

Background

Water management issues have fundamentally shaped the laws, political and social institutions, economy, and culture of the residents of the Denver area since its beginnings. Denver was founded where two streams meet in the semiarid South Platte valley and has since spread outward to occupy the juncture between the water-scarce High Plains to the east and the Front Range of the Rocky Mountains to the west.

Since permanent settlement began in 1859, the successive waves of occupants – miners, irrigation farmers, and later residents of a complex urban service and trade center – have fought to conquer and shape the natural environment to their needs. Land, water, and minerals have been taken where they could be found and exploited to their full, and new resources sought in more distant searches. The values of this society have been those of achievement through aggressive struggle to create and mold an improved environment that compensated for the inadequate environment that nature provided. The society has grown and flourished under an entrenched system of law, custom, and institutional culture that was formed by, and in turn protects, traditional values.

These values are now being challenged by continued growth. Many people now recognize that the growth that has sustained economic vitality is increasingly costly to support because the resources that fuel growth are scarcer and more distant. Another challenge is an awareness that growth itself reduces amenities of life and fouls the environment.

As dissatisfaction intensifies over the conditions of life in the Denver area, an area still considered attractive although no longer ideal, the values that have sustained growth and prosperity are increasingly attacked. The major focus of controversy is the region's system of water management, which was fundamental to its growth. Traditionalists fear that tampering with the water management system will jeopardize the economic prosperity and thus reduce the ability to conquer society's problems. Reformers state that the water management system must be changed to eliminate systemic inefficiencies and anomalies and to recognize new public interest values, notably the value of managed growth. Neither side is winning. Rather, the growing tactical skill of the reformers has caused a stalemate.

The setting

Concern over water issues existed before Denver's founding, before Colorado was named and its rectangular boundary severed from the Kansas Territory. Unlike the lands of southern Colorado, where Hispanic settlers on land granted by the king of Spain had built irrigation works to manage water supplies some decades before Denver's permanent settlement, the Denver area had no permanent farmers and ranchers. Early occupants included roaming tribes of native Americans – who followed the deer, antelope, and bison in their search for pasture and water – and some trappers and mountain men. These people did not manage water; water managed them. Thomas Hornsby Ferril, whose poem is inscribed in the rotunda of Colorado's Capitol, wrote of them:

Here is a land where life is written in water,
The West is where the water was and is,
Father and son of old, mother and daughter,
Following rivers up immensities
Of range and desert, thirsting the sundown ever. . . .

The first permanent settlers of the Denver area were the argonauts of 1859, disappointed gold seekers returning to the east after the California gold rush had ended. Gold was found in the sandy stream beds at the confluence of the South Platte River and Cherry Creek, and a frontier boom town, later to become Denver, was settled.

Perhaps regrettably, the settlement lacked city planners. For all its scenic and climatic assets, the site chosen for settlement lies in a shallow valley at the edge of the arid High Plains near the foot of the Front Range of the Rocky Mountains. The various ranges of high mountains, culminating in the Continental Divide some 35 miles west of Denver, effectively block many of the moisture-laden clouds that carry Pacific moisture to the midwestern states. Much of the precipitation falls as rain or snow in the mountains and in the high mountain valleys of the Western Slope – the lands west of the divide that gradually decline in elevation as the major tributaries and main stem of the Colorado River carry snowmelt to the Utah border.

Some rivers and streams east of the Rocky Mountains are ephemeral, and others have flows that vary greatly from season to season and from year to year. Flows peak as the snow melts in spring, and the flow variations, from flood to drought, are much wider than in most parts of the nation.

Origins of Colorado water law

The early gold miners developed a system of water management that was simple and useful for its day and for over a century thereafter. But, recently, this system has come under increasing criticism. "First in time, first in right," the doctrine of prior appropriation, was the rough-and-ready rule for gaining permanent possession of both land – the mining claim – and the water used in placer mining operations. This frontier miners' law was adopted by the farmers and ranchers who soon brought a greater economic diversity to the Denver area. Each water user could take as much unclaimed water as he could use and could retain this right against the claims of those who came later, whenever the recurring droughts reduced the streamflow below the amount needed to satisfy all user needs. Using previously unused and unclaimed water gave the user a property right to the water, which he could keep as long as his use continued. This right could be sold to another user or be lost through abandonment, and common law prohibited waste or storage if such storage kept downstream users from using the water.

Water law in Colorado, which is similar but not identical to the law in other western states, predates Colorado's state constitution. A famous early case (*Coffin* v. *The Left Hand Ditch Co.*, 6 Colo. 443, 1882), which upheld the right to take water and use it where needed, even if in another river basin, stated:

> The right to water in this country, by priority of appropriation thereof, we think it is, and always has been, the duty of the national and state governments to protect. The right itself, and the obligation to protect it, existed prior to legislation on the subject of irrigation.

Colorado's constitution formally recognized the doctrine of prior appropriation, stating:

> The water of every natural stream, not heretofore appropriated, within the state of Colorado, is hereby declared to be the property of the public, and the same is dedicated to the use of the people of the state, subject to appropriation as hereby provided (art. XVI, sec. 5).

> The right to divert the unappropriated waters of any natural stream to beneficial uses shall never be denied. Priority of appropriation shall give the better right as between those using the water for the same purpose; but when the waters of any natural stream are not sufficient for the service of all those desiring the use of

the same, those using the water for domestic purposes shall have the preference over those claiming for any other purpose. . . . (art. XVI, sec. 6).

For over 110 years, since statehood, water law has been based on prior appropriation. It has become increasingly complex to administer as demand for competing uses has grown, with many water resource allocation decisions made adversarially in water courts, rather than primarily through administrative decisions of the state engineers, as is done in neighboring states. In Colorado, anyone who wants to create a surface water right or to modify it (for example, through change in use after purchase) must apply to the water court. The court gives public notice, and a court referee investigates and rules on the change. Anyone who claims to be damaged by the ruling may apply to the court for a hearing in which the applicant must prove that the new or modified right does not harm the protester. An appropriator of surface water can file for a conditional right and, if granted, must exercise reasonable diligence to put the water to beneficial use, reporting every four years on actions taken until the facilities are completed and used. Once the beneficial use is established, the user can file an application for a perfected water right.

Colorado, the only state with a system of water courts, is allegedly the home of half the water lawyers in the West. Defenders of the system claim that it has maintained consistent rules to govern a complex process and that it responds to society's changing needs and values. As an example of the system's adaptability, they point to the 1973 legislation that for the first time established instream flow to support fish and wildlife as a beneficial use.

Agitation for reform in the water management system

Critics of Colorado's water management system note its anomalies, many of them perpetuated by the doctrine of prior appropriation: (1) the lack of incentives to promote greater efficiency of water use and the failure to recognize inefficient use as water waste while total demand for water far exceeds supply; (2) the slow, expensive, and uncertain method of reallocating water from low-value uses to higher-value uses, requiring compensation to the person making low-value use of the property of the public and legal action through the water courts (and perhaps also the civil courts if the right is taken through condemnation); (3) continued failure to recognize certain uses of water as beneficial and thus lawful (for example, the retention of water flow in natural streams as a recreational and scenic amenity, one that supports Colorado's second most important industry, tourism); (4) the traditional encouragement of water use rather than water conservation and preservation, which is exacerbated by government subsidies for development of water use facilities; (5) the lack of public interest provisions in water appropriations and transfers, which are managed as a form of private property transaction (although water courts protect rights of other water users, third parties affected in indirect ways – for example, through loss of job opportunities – are not protected); (6) the separation of water quantity and water quality management and the historical lack of recognition of water quality values prior to the federal legislation of recent years; and (7) the absence of requirements to mitigate or compensate for damages to the basin of origin caused by interbasin water transfers.

Water management has been controversial in Colorado since settlement began. But agitation for reform is clearly growing, albeit over the protests of those who have played by the rules of prior appropriation, diligently followed the existing law, and invested capital in water system development. Opponents of reform resent new rules and defend the system of water law that has avoided chaos and violence in water use for more than a century. They resist new values, much as an experienced poker player might resist a call for deuces and treys wild after being dealt a straight flush.

The controversy is neither academic nor one that can be ignored. Those calling for reform to reduce inefficient use of water, increase economic efficiency through reallocation, and better protect the public interest have enough power to force a stalemate. In recent years, rights to divert the unappropriated waters of any natural stream have been denied regularly and systematically by proponents of system reform – including environmental advocates, basin-of-origin interests, and opponents of further urban growth – through political action and litigation. The subsequent delay and cost escalation have created a deadlock.

Is the stalemate to end? If so, how? This is the preeminent water management issue facing the Denver area, and its resolution seems likely to reshape Colorado's water management system for the foreseeable future.

Evolution of the Denver metropolitan area water system

*Growth of the core city system**

The water supply system in the Denver metropolitan area (see Figure 8.1) comprises a system that began in the core city and grew

*This section draws many facts from Cox, 1967, chaps. IV and V.

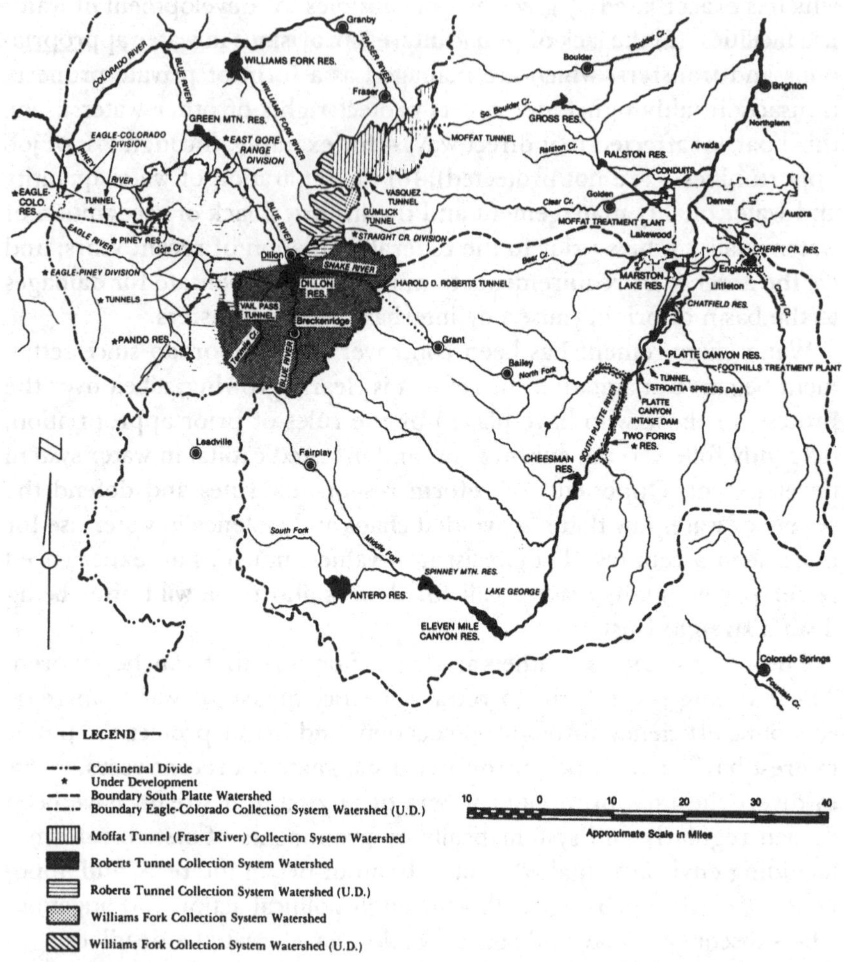

Figure 8.1. Denver metropolitan water supply system (adapted from a map by the Denver Board of Water Commissioners; courtesy of Denver Water Department).

outward, plus a group of town systems that served Golden, Boulder, Littleton, and other distant localities that are now a continuous urban metropolitan area. They have not grown together, or accommodated new political jurisdictions, in harmony.

From 1859 to 1872, Denver residents obtained water from private wells or from streams. In 1872 the first private water company began serving Denver residents. As Denver grew over the next 20 years, several competitive private companies were formed, diverting water from the South Platte River and some streams. Competition became so great that one company provided free water during 1889 and 1890. After some consolidations and bankruptcies, the Denver Union Water Company was formed, incorporating its competitors. Because of its size and monopoly control of water service, the new company succeeded in increasing supplies and storage, treatment, and distribution facilities. In 1905 it built Cheesman Dam and Reservoir 40 miles upstream on the South Platte, a facility that is still an integral part of the Denver water system. Most of the water still comes from surface supplies. Some water came from wells, primarily alluvial wells, but some also come from the deep aquifers underlying Denver. This good-quality deep groundwater is being mined because geological strata make natural recharge essentially impossible.

The Denver Union Water Company obtained a 20-year franchise from Denver in 1880, and during subsequent years it also served the adjacent suburban cities of Englewood and Aurora. In 1918 a Denver bond issue provided funds to buy the Denver Union Water Company and form the Denver Water Department. This system still serves Denver and various suburban customers, although some suburban cities and towns began with their own systems and later bought Denver water, and others broke away from Denver after disputes.

For three decades after 1918, the Denver Water Department was the predominant water supplier to the area. Several cities separated from Denver by several miles of undeveloped land – Golden, Boulder, Littleton, and Brighton – had their own systems. Denver supplied the adjacent suburbs as a matter of course, anticipating eventual annexation of the unincorporated areas to the city and county of Denver.

The rapid growth of the late 1940s and early 1950s strained Denver's water supply capacity. The Denver Water Board, faced with unusual growth in Denver itself, chose not to expand the system to serve accelerating suburban growth. This position was reinforced by a city charter that made in-city service obligatory and outside-city service subject to the availability of surplus water supplies. In addition, expansion of the water system would require a bond election by Denver residents. The board instead drew a blue-line boundary beyond which it would not serve water customers and maintained this service limit from the early to the late 1950s. This restriction was coupled with a city charter provision that limited outside-Denver service to contracts of no more than one year, with uncertain annual renewals, until the charter was revised in 1959.

The board's policy of financial conservatism avoided some political risks of rejection of new water bond issues by Denver voters. However, the unexpected reversal of Denver's traditional policy of suburban service has forced financially strained suburbs to develop new water supplies in an environment of engineering and economic uncertainty. It has led to fragmentation of the metropolitan water supply system, competition for water supply sources, competitive pressures to serve new customers, unwise suburban growth dependent on water supplies of dubious reliability, and destructive political feuds between Denver and its suburbs. In hindsight, had the Denver Water Board taken the political risks of leadership in metropolitan water management 35 years ago, the Denver metropolitan area would have a better-planned infrastructure of public services and probably a more unified, rational political structure able to respond to the challenges of the last half of the twentieth century.

Creation of suburban water systems

The Denver Water Department supplies water to about 57 percent of the metropolitan area residents. Nearly half of its customers live outside the city and county. Major suburban municipalities supplying their residents and some outside-city customers include Aurora, Englewood, Boulder, Golden, Morrison, Northglenn, Westminster, and Thornton. Arvada buys raw water from Denver at cost, treats it, and distributes it through its own system.

Thornton, Northglenn, and South Adams are more recently founded northern suburbs that developed or purchased their own supplies. Littleton, an early suburb south of Englewood, took its water from shallow wells along the South Platte until growth exhausted this source in 1966 and required Littleton to contract with Denver for water service.

Westminster, a northwestern suburb, tried unsuccessfully to obtain Denver water during a decade of serious water problems between 1955 and 1964. Not only was Westminster water-short, but its supply was badly polluted with untreated sewage from Arvada. Although chemically treated to avoid serious health effects, the water was odorous, discolored, and disagreeable to taste. Denver refused to help during the 1950s because Westminster was outside the blue line, causing Westminster to make expensive but inadequate improvements in its water system. When the water supply again became critically short in 1962 and 1963, Westminster again appealed to Denver but was offered unacceptable terms for service: the abandonment of Westminster's entire water system, recently expanded at the cost of a substantial bonded debt, and purchase of all water from Denver at outside-city rates. Westminster voted narrowly to improve its own system, although its water consultant warned that the system would be inadequate again within 20 years and that merger with Denver or another system would become necessary.

The dispute that led Englewood to separate angrily from Denver and develop its own system contains an element of farce. Although Denver did not require new in-city homes to install water meters to eliminate water waste through volume-of-use pricing until 1957 ("Water Meter Work," 1986), and although pre-1957 homes continued on flat-rate service until a dramatic 1986 policy change, all outside-city customers have been required to have meters since 1957. In 1948, appalled by the evident waste of water by flat-rate customers in Englewood, the water board sanctimoniously raised water rates in Englewood 30 percent until customers there installed water meters. Enraged by this high-handedness, Englewood sued Denver to enjoin collection of the new rates and petitioned the courts to place the Denver Water Board under Public Utilities Commission control. Losing in the district and supreme courts, Englewood decided to develop its own water system; it acquired some South Platte River rights and by 1955 had developed substantial water rights on the Western Slope. Unfortunately, Englewood had no means of bringing the transmountain water to its reservoirs until, after further disputes with Denver, an exchange agreement was signed with Denver to trade the Western Slope water for some of Denver's South Platte rights. Englewood finally had an independent supply ample enough to serve itself and outside-city customers. Englewood continues to serve some 80 percent of its in-city customers at flat rates but requires water meters of all outside-city customers.

Aurora, Denver's eastern suburb, obtained most of its water from the Denver Union Water Company and subsequently from the Denver Water Department until the early 1950s, when Denver established the blue line. Most of Aurora, the largest suburb, was outside the blue line. This restriction, a 50 percent rate differential for water and no limiting clause on rate escalation, and the insecurity of only a year-to-year contract for water service caused Aurora to develop its own system. Aurora developed groundwater supplies, diverted some South Platte River water, purchased agricultural water supplies from the Arkansas River Basin in southeastern Colorado, and embarked on an ambitious transmountain diversion project in partnership with Colorado Springs. In April 1986, Aurora filed claims on up to 108,000 acre-feet per year of water rights from the Upper Gunnison Basin, another transmountain diversion planned for 20 years in the future. Aurora is now growing rapidly, is annexing suburban land, and is aggressively competing with Denver to provide water service to new suburban developments. The Aurora City Council's goal appears to be to attract new growth and to seek water to support it.

Transmountain diversions from the Western to the Eastern Slope

For 70 years, the growth of Denver and suburban water systems has relied increasingly on transmountain diversions of water from western Colorado through the mountain ranges of the Continental Divide as eastern Colorado water resources have been appropriated and developed. The Colorado constitution designates any unappropriated water in the state as public property, so water may be taken and used anywhere that it is technically and economically feasible to do so within the state.

Engineering consultants recommended that Denver develop a water supply from the Western Slope and identified the Fraser, Williams Fork, and Blue rivers as the most promising sources. Following preliminary surveys, Denver filed for water rights on the Fraser and Williams Fork rivers in 1921 and on the Blue River in 1923. Although Denver voters had rejected a $5 million bond issue to pay for badly needed water system improvements in 1921, the next year a comprehensive engineering report (Cox, 1967, quoting Mosely, 1957, p. 253) cautioned residents about future expansion:

> It would be a very short-sighted policy to limit the growth of Denver by failing to provide a sufficient water supply for the future. The supply available from the Eastern Slope is inadequate for the future Denver. Provisions should be made at once to secure a perpetual right to all possible diversion from the Western Slope.

Although Denver was planning for Western Slope water diversions, it first needed to establish a justification by developing its Eastern Slope supplies fully. By 1932, when Eleven Mile Canyon Reservoir was constructed, there was almost no unappropriated water in the South Platte River Basin (Cox, 1967, pp. 58–59).

In 1922 the Colorado legislature had authorized the Moffat Tunnel through the Continental Divide, primarily to facilitate a transcontinental railroad route passing through Denver but also to construct a second tunnel for conveying water from the Western Slope. The tunnel was completed, and in the dust bowl year of 1936 delivered the first water from the Fraser River and its tributaries to the Eastern Slope (Cox, 1967, p. 61). Denver completed another collection system on the Williams Fork River, diverting water through the Continental Divide to Clear Creek on the Eastern Slope in 1940. In 1959, in order to use Williams Fork water more efficiently, Denver brought more of it through the Jones Pass Tunnel and diverted it from upper Clear Creek into the Fraser River collection system. The water then flows into the Moffat Tunnel for direct use by Denver.

Although Colorado law does not require municipalities or individuals to pay the basin of origin or to mitigate stream-depletion effects, it does require that downstream water rights holders continue to obtain water. Thus, Denver constructed Williams Fork Reservoir on the Western Slope for replacement storage to be released into the Colorado River to compensate downstream users for water diverted from the Fraser and Williams Fork to the Eastern Slope. This replacement water comes from unappropriated water in the Colorado River system.

By 1963, just before completion of an even larger project to divert Western Slope water to Denver, the Fraser and Williams Fork systems delivered nearly half Denver's municipal water supply. In 1984 the two systems supplied 50,316 acre-feet of water, one-sixth of the (then much larger) municipal supply (Denver Water Department [1985], p. 16).

The Blue River–South Platte diversion project is the largest and most costly of Denver's transmountain diversions. The Denver Water Department filed for Blue River water rights in 1923. Planning efforts by the Denver Water Department and the Bureau of Reclamation continued until 1946, when they approved a transmountain diversion tunnel into the upper South Platte River. A bitter legal controversy between Western Slope and Denver interests was not settled until 1955, when the Western Slope antagonists accepted a compromise stipulation by Denver, reportedly after President Eisenhower proposed an out-of-court settlement to avert an even lengthier appeal to the Supreme Court. The Blue River stipulation obligated Denver to protect the rights of downstream water users in western Colorado for existing and some future uses, to stop diversion when needed to satisfy Colorado's delivery obligations under the Colorado River Compact, and to reimburse the Bureau of Reclamation for losses in hydropower generation at Green Mountain Reservoir, a feature of the Colorado–Big Thompson Reclamation Project (Cox, 1967, pp. 74–75).

Denver then began building the Blue River diversion project and passed two bond issues totaling $115 million to pay for it. The project required relocating the town of Dillon, completing Dillon Dam and Reservoir, relocating highways and power lines, and constructing the 23.3-mile Roberts Tunnel through the Continental Divide. The Roberts Tunnel was completed in 1962, and Dillon Dam was closed in 1963. Dillon Reservoir stores 254,036 acre-feet, half the total storage capacity of the

Denver Water Department system. The Blue River system supplies varying amounts of Denver's water needs and protects against sustained drought on the Eastern Slope. In 1982 it supplied 37 percent of Denver's water. In 1984, a wet year, it supplied only 6 percent, although Dillon Reservoir was kept full for future dry year needs (Denver Water Department [1985]).

The time required to complete a transmountain diversion project, from filing for the water rights to delivering the water, is growing. The Fraser River–Moffat Tunnel diversion took 15 years, and the Blue River–Roberts Tunnel 42 years, even with presidential intervention. Opposition by environmental and western Colorado interests is steadfast, permitting regulations are increasing complex, and delays in the court system grow more complex. With Colorado's water management system in an unprecedented stalemate between water development proponents and opponents, the timing of any new transmountain diversion is open to question.

Aurora and Colorado Springs are proposing a second major transmountain diversion, Homestake II, and Aurora is planning a Gunnison River diversion in the twenty-first century. Future transmountain diversions planned by the Denver Water Department, bitterly opposed by a coalition of Western Slope and environmental preservation interests, include extensions to the Williams Fork collection system to augment Moffat Tunnel flow, and possible extensions of the Roberts Tunnel collection systems, designated as Eagle-Piney, Eagle-Colorado, East Gore, and Straight Creek. If trends continue, the likelihood of a future transmountain diversion project is slim.

Water supply and demand issues

Denver-suburban political disputes

The political rancor that began with water disputes between Denver and its suburbs in the 1940s and 1950s continues to fragment water policy in the metropolitan area and to hamper formation of an effective regional growth management system.

After the Denver Water Board (an independent body appointed by the mayor of Denver without city council approval) lifted the blue-line service area limit in January 1960, it again began supplying water to suburban areas that could be served economically. At the same time, the city council shifted from a no-annexation policy to one that sought to bring adjacent areas into the city and county of Denver. Although the annexation of suburban areas was not required, it was widely perceived by suburban officials that Denver was using the power of its water supply to force annexation of suburban areas.

During the 1960s, Denver annexed land in the surrounding unincorporated suburban counties, including Adams, Arapahoe, and Jefferson. Because Denver is the only combined city and county in Colorado, its annexed land was automatically de-annexed from the surrounding counties and from county school districts, wreaking financial havoc on suburban school systems that unexpectedly lost students and tax base to Denver and sometimes found their school locations ill-suited for their surviving student population. To a less degree, the suburban county governments felt the loss of property and sales taxes generated in growth areas that were newly annexed to Denver. Some suburban residents were also apprehensive because they preferred the less structured government of an unincorporated suburb. Others disliked the Denver public school system's court-ordered policy of busing students to achieve ethnic and racial balance in schools.

Growing suburban hostility to Denver's annexations led in 1974 to passage of a constitutional amendment effectively barring further annexations by Denver. As a result, Denver is locked in a suburban steel ring, unable to expand its boundaries and tax base to meet the growing financial demands faced by central cities. Denver's population has dropped since the 1974 amendment, and in 1985 it was lower than in 1970. All metropolitan growth occurs in the suburbs ("Metro Growth," 1985; Bureau of the Census, 1973).

Denver's political power and legislative representation have been significantly eroded by the growth imbalance. Denver's municipal officials resent the shift in power and affluence to the suburbs, particularly because Denver provides and financially supports facilities that serve the entire metropolitan area – the major airport, convention center, general hospital, symphony and performing arts center, art museum, zoo, and athletic stadiums.

Since the mid-1970s, Denver has not aggressively sought new suburban customers. Augmenting its water supply in the face of opposition from Western Slope interests and environmentalists is too difficult, and any such augmentation would be for suburban service rather than for Denver itself. Rather, Denver has limited the number of new taps in its total service area (now at 7,000 taps per year) to conserve available supplies ("Prolonged Water Curbs," 1977; *Denver Post*, June 21, 1979). Some recent service additions are in response to appeals from suburbs facing severe water supply problems, such as Westminster and Broomfield – a more neighborly policy than Denver exhibited during the 1950s.

In the absence of service by the Denver Water Department, some suburban growth has been risked despite marginal supplies of water (Pokorney, 1982, pp. 66, 67, 69). Some development has occurred in areas dependent on nonrenewable groundwater that is rapidly being depleted. Some individual residential wells failed, causing pleas for Denver Water Department service as hardship cases; some large tract developments rely entirely on supplies from well fields in rapidly depleting aquifers. In recent years, the Colorado state engineer has controlled use of groundwater supplies more severely; fewer wells may be drilled, and pumping volumes are lower. Nevertheless, several large suburban developments in unincorporated areas and municipalities have been built without a water supply that is reliable for even a decade or two at current rates of use. Pro-growth Aurora is notoriously eager to annex new areas and is willing to rely on groundwater during dry years to augment its none-too-ample supply of water. According to several informed water supply planners, Aurora is prepared to gamble with development that relies on the city's average yield of water rather than on a safe yield (that would also cover droughts).

Opinions differ on whether development that relies on marginal supplies is foolhardy. Those willing to tolerate risk can argue that so long as a supply agency develops adequate long-term supplies, a temporary source of new supply or short-term demand reduction (through, say, use limitations) can insure against drought. So long as there is a major water supplier in the area with a reliable supply, other water agencies can take greater risks, knowing that social pressure or financial reward will force the supplier to share.

A multijurisdictional water supply system

In 1986 there was enough municipal water to serve the total Denver metropolitan area, barring a severe drought. However, jurisdictional fragmentation intensifies the area's water supply problems. Paradoxically, several supply agencies face water constraints while other municipalities or special districts have more than their service areas need. If there were a single metropolitan water authority or if an effective market existed to sell surplus rights or to arrange short-term water leases, allocation problems could be readily solved. But the fragmented multijurisdictional water systems serving the Denver urban area do not readily respond to such problems. Uncertainty about each jurisdiction's future needs in a dynamic and competitive metropolitan area is one problem. Although suppliers unwilling to sell surplus water rights because of future uncertainty should have little concern over leasing unneeded supplies for a few years to other suppliers with near-term shortages, only rarely do they do so. As a result, the Denver Water Department, which since 1977 has limited the issuance of new taps to a quota, now 7,000 per year, announced that it will allow no new taps in its extensive suburban service area after 1988 until its safe yield supply is increased – probably for seven years and possibly longer ("Denver Extends Tap Allocation," 1986).

Why other suppliers with excess water are unwilling to lease water to Denver to permit lifting the announced tap moratorium remains unclear. One possibility is that those with a surplus (Englewood, Aurora, Thornton, for example) wish to attract new growth to the areas they serve or to add portions of the Denver-supplied suburbs to their own service areas when Denver cuts new taps off entirely. Another possibility is that traditional political hostility with Denver influences their refusal.

Environmental battles over metropolitan water system expansion

The Denver area water supply agencies seeking to expand their systems' supplies have for decades faced opposition from various organized groups. Besides the long controversy with Western Slope interests over transmountain diversions, other opposition from environmental groups has become well-organized, politically skilled, and effective in the past 15 years.

The environmental opposition to traditional structural development of water resources for metropolitan supplies is based on the preservationist conservation ethic, not the utilitarian conservation ethic, as distinguished by Martin et al. (1984). Opposition is based on one or more of these reasons (see Weaver, 1985):

- More water supports more urban growth, which reduces the quality of life for all urban residents save those who profit financially from urban land development.
- If some urban growth is inevitable, the necessary water can and should be obtained through water conservation programs and new supplies developed in environmentally sensitive ways.
- Developing surface water supplies requires constructing dams, reservoirs, and aqueducts, which are unnatural and unsightly intrusions in areas of natural beauty that should be preserved.
- Water resource developments destroy fish and wildlife habitat, dewater streams, and ruin wilderness and recreational areas, where fishing, hiking, and whitewater rafting would otherwise occur.

- Water for metropolitan supplies is taken from other valuable uses, such as tourism, agriculture, and potentially industry, that will occur in rural and economically underdeveloped regions in future years.

Not all who oppose new dams and reservoirs agree. The Colorado Environmental Caucus, for example, considers water an ineffective growth control mechanism. But such arguments have been heard for a decade in political battles with those who would develop additional water supplies in Denver and in much of the west.

The pro-development faction responds with its own arguments and counterarguments:

- Population growth will occur when economic conditions and perceived attractiveness cause in-migration; thus, lack of an adequate water supply simply makes urban life less comfortable rather than inhibiting population growth.
- Water is necessary to continue new construction and economic growth, however.
- Some methods of water conservation are cost effective and useful, and others are not useful, desirable, or necessary; in any case, extreme conservation measures should be reserved for extreme droughts or sudden emergencies rather than used prematurely as a cheap substitute for long-range investment in needed supply facilities.
- Some physical losses of the natural environment of streams and scenic resources are inevitable but necessary; yet wise planning can reduce or offset undesired impacts.
- A water supply adequate for reasonable needs of an urban population is a high-value beneficial use of water that provides amenities and avoids constraints on healthy growth of the urban economy; such use has a higher economic value than use of the same amount of water in agriculture, or in maintaining natural river flows for habitat, recreational use, or scenery, or in preservation for uncertain future development in rural areas.

In 1974, Governor Richard Lamm was elected on a slowing-of-growth platform, and Colorado voters refused to provide financial support for the hard-won 1976 Winter Olympic Games, largely for fear that unwanted growth would occur. Among policies promoted by Lamm's administration were population dispersion from the fast-growing Front Range to the less-developed rural regions of Colorado and the preservation of the state's agricultural sector whose irrigation water is sought by cities and energy resource developers (Milliken et al., 1981, pp. 250–253).

Imaginative leadership alone, including decentralization of state government offices from Denver to rural areas, has not reversed trends based on economic realities. Colorado's agricultural sector, crippled by low prices for wheat and beef and by the high interest rates of the past decade, faces foreclosures and bankruptcies. In these hard times, the sale of water rights to cities is increasingly seen as the only way to realize a return on years of investment. (A 1984 study found that 35.4 percent of Colorado's farmers expect to quit farming within five years, 16.5 percent within one year ["Water War," 1984].)

Just as Los Angeles purchased land and water rights in California's Owens Valley several decades ago and as Tucson has retired Avra Valley farms and transported their water over the Tucson Mountains for municipal use, so have some Colorado Front Range cities raided High Plains farms. Aurora and Colorado Springs have aggressively bought water rights in the Arkansas River Valley since 1973 – Aurora to serve a 117-square-mile area primed for development and Colorado Springs to supply development around its Space Center. The result is dried-up cropland that may stretch eastward from Pueblo County 110 miles to the Kansas border, leaving barren land and depopulated towns in the formerly rich vegetable farming area and "melon capital of the world" ("Water War," 1984).

Forecasts of continued metropolitan Denver growth

Population of the Denver metropolitan area has grown rapidly since World War II, accelerating from 1965 to 1977, slowing until 1980, then rising again. Most growth has come from in-migration, lured by climate, amenities, and high employment demand in energy, high technology, and service industries. Metropolitan area population reached 1.8 million in 1985 ("Metro Growth," 1985). Demographers differ in their assumptions about estimates of future growth. The Denver Regional Council of Governments, the metropolitan planning agency, plans infrastructure using its population forecasts of 2,629,000 in 2010 for a six-county metropolitan area (Denver Regional Council of Governments, 1986). This estimate, reduced from earlier projections, reflects labor demand forecasts and the assumption that interregional migration will continue to provide workers to meet that labor demand.

A systemwide environmental impact statement (SEIS) of the metropolitan Denver water supply is being prepared by the Army Corps of Engineers to help determine the need for new water supply and storage facilities, based on forecasts of Denver metropolitan area growth to 2035. The assumption here is that in-migration to the Denver area will slow, then stop, as the relative attractiveness of the Denver area diminishes and reaches equilibrium with the rest of the United States by 2035. Population forecasts used in the SEIS for a planning area roughly a tenth smaller than the Denver Regional Council's area are 2,581,320 in 2010 and 3,302,000 in 2035 (Corps of Engineers, 1985, app. 5, p. I–25). The draft SEIS, which incorporates the lowest of several population forecasts developed using differing in-migration assumptions, forecasts an annual metropolitan area demand (safe yield) increase from 364,000 acre-feet in 1985 to 703,000 acre-feet in 2035. The total available safe yield, although not efficiently distributed throughout the metropolitan area, is 418,000 acre feet, indicating a shortage that will begin in 1989 and a deficit of 288,000 acre-feet by 2035 (Corps of Engineers, 1985, app. 5, fact sheet, p. 10).

Controversy surrounds the discrepancies between the forecasts and their underlying assumptions. Development opponents consider forecasts based on past growth trends unrealistically high thanks to the suburban growth boosters who lobby the Council of Governments board. They point to Colorado's failure to meet growth predictions during the 1980s, when the energy boom collapsed and the electronics industry declined. They also question the assumption that cities continue to grow at historical rates, citing evidence that natural forces cause metropolitan area growth to slow after a few decades (Morris and Luecke, 1985). These analysts sharply criticize plans for additional structural water development, with its environmental and economic costs, to meet inflated future needs.

Proponents of water development, notably the Metropolitan Water Providers (an organization of suburban water suppliers), believe that Denver metropolitan growth will continue, particularly because no regional growth management plan exists and some metropolitan jurisdictions actively promote growth. This faction considers it irresponsibly risky to rely on predictions that future growth will spontaneously slow despite a lack of growth controls, given the area's history of rapid growth over several decades. Even though conservation programs are expected to reduce per capita water use, the supply agencies forecast a major increase in total water demand over the next 50 years, far exceeding the present supply.

Even the smallest of the SEIS forecasts represents major population growth that challenges planners of public service facilities. Demand is expected to exceed the developed safe yield by about 1989, and 69 percent more water will be required in 2035 (Corps of Engineers, 1985, app. 5, fact sheet, p. 10). Because so much lead time is needed to expand water supply facilities, the challenge is more severe for water supply agencies and the risks of forecast errors are the greatest.

To meet the expected water needs, metropolitan area water agencies until recent years relied exclusively on augmenting supplies, saving demand modification for times of severe drought, failure of system facilities, or some other emergency. Groups opposed to the environmental costs and economic inefficiencies of traditional water supply agency practice have long proposed alternative management practices. They would maximize conservation, price water to promote efficient allocation and use, recycle and reuse water supplies, and augment supplies where necessary in the least damaging ways, such as using groundwater and water saved from inefficient agricultural use.

When these exhortations were ignored by water supply agencies during the 1960s and 1970s, opponents to traditional water supply development resorted to a tougher and more resourceful strategy. They sued the Denver Water Board to block its plans to perfect new water rights on the Western Slope and to construct new collection, diversion, storage, and treatment facilities. Similar lawsuits were filed by Western Slope groups to defend their water supplies for future need. The lawsuits relied on the complexities of Colorado water law and also made full use of the federal laws dealing with environmental protection, wilderness, endangered species, and preservation of fish and wildlife habitat. Even if the lawsuits are ultimately lost, they cause extensive delay and greater expense to the water developers and thus discourage construction. The Denver Water Board has vigorously defended the lawsuits and even filed its own suit against environmental groups, perhaps as an attempt at deterrence.

The water development-environment controversy focused on five system features: perfection of water rights on the Upper Colorado River and its tributaries, expansion of the Eagles Nest Wilderness, expansion of the Williams Fork collection system, the Foothills Treatment Plant, and the Two Forks Dam and Reservoir.

Upper Colorado River water rights

The Denver Water Department's long-range plan for water supply in the twenty-first century relies on an extensive amount of transmountain diversion from the Upper Colorado River and its tributaries. In 1956, the Water Department filed claims for water rights on the Eagle and Piney rivers and their tributaries near the area that later became the resort town of Vail and on tributaries of the Blue River near Dillon – Straight Creek and several streams diverted by the proposed East Gore Canal. In 1971, it also filed claims on waters tributary to the Eagle and Colorado rivers. Altogether, these claims involve an average supply of 264,000 acre-feet and about equal the Denver Water Department's average demand in 1980 (Pokorney, 1982, p. 74). How much these claims will increase the department's safe annual yield cannot be determined until it is known what storage increases will be made in the system.

Denver's claims have been denied pending appeal to the Colorado Supreme Court. Two Western Slope water conservancy districts, the U.S. government, and the cities of Aurora and Colorado Springs sued the Denver Water Department to deny those water claims. Although Denver expects the Supreme Court to uphold its claims, the decision adds delay, expense, and uncertainty to long-range planning (Marston, 1982).

Eagles Nest Wilderness

When the Denver Water Department filed claims for water rights in the Eagle-Piney, East Gore, and Eagle-Colorado areas, it avoided encroachment on the 62,125-acre Gore Range–Eagles Nest Primitive Area. In 1972 the U.S. Forest Service recommended boundaries for a proposed Eagles Nest Wilderness of 87,755 acres, somewhat larger than the Primitive Area, but one that would not interfere with the Denver Water Department's plans to divert water by gravity flow to Dillon Reservoir and from there to Denver. Subsequently, in large part to block the transmountain diversion plans, a coalition of Western Slope and environmental interests lobbied Congress to expand the wilderness boundaries by adding nonwilderness land so that Denver's structures would need to be moved to lower, more distant locations. This change threatened to make construction and operation much more expensive and to require costly pumping instead of gravity flow. Aided by a strategic error by the Denver Water Department – refusal to compromise short of absolute victory – the Western Slope–environmental coalition won. In 1976, a 134,000-acre Eagles Nest Wilderness was approved, adding major construction and operational costs to the already very costly plans for the Eagle-Piney collection system (Pokorney, 1982, pp. 81–92).

Williams Fork expansion

The Williams Fork diversion system collects water from tributaries of the Williams Fork River and diverts it to the Eastern Slope. How much is diverted depends on demand and climatic conditions. In 1984 the total was down from 9,418 acre-feet in 1981 to 2,814 acre-feet (Denver Water Department [1985], p. 16). Denver has had water rights on Williams Fork since 1921 and a federally approved right-of-way since 1924.

In 1978, the Denver Water Department began expanding the Williams Fork collection system to obtain another 12,000 acre-feet annually. The Forest Service stopped the project, maintaining that the Denver Water Department had changed the structural form of its construction and moved it beyond the 1924 right-of-way, thus requiring both an amendment to the federal right-of-way approval and an environmental assessment. In 1981 the court held that the National Environmental Policy Act (NEPA) applies to the project, requiring an environmental assessment and possibly an environmental impact statement, and that the Grand County land use regulations also apply, even to projects (such as this) that are wholly on federal land within the county. Denver has appealed the ruling, which puts any future Colorado transmountain diversion project in jeopardy (Pokorney, 1982, pp. 78–80).

Foothills Treatment Plant

In 1973, faced with an impending shortage of water treatment capacity to meet summertime peak demands, the Denver Water Department proposed a $160 million bond issue that would, in part, provide a 500-million-gallon-per-day (mgd) treatment plant on land it owned in the foothills of the South Platte River south of Denver. The bond issue passed, and $70 million was allocated to build the Foothills Treatment Plant and two companion features, the Strontia Springs diversion dam on the South Platte and a tunnel from the dam to the treatment plant ("Denver's Water Dilemma," 1977). Thus began a bitter six-year controversy whose settlement was compared by Governor Lamm to the historic diplomatic recognition by the United States of the People's Republic of China ("Corps, EPA Agree," 1978).

A broad coalition of interests opposed to Denver's growth and to the environmental effects of further water development organized to defeat Foothills. Although Foothills was to be built in four stages of 125 mgd each, the dam and tunnel were sized for the complete project as engineering prudence dictates. Foothills opponents took this information as proof that Foothills was but the first part of an integrated Denver Water Board plan for massive development involving Foothills, Two Forks Dam and Reservoir, and enlargement of the Water Department's collection and diversion scheme to bring more water from the Western Slope to Denver. Not only did a broad front of environmental and outdoor recreation groups oppose Foothills; the EPA regional director also opposed it strongly as a first step toward undesirable metropolitan growth ("Regional EPA Chief," 1977).

In the hot, dry summer of 1977, the Denver Water Department restricted outdoor water use, limiting lawn and garden watering to every third day to conserve treated water ("Hot Dry Summer," 1977). These restrictions marshaled public opinion in support of Foothills, and later that year some Denver suburbs began backing Foothills and offered to help pay for it.

Ultimately, after a complex series of lawsuits, the Foothills controversy had a political settlement engineered primarily by Congressman Tim Wirth. The Denver Water Department was allowed to build the first stages of Foothills, including Strontia Springs Dam and the tunnel. EPA received the Water Department's commitment to a strict water conservation program designed to reduce per capita water consumption 20 percent by the end of the century. The Denver Water Department also agreed to minimum streamflow releases to protect fish in the South Platte below Strontia Springs, to pay the legal fees for some environmental groups involved in the lawsuit – other groups refused to receive such payment – and to appoint a citizens' advisory committee with environmental participation to help the board make decisions ("All Sides," 1979).

Two Forks Dam and Reservoir

With the Foothills battle resolved by armistice, the controversy has moved to Two Forks Dam and Reservoir. Two Forks represents a major escalation of the dispute between environmental advocates and traditional water developers. Foothills merely allowed filtration and treatment of existing raw water supplies to make them potable; its impact on the flow of the South Platte River and its fish habitat was limited. These issues were effectively and reasonably resolved in the hard-won compromise (which took as long to resolve as World War II), even if not to the satisfaction of the combatants. In contrast, Two Forks will store massive amounts of water, usually estimated at 860,000 acre-feet but with a potential design that would hold 1.1 maf, making it Colorado's largest water storage facility. Through that storage, Two Forks would increase the firm yield of the Denver Water Department system from 88,000 to 98,000 acre-feet per year, depending on the dam's height. Lack of storage capacity limits the water department's firm yield, and the free water available in wet years on both eastern and western slopes is lost. The additional storage in Two Forks would permit the Denver Water Department to maximize its yield from existing supplies in the Roberts tunnel system, although at a cost of $337 to $392 million (Denver Water Department, 1986).

Two Forks seems a true disaster to preservationists and river recreation groups. A prominent attorney and historian summarized several of the opponents' objections over a decade ago (Hart, 1975):

- It would ruin the Denver metropolitan area by encouraging excessive increase in population rather than reasonable, planned, and controlled growth.
- It would deprive water users in western Colorado, on the Arkansas and possibly the South Platte, of the use of rights junior to Denver's conditional decrees for agriculture and future resource development.
- It would deprive residents and visitors to Colorado of its natural setting, streams, meadows, and agricultural greenbelts and would destroy wildlife, flowers, shrubs, and trees along natural streams.
- It would deprive residents and visitors of the outdoor recreational activities of fishing, boating, hiking, hunting, climbing, camping, bird watching, and touring, replaced by a fluctuating reservoir.
- It would threaten the habitat of birds and animals, some endangered.

Environmental opponents to Two Forks who discount the slow-growth argument (as does the Colorado Environmental Caucus) raise further objections. They claim that enough water can be supplied to meet any possible growth needs by far less environmentally damaging means than Two Forks. They say that construction of Two Forks shows Colorado's lack of concern for the environment, which has traditionally been the state's major asset. Still another objection is to the economic cost of Two Forks – more than a third of a billion dollars, which will place a heavy debt on the Denver metropolitan area, an area already suffering from economic malaise and needing expensive highways and other infrastructure.

The systemwide environmental impact statement of the Denver Metropolitan Water Supply presents seven scenarios involving various combinations of structural (dams, reservoirs, aqueducts) and nonstructural (water conservation) methods to meet the future water demand in the study area (U.S. Army Corps of Engineers, 1985, app. 5, fact sheet, pp. 13–46). Six of the scenarios involve building a regulatory and storage reservoir on the South Platte River upstream from Denver, Two Forks Dam and Reservoir figure in four scenarios, and other structural scenarios feature alternative sites. A seventh scenario, the no-federal-action alternative, does not involve reservoir construction.

The Denver Water Department and most of the suburban suppliers, which are paying the costs of the SEIS, are determined to use traditional structural methods to develop the new water supplies that, even with conservation, will be increasingly needed from the mid-1990s for at least 40 years. Despite differences of opinion on the rate of population growth in the area and widespread concern over its undesirable consequences, substantial population growth is viewed as inevitable unless a regional growth-management mechanism is adopted.

The Corps' SEIS does not ignore the potential effects of such demand-reducing policies as residential metering, innovative conservation techniques, and regulation. Pricing water to reduce demand is not addressed, but in its scenario the Corps treats such potential demand reduction as an equivalent increase in supply. The magnitude of demand reduction by 2035 ranges from 18,600 to 119,500 acre-feet per year among the six structural scenarios; in the no-federal-action-alternative it is 125,000 acre-feet per year. As demand reduction increases, reliance on structural features is reduced accordingly.

The seven scenarios are based upon the best available technology and upon current law and institutions. Opinions differ on whether current water law and institutions are the best available. Voices calling for changes in the rules are becoming more evident, but change will not come easily. Responding at a 1985 water conference to numerous proposals to improve water management by changing water law and by overcoming institutional barriers to certain innovative means to improve allocation efficiency, a prominent Colorado state senator commented: "*I* am an institutional barrier! I took an oath to uphold the Constitution of the State of Colorado, and I will do so" (McCormick, 1985).

Alternatives in water supply and demand management

Water supply can be augmented or water demand managed to alleviate the large water shortage facing the Denver metropolitan area during the next half-century in various ways. Legal and institutional barriers to water management innovations are especially important because they can resolve the divisive issues facing those who care deeply about the future of the Denver area.

Transmountain diversion

The future transmountain diversion projects planned by the Front Range cities of Denver, Aurora, and Colorado Springs cost far more than any past diversions in Colorado. Their construction is less certain because of the legal entanglements over perfection of conditional water rights, wilderness intrusion, and NEPA and land use regulations. One scenario that includes Two Forks Dam, Williams Fork Gravity, Straight Creek, East Gore Canal, and ultimately Eagle-Piney and Eagle-

Colorado has capital costs of $2.7 billion in 1985 dollars, or about $10,000 per acre-foot of safe yield from its structural features. Long-term capital costs per new household are estimated at $11,000 (in 1985 dollars), in addition to a share of the amortized cost of the existing water system (U.S. Army Corps of Engineers, 1985, app. 5, fact sheet, p. 25).

Besides these direct costs to the diverter are the social and external costs to the basin of origin and to downstream water users. The social costs of diverting water are the values of alternative uses of the same capital resources, of natural resources lost or transformed, and of what could have been produced using the undiverted water. External costs imposed on parties distant from the project site include increased degradation and salinity in downstream water resulting from an upstream diversion of fresher water. Conceivably, the loss of water may interfere with the future development of natural resources and resulting employment in the basin of origin. These costs are difficult to estimate, particularly when the basin of origin is undeveloped, is not putting all of its water to beneficial use, and possibly will never do so. Further, neither Colorado water law nor custom requires the diverter to pay these costs, even though the National Water Commission has stated the now-conventional economic wisdom:

> [A]n interbasin transfer should produce benefits from the new uses of water that exceed the losses from the present and foregone uses in the area of origin and that exceed the costs of the project as well. Applying this criterion may be difficult, because the foregone uses include both present and prospective uses whose value must often be simulated, since they are not priced (1973, p. 321; see also Howe and Easter, 1971, p. 105).

Colorado law does not require an individual or a municipality that diverts water between river basins to compensate or mitigate losses to a basin of origin. However, a 1937 law does require Water Conservancy Districts to pay to construct compensatory storage reservoirs in the basin of origin so that present and future appropriations of water will not be impaired or cost more (Colo. Rev. Stat., sec. 37-45-118(b)(IV), 1973).

Western Slope officials are currently agitating for a new state law that requires cities and other diverters to pay for compensatory storage and perhaps for other forms of mitigation. They offer to Front Range diverters, in return, the promise of reducing the cost of delay and litigation and of permitting certain diversions to go forward if adequate compensation is paid (Fischer, 1985). Such a proposal is hard for elected or appointed water officials in Front Range cities to accept because the law does not require such compensation and paying it could easily be considered a breach of fiduciary responsibility. Passing legal and inflationary costs along in water rates and tap fees is unpopular but lawful; passing along discretionary costs paid to build compensatory storage is risky in Colorado's litigious society. Furthermore, the proposal has overtones that resemble the offers of the Capone gang to sell protection to owners of Chicago saloons: only the bravest and most foolhardy saloon keepers refused to pay, usually to their regret. Some water agencies may in fact support a compromise that results in a legal obligation to pay for reasonable compensatory storage if this provision will avoid the seemingly endless delay and costly opposition to diversion projects.

An imaginative alternative way to compensate the basin of origin was proposed in 1985 by Governor Lamm and subsequently analyzed in a report by University of Colorado experts in law and economics (MacDonnell et al., 1985). The governor proposed an export fee, set by the legislature, to be paid by the importer annually for each acre-foot of water diverted outside a primary river basin. Fees would be held in trust by the state and authorized by the Colorado Water Conservation Board for spending on projects proposed by public entities in the basin of origin: new storage projects or repair of existing ones; municipal water systems; improvement of irrigation systems, on-farm water efficiency measures, and water-based recreation facilities; and securing in-stream flows (MacDonnell et al., 1985, pp. 36–40). Lamm's concept reflects the recognition that new forms of compensation may well be more valuable to a basin of origin than compensatory storage alone.

Reallocation of agricultural water supplies

Most Colorado water is used and reused in irrigated agriculture until it flows out of the state. About 85 percent of the water depletions in the South Platte system and 89 percent of the depletions in the Arkansas system are for irrigation (Milliken et al., 1981, p. 51, quoting the 1974 state water plan). Although irrigators value water highly, some is used inefficiently because economics does not permit farmers to invest heavily in the high-technology sensors and drip-irrigation equipment needed to boost efficiency. Farm economics in recent years makes it challenging enough to survive and to pay interest on bank loans in eastern Colorado. Then, too, much water wasted through excess runoff or loss in unlined canals is merely transferred to the next users; only consumptive loss, as through evaporation, is lost to the river system. Nevertheless, for the same financial investment, more water can be saved from irrigated agriculture than from municipal conservation. Because all water in the South Platte and Arkansas river systems is in beneficial use within Colorado and some flows into Nebraska and Kansas to meet compact obligations, water is reallocated from agricultural to municipal use through the market system. (It is purchased, and the water courts confirm the transfer.) Although Colorado's constitution gives preference to municipal purposes over agricultural uses, thus permitting cities to condemn water rights under the power of eminent domain, such taking is rare because there are many willing sellers.

The direct monetary costs of reallocation are less than those of other water supply sources because the value of water used in irrigation is lower than that for municipal purposes. Indeed, the direct costs of buying the water rights, and sometimes also the land that the water has irrigated, and of constructing diversion and conveyance facilities are quite reasonable compared with those of transmountain diversion. However, indirect costs (including the loss of agricultural production) are apt to be high and to fall on a relatively small and localized part of the population: agricultural supply and processing businesses and rural towns. When water rights are taken for municipal use, the water rights holder is compensated. The tenant farmers and farm workers receive no compensation; nor do those who lose income as agriculture-related enterprises decline.

Where maintenance of a healthy agricultural sector is a goal of state policy, as in Colorado, continued reallocation of agricultural water is viewed with increasing concern. Purchase of agricultural water from the Arkansas Valley by Aurora and Colorado Springs has caused a 35 percent decline in irrigated farmland over the past 24 years in Crowley, sometimes called "the county that sold its birthright" ("Water War," 1984). In 1985 the governor reversed his long-established opposition to transmountain water diversions, stating that without some transmountain diversion, cities along the Front Range would be forced to take over agricultural water rights to meet growing demand ("'Reality' of Growth," 1985).

Proposals have been advanced to keep the best of both worlds by having the cities, which are richer, pay for such conservation techniques as canal lining and drip irrigation in order to increase the efficiency of water use in irrigated agriculture and to allow the same crops to be produced with less water. The cities could then receive the water relatively cheaply (Weaver, 1985, pp. 12–14). A similar proposal has received favorable attention in southern California; the Environmental Defense Fund suggests that the Imperial Irrigation District and other large agricultural districts pay for conservation improvements and sell the saved water at a profit to San Diego and other cities ("How to Find Western Water," 1985). The major problem with this plan in Colorado is that it is unlawful. Although Colorado water law prohibits waste, it also prohibits spreading – salvaging water through some conservation method that improves efficiency and results in using salvaged water to irrigate more land. No one can acquire an independent right to water unless, among other proofs, it can be proved that the water was not part of the stream system when senior rights attached (Martz, 1975, p. II-27). Water can be conserved, but in Colorado the saved water belongs to the next-most-senior water rights holder, not to the conserver, unless the conserver can prove that the savings come from lower consumptive use and that the return flow to subsequent users is not reduced (Danielson, 1986). This anomaly seems to remove the incentive to conserve, but a case can be made that in its absence an incentive would exist to claim water in excess of that needed for beneficial use and then, after reducing use through conservation, to profit by selling the excess. In jurisdictions other than Colorado, the legal principle that a party salvaging water has the proprietary right thereof is virtually unchallenged (Martz, 1975, pp. II-28 to II-29).

Using groundwater

Colorado has large supplies of groundwater. Nearly 22 maf is available in alluvial aquifers hydraulically integrated with the surface water system and is recoverable, and another 909 maf is in deep, confined aquifers not tributary to streams (Milliken et al., 1981). (These amounts are of recoverable water only.) Groundwater has been used mostly for agriculture in eastern Colorado and for rural domestic and supplementary municipal uses. Greater use of groundwater has been proposed to avoid the problems apparent in obtaining municipal water from transmountain sources.

The major advantages of using groundwater as a municipal supply, as Tucson, Arizona, does, are that the total volume available can be enormous, storage costs nothing, no environmental damage from reservoir construction is involved, and the supply is comparatively immune to pollution and to such natural disasters as earthquakes. On the other hand, response is slow, the source is relatively uncontrollable, groundwater is mined when withdrawal exceeds recharge, land can subside after groundwater is removed, energy costs for pumping increase as groundwater levels fall, and treating or eliminating pollution is difficult once it has occurred.

In the Denver metropolitan area, little use has been made of groundwater when more reliable sources have been available. A number of deep wells supply some suburban developments, although the reliability of such supplies is dubious. Other deep wells are used to water parks and golf courses. Denver's deep groundwater is essentially nonrenewable because it is confined by an impervious rock layer between the aquifer and the land surface. Natural recharge occurs only in limited outcrops by the foothills, so the water table is dropping without hope of natural recharge. The water quality is considered good, although in the metropolitan area toxic waste disposal has been extremely careless. Nerve gas and other dread poisons from the Rocky Mountain Arsenal were injected into deep wells during the 1950s and 1960s, and regional chemical wastes have been dumped in Lowry landfill, where an unlined pond has allowed access to groundwater migrating toward urban areas with well systems.

In its SEIS water supply scenarios, the Corps of Engineers estimates that pumping groundwater from under municipal boundaries could yield 20,000–87,000 acre-feet per year. Such groundwater pumping would have a capital cost of about $1,000 per acre-foot of yield and operation and maintenance costs of $75 per acre-foot per year (both in 1985 dollars). For groundwater collected in a satellite well field outside municipal boundaries, the capital cost would be $5,000 per acre-foot of yield, and operation and maintenance costs would be $200 per acre-foot per year. A satellite well field could yield up to 28,000 acre-feet per year (U.S. Army Corps of Engineers, 1985, app. 5, pp. 2–49 to 2–50).

Colorado law places nontributary water under the jurisdiction of the state engineer, who may grant permits to withdraw it from wells so long as unappropriated water exists under the land of the applicant and other consenting parties and no more than 1 percent of the estimated water will be pumped each year (Martz, 1975, pp. II-25 to II-26). The city of Aurora has proposed using groundwater under new lands to be annexed as a supplementary supply during dry years when surface supplies are inadequate. This proposal is similar to one made by the Colorado Environmental Caucus, one of the groups participating in the Governor's Metropolitan Water Roundtable, which is the main forum for water issue debate. The Environmental Caucus proposes new institutional and legal arrangements to integrate nontributary groundwater supplies with surface water systems so that groundwater would augment surface supplies in dry years and surface supplies would be used to replenish groundwater when surface water flow is adequate (Weaver, 1985, pp. 10–12). How the institutional and legal barriers to this proposal might be overcome is unclear, but broad and fundamental changes would seem to be needed in the doctrine of prior appropriation and in the 1965 Colorado Ground Water Management Act, as amended

(Martz, 1975, pp. II-23 to II-26). Even harder to overcome may be the technical geologic and hydrologic barriers to the recharge of groundwater aquifers through impervious rock strata from surface water supplies during above-average runoff years.

Still other means of augmenting water supplies have been proposed, most of them potentially applicable to the Denver metropolitan area (Milliken et al,, 1977; Milliken and Taylor, 1981):

- Using watershed land management techniques (for example, enhancing the snowpack, modifying vegetation within watersheds, and eliminating riparian vegetation along rivers and streams;
- managing the water system to increase yield (for example, by adding storage, particularly at elevations with lower evaporation rates);
- augmenting precipitation (for example, through cloud seeding);
- desalinizing brackish or saltwater; and
- reusing municipal wastewater.

Except for desalination, all these techniques may be useful to the Denver metropolitan area. As noted above, only improved water system management has been actively proposed, and it remains controversial. As for reusing municipal waste water, Denver's pioneering 1-mgd pilot plant is under development, but recycling wastewater is several years away and is not yet an issue (Lohman and Milliken, 1985). Cloud seeding continues experimentally, but it will not become an issue until cause and effect are proved. (The legal issues underlying these methods for increasing municipal water supplies and reducing water demand – conservation and pricing – have been definitively discussed by Martz [1975].)

Conservation measures

The desirability of water conservation in the Denver metropolitan area is almost universally accepted. But there are questions about the reliability of conservation as a water management tool and the degree to which conservation should substitute for development of additional supplies during nondrought periods. The issue of metering water use in 88,000 Denver households was resolved by the Denver Water Board's 1986 decision to require universal metering. Englewood, however, has not taken action on its 8,000 unmetered homes.

Numerous water conservation measures have been proposed, and many have been implemented in whole or in part: universal metering of homes in the metropolitan area to encourage water saving in the 20 percent of homes on flat rates; the well-received ET (evapotranspiration) program, through which the news media daily advise residents on the amount of water needed to keep a lawn healthy; "xeriscape" programs promoting landscaping with plants that thrive in semiarid climates; local ordinances restricting lawn sizes and requiring soil preparation when planting new lawns; the promotion of energy-efficient retrofitting and the installation of water-saving appliances or devices; and leak-detection programs in water systems ("Conservation: An Attractive Option, But . . .," 1985).

In its SEIS, the Corps of Engineers estimates four levels of demand reduction based on various assumptions about which conservation programs will be adopted, how intensive the public awareness program is, and whether certain programs are mandated or left voluntary. Demand reduction would be lowest with a low-penetration program involving natural retrofit of plumbing devices in existing homes (3,800 acre-feet per year), efficient plumbing in new homes (11,100 acre-feet per year), an enhanced xeriscape program (1,700 acre-feet per year), and an ET program (2,000 acre-feet per year). The plumbing costs would be borne by homeowners, and the xeriscape/ET program costs would be $76,000 per year (in 1985 dollars), or $20 per acre-foot of yield. The highest level of demand reduction would require a high-penetration public awareness program, mandatory universal metering, and low water use for landscaping and plumbing devices in new homes. Estimated savings would be 21,000 acre-feet per year for the retrofit of existing homes, 26,700 acre-feet per year for efficient plumbing in new homes, 13,700 acre-feet per year in 2035 for universal metering, 1,700 acre-feet per year for xeriscape, 6,500 acre-feet per year for low water use for landscaping, 13,300 acre-feet per year for the ET program, 11,200 acre-feet per year for the retrofit of commercial and industrial buildings, 26,300 acre-feet per year for a leak-detection program, and 4,800 acre-feet per year for minor household conservation items, such as low-flow shower heads. The capital cost of metering is estimated at $17,237,000 (in 1985 dollars) or $60 per acre-foot per year. At these estimated costs, metering would provide conserved water for $1,075–$2,150 per acre-foot, which is competitive with water from other sources (U.S. Army Corps of Engineers, 1985, app. 5, chap. 2).

The effectiveness of conservation measures proposed for the Denver metropolitan area is a topic of active contention, as the Corps of Engineers' SEIS nears completion. The Corps' consultant estimates a savings range of 25,000–60,000 acre-feet by 2010. The Metropolitan Water Providers, representing suburban water agencies, estimate 14,000–28,000 acre-feet saved. The Governor's Metropolitan Water Roundtable, which represents all factions, estimates 30,000–70,000 acre-feet. The Environmental Defense Fund, which funded a 1980 study of alternative means of meeting Denver's water needs, estimates 100,000 acre-feet (Morris and Jones, 1980, pp. 108–112). These wide-ranging estimates reflect various assumptions about program components and the years in which the savings would be realized.

The philosophic issue of water management in Denver's future is closely tied to conservation. The water development proponents take the following view (Metropolitan Water Roundtable, 1985):

> Estimated conservation savings should be established and every effort made to achieve these goals but they should not be used as a source of water supply in planning water projects which are presently proposed to meet future needs. At such time as history has shown that they are realistically achievable, conservation program results can then be taken into account in planning future water projects and diversions.

Conservation and environmental proponents support a different view:

> Estimated conservation savings goals should be counted as a source of water supply in planning future water projects and diversions since data and studies for Denver and other western urban areas strongly suggest that opportunities exist for making water use more efficient.

The issue remains divisive, pitting those who see conservation as a rational, inexpensive, and environmentally undamaging alternative against those who consider it risky, unproven in reliability, and best reserved for emergencies or severe droughts.

Water pricing to reduce demand

Water is still a relatively inexpensive commodity in the Denver area, but its cost is rising, and the costs of new water taps are quite high compared to those in other U.S. cities. Metered customers in the city and county of Denver pay about $176 per year for water, following a 35.3 percent raise in April 1980 and an 8.7 percent raise in February 1982 ("Water Board," 1984). Bimonthly rates in the city are $0.74 per 1,000 gallons for the first 30,000 gallons, then $0.60 per 1,000 gallons for the next 70,000 gallons, then $0.46 cents per 1,000 gallons, plus a $5.30 bimonthly service charge. Total-service suburban customers pay double these amounts, averaging $350 per year (Denver Water Department [1985], pp. 60–61; "Water Board," 1984). In early 1986, rates were increased 5 percent; a new rate study is underway (Brimberg, 1985). Neither Denver nor the major suburban suppliers use inverted block rates to reduce demand, as Tucson does (Martin et al., 1984), nor has peak demand pricing been used in the Denver metropolitan area. Marginal cost pricing has been discussed by Milliken (Milliken et al., 1977;

Cristiano and Milliken, 1977) and in more detail by Morris and Jones (1980, pp. 117–120). Evidently, it has not been considered seriously by any Denver area water agency, probably because the true marginal cost of water is many times above the price now charged.

After some 70 years of increases in daily per capita water demand by Denver Water Department customers (Milliken et al., 1977, pp. VI-23 to VI-25), demand for treated water peaked about 1980 at 244 gallons per capita per day (gpcd) and has declined since. From 1981 to 1985, it averaged 217 gpcd (Denver Water Department [1985], pp. 8–9). A similar drop in demand occurred throughout the Denver metropolitan area, where demand peaked in 1974 at 231 gpcd and averaged 192 gpcd between 1974 and 1982 (U.S. Army Corps of Engineers, 1985, app. 5, pp. 22–24). No convincing evidence explains this welcome reduction in demand, although the differential between the lower per capita demand in the suburbs and the demand in the city and county points up the effect of the higher pricing of suburban water. The entire metropolitan area receives the benefits of the ET program and of the voluntary every-third-day outdoor water use restriction program in effect from June 1 through September 30.

Although water tap fees tend to deter new construction rather than per capita water use, tap fees have increased rapidly in the Denver metropolitan area in the last decade through a combination of events and policy that places much of the water system development cost on new growth. In 1974 the Denver Water Department placed a system development charge of $700 on applicants for a three-quarter-inch water tap (typical for a single-family residence) outside the city and county, with a smaller charge in the city. In February 1982 the system development charge for a three-quarter-inch tap was $2,550 in Denver and $3,570 outside (Denver Water Department [1985], pp. 8–9). A 15 percent increase in these charges took effect early in 1986, along with a 5 percent increase in water rates.

In 1982, the Denver Water Department entered into an unprecedented agreement with suburban cities, counties, and special water districts to sell them 80 percent of the present and future water facilities that it is building to expand its water system. As a consequence of the agreement, suburban water providers have contracted to pay a large but as yet unknown sum to the Denver Water Department. Some municipalities, such as Littleton, have thus far paid most of these participation costs from general revenues, although Littleton recently announced an increase from $1,300 to $3,000 in city tap fees. Others, notably special water districts, pass these costs along to new homes rather than raise the general property taxes or water rates for all residents. Thus, district tap fees have been superimposed on DWD system development charges. Combined water and sewer tap fees average $6,370 in the Denver metropolitan area and exceed $10,000 in Cherry Hills Village (Home Builders Association, 1985), compared with $1,200 for Phoenix, $1,074 for Tucson, $1,712 for Albuquerque, $1,522 for Salt Lake City, and $1,093 for Dallas ("Denver Outpaces Counterparts," 1985). It is too soon to tell how steep tap fees will affect growth in the area, particularly in view of the fact that the Denver economy has slowed and some 30,000 completed residential units remain unsold. But, since 1979, the number of new taps sold and put to use in the Denver Water Department system has been smaller than the quota (now 7,000 per year) that the department allocated during its 1977–1986 tap-allocation program. If tap costs were uniformly high through the metropolitan area, rather than varying widely among jurisdictions, better evidence might be found on how growth paying its own costs affects the efficiency of water allocation and use.

*A regional growth management plan**

Will the Denver metropolitan area ever free itself from the high costs of fueling continual growth and from the loss of traditional amenities – clean air, open space, convenient driving, efficient local government services? Can the area end its ceaseless efforts to obtain more and more water (efforts that remind us of Disney's sorcerer's apprentice) and turn its attention to other needs?

Achieving this goal will require an integrated growth management system supported by a comprehensive planning process and a substantial political consensus. It will not come about as easily as some proponents of slowed growth believe, through curtailment of structural measures to obtain new water supplies. Such simplistic strategies are ineffective. If growth is desired, evidence attests, water will be found to support it. As for those trying to constrain growth, their goals are sensible and are widely shared among area residents. However, they propose to achieve these goals by sacrificing the adequacy and integrity of the metropolitan water system – an unwise and unsuccessful strategy for pursuing a wise goal.

How large a role does water play in development? In a semiarid region, a reasonably secure supply of potable water is no doubt a major

*This section draws heavily on Milliken, Lohman, and Gougeon, 1985, and on Gougeon, 1978.

prerequisite of land development. Residential, commercial, and industrial land uses all require water. But the availability of water is not the sole factor influencing the pace and pattern of urban development. Additional determinants include:

- Cost and location of raw land;
- suitability of available land (in terms of slope, drainage, soil characteristics, etc.);
- local government development policies and approval requirements;
- availability of other utilities (for example, electricity and natural gas) and services;
- transportation access;
- social and physical amenities (site-specific and regional);
- demand conditions in the local market (influenced by population growth and other demographic factors); and
- local, regional, and national economic factors (for example, interest rates).

On balance, the availability of water permits development to take place once the factors listed above have exerted their influence, but demand for urban land is driven first by natural population growth, in-migration, employment growth, household formation, technological change, and other forces. By itself, the availability of water does not create a demand for it.

For managing growth through the provision of water service, the three most commonly used techniques are: urban service boundaries, public facilities requirements, and rationing systems. These techniques can be used individually or in combination.

Urban service boundaries. Urban service areas (or limit lines) have been adopted by numerous jurisdictions to manage growth. Under such a system, the local government identifies the geographic area slated to receive urban services during some specified period (typically five to 15 years). The exact boundary is usually linked to the provision of utility services, such as water. Land lying beyond the designated urban service boundary cannot be developed until the boundary is extended. Care must be taken to ensure that the amount of developable land within the service area can meet the level of demand that can reasonably be expected. If the supply is not adequate, raw land prices in the service area will skyrocket until the service boundary is next adjusted. Urban limit lines are typically established by local governments during planning and capital programming.

In some respects, the Denver Water Department blue line served as an urban service boundary from 1951 to 1960. The blue line, however, was not a true urban service boundary for several reasons:

- Its implementation was motivated strictly by water supply rather than by growth management considerations.
- The Denver Water Department was not the exclusive supplier of water service in the growth area (a number of independent systems were formed during this time that provided water to areas beyond the blue line).
- The Denver Water Department did not have jurisdiction over the entire growth area or over other services (sewer hookups, natural gas taps, etc.) (*Englewood* v. *Denver*, 123 Colo. 290, 1951).

In 1951, the Colorado District Court decided that the Denver Water Department does not function as a public utility. A contrary Colorado District Court decision in 1982 was reversed on appeal in 1986 by the Colorado Supreme Court ("Denver Water Board Wins," 1986). Any change that would bring the Denver Water Department under the jurisdiction of the Public Utilities Commission could bring the *Robinson* decision into play. In *Robinson* v. *Boulder* (547 P.2d 228, 1976), the court found that the city of Boulder was acting as a public utility and could not legally deny water service outside its municipal boundaries if it had the physical capacity to do so, despite its growth management policy. As a public utility, it probably would be impossible to create an urban service boundary for nonutility reasons even if other conditions were favorable.

Public facilities requirements. The term "public facilities requirements" describes techniques that make approval of development conditional on the existence or provision of adequate public facilities (water, sewer, roads, schools, etc.). Public facilities requirements can affect both the location and the rate of growth. Local governments can direct the siting and timing of development through decisions on capital investments in public facilities, access to existing facilities, and provision of services.

It is particularly important that the local jurisdiction involved exhibit a strong commitment to providing programmed service improvements. As some observers have noted, a conflict can arise between local desires to manage growth and the responsibility to provide the highest-quality utility service achievable. For example, the Board of Supervisors of Fairfax County, Virginia, was widely accused in the mid-1970s of deliberately allowing a county sewer system to become inadequate in order to justify limitations on urban growth (Rivkin, 1975, pp. 473–482). The quality of sewer service in the county declined, and a large number of septic tank systems were required, increasing the potential for significant water pollution. Clearly, the integrity of publicly operated utility systems should not, and need not, be sacrificed to achieve legitimate growth management objectives.

As currently organized, the Denver Water Department is not an appropriate body to implement any type of public facilities requirement. It lacks jurisdiction over the metropolitan area, it is not responsible for issuing development approvals, and it exercises no control over the provision of other public facilities and services. At present, the department's role would be limited to coordinating its service and expansion plans with some larger, more appropriately structured entity empowered to implement a public facilities requirement. But that entity does not exist; nor does a regional plan.

Rationing systems. An increasingly popular growth management technique is rationing. Generally, rationing systems allocate a fixed number of building permits or development approvals. The city of Boulder has operated such a system since 1976; following a study of its impacts, the system was made permanent in early 1985 (City of Boulder Ordinance No. 4877, 1985). Access to utilities can also be rationed. Rationing systems influence the rate of growth, but they can also influence location of development.

The Denver Water Department's tap-allocation program resembles a rationing system in many ways. It differs only in that the allocation decisions are not directly related to land use factors, and the limits and allocations adopted are not the result of any comprehensive planning process. Rather, they are an attempt to distribute limited water supplies fairly, evenly, and equitably throughout the service area. In addition, because of the department's city charter requirements and department policy (the imposition of the blue line and continued use of the tap-allocation program), Denver proper is not subject to any limitations on the number of taps it can receive annually. Still, the Denver Water Department has not yet been involved in true growth management and as now structured is not an appropriate tool for managing urban growth in the metropolitan area; nor does it wish to be.

At least two significant and complex changes in the existing structure are required for water service to be used effectively in managing growth in the Denver area. First, the nature of the several water service agencies would need to change dramatically. Establishing some form of metropolitan water agency could, if properly designed, achieve geographic congruence between the service agency's boundaries and the growth management area, provide political accountability, and overcome any potential obstacles created by state utility law and local charter provisions. Second, some form of metropolitan planning mechanism for coordinating service and developing an enforceable growth management plan would have to be instituted. Water service can be used effectively and appropriately as a growth management tool only when it is but one element of a complete system and when the system is supported by a comprehensive planning process and a substantial political consensus.

Achieving the necessary consensus over a regional plan could be difficult. Creating a metropolitan water agency with political accountability to voters throughout the metropolitan area would be welcomed by many suburban jurisdictions that dislike control exercised by the Denver Water Board, but Denver would certainly expect substantial concessions in return. It is unlikely that local governments in the area would be willing to sacrifice any autonomy to create an enforceable growth management mechanism.

Even with these difficulties, there is hope for a growth management system. With the type of leadership and effective educational campaign that led to Colorado's 1974 vote to reject the award, already granted, to host the 1976 Winter Olympics, the voters could create a system for effective regional growth management through a referendum. This approach would involve establishing a regional planning authority with directors accountable to the voters, probably elected partly by county or election district and partly at large. The planning authority would develop and enforce a regional growth management plan using some combination of rationing and urban service boundaries. Within these constraints, local planning and zoning decisions could be left to local governments. The regional planning authority would also set policy on the extension of service to undeveloped land, on water pricing, and on water conservation regulations. The people *can* decide, and the fragmented local governments must then follow.

New rules versus new alliances

In 1985 two dominant advocacy groups mobilized campaigns to shape the direction of water management in the Denver metropolitan area. Each group has an ideology, experience, and allies. For lack of better terminology, one can be called the water system developers, and the other can be called the water system management reformers. Although they are now deadlocked, events will probably break the impasse in the next two or three years, when the systemwide environmental impact statement on the metropolitan Denver water supply system is complete and the scenario decided upon to meet the Denver metropolitan area's water needs for the next 50 years.

The new alliance: the suburban water agencies join Denver

More than three decades of bickering, during which Denver antagonized Englewood, Aurora, Westminster, and its three surrounding counties and was in turn punished by them, suddenly ended in 1982. The Denver Water Board and the suburban governments and water agencies agreed to negotiate, and within a few months had signed the innovative, if not historic, Metropolitan Water Development Agreement. It binds Denver and suburban jurisdictions in Adams, Arapahoe, Boulder, Douglas, and Jefferson counties to mutual development of water for metropolitan use. They pledged "to cooperate among themselves to supply the water supply facilities . . . they all will need in effectuating a more economic use of water resources as well as a more cooperative input to the decision-making processes" (Metropolitan Water Development Agreement, 1982, p. 2). In other words, they chose to unite in pursuit of common water goals.

The first phase of the agreement obligated the providers (county governments, municipalities, special water districts, and other suburban signatories to the agreement) to purchase shares totaling 80 percent of the development costs in return for future raw water supplies. Denver kept 20 percent. The original agreement was oversubscribed (182 percent of the 80 percent share was sought, but it was reduced by negotiation). Not all suburban jurisdictions chose to join, some because of risk aversion or indecision, others because they were too small to subscribe the minimum 0.5 percent of the 80 percent suburban share, and others because they anticipated little future growth.

The original 47 providers agreed to pay 80 percent of the rapidly escalating costs of the SEIS as a form of ante, which permits them to participate in future project-participation agreements to develop new water supplies. A 1984 agreement covering participation in Two Forks Dam and Reservoir, a Colorado River joint use reservoir and related facilities, was signed by Denver and 44 providers (Platte and Colorado River Storage Projects Participation Agreement, Jan. 24, 1984). The ultimate costs of participation are not yet known, but participation will be quite expensive. Not only the costs of developing the two specified reservoirs will be included, but also those of developing the Blue River diversion, Roberts Tunnel, and Dillon Dam and Reservoir (which are "attributable" to the Two Forks project) plus those of other ancillary facilities. Politically, the providers have organized to unite suburban opinion in favor of development, to scrutinize the SEIS critically, and to promote an integrated plan for future water resource development for the Denver metropolitan area. The providers' preferred scenario would be a largely traditional structural form of development involving Two Forks Dam and Reservoir and future transmountain diversion. Conservation is considered a useful goal but one that is not yet reliable enough to substitute for structural features.

The water management system reformers

The reformers' strategy has become more sophisticated during the years of battling water development in the Denver area. They found in the Foothills agreement that they can obtain many, if not all, of their high-priority goals by negotiation. Through participation in the Colorado Environmental Caucus in the Governor's Metropolitan Water Roundtable and the Denver Water Board's citizens' advisory committee, they now influence ongoing water decisions greatly. Through research, they (in particular, the Environmental Defense Fund) found blunders in the population forecast method for the SEIS and forced its revision.

Further, the reformers identified anomalies and inefficiencies in the archaic water management system that was shaped by nineteenth-century cultural values of struggle and conquest of nature. The reformist faction has proposed some specific and desirable improvements to the system, recognized the legal and institutional barriers to reform, and is actively recommending changes to the state legislature.

Tradition versus change and persistence versus reform are the water management issues facing Denver. A related issue may be continued deadlock versus negotiation and compromise. Much that is vital to the Denver area's future environment, culture, and economy will be determined by how these issues are resolved. As the Chinese curse puts it: "May you live in interesting times!"

References

"All Sides in Foothills Project Come up Winners." 1979. *Rocky Mountain News*. Jan. 8.

Brimberg, Judith. 1985. "5% Water Rate Increase, 15% for Tap Fees Planned." *Denver Post*, Nov. 23, p. 1A.

"Conservation: An Attractive Option, But . . ." 1985. *Metropolitan Water Providers Newsletter*, *2*(10).

"Corps, EPA Agree on Foothills Project." 1978. *Denver Post*. Dec. 17.

Cox, James L. 1967. *Metropolitan Water Supply: The Denver Experience.* Boulder: University of Colorado, Bureau of Governmental Research and Service.

Cristiano, Carole R., and Milliken, J. Gordon. 1977. "Demand Rates for Water: Special Problems of the Southwest." Paper 23-b. Proceedings of the 97th Annual Conference, American Water Works Association, Anaheim, California, May 8–13.

Danielson, Jeris. Colorado State Engineer. 1986. World Resources Institute Workshop on Water and the Arid Lands of the West, Tucson, Arizona. Feb. 20.

"Denver Extends Its Tap Allocation Program." 1986. *Denver Post.* June 11, p. 3B.

"Denver Outpaces Counterparts on Water Tap Fees." 1985. *Denver Post.* May 19.

Denver Post. 1979. June 21.

Denver Regional Council of Governments. 1986. *Population and Employment Distributions for 1990, 2000, and 2010.* Denver.

"Denver Water Board Wins 13-Year Fight." 1986. *Rocky Mountain News,* Apr. 8, p. 6.

Denver Water Department. [1985.] *1984 Annual Report.*

Denver Water Department. 1986. *Water News.* May-June.

"Denver's Water Dilemma: How Foothills Became Stalled." 1977. *Denver Post.* Dec. 4.

Fischer, Roland C. 1985. "The Resource and the Injury." Paper presented at the Tenth Colorado Water Workshop, Gunnison, Colorado. Aug. 1.

Gougeon, Thomas A. 1978. "Urban Growth Management and Local Government." Unpublished honors thesis, University of Denver.

Hart, Stephen H. 1975. "Two Forks: A Symbol of Unwanted Growth." *Denver Post.* Feb. 9.

Home Builders Association of Metropolitan Denver. 1985. *1985 Water and Sewer Tap Fee Survey and Supplement.* Denver.

"Hot, Dry Summer Could Boost Water Board Plans." 1977. *Rocky Mountain News.* June 12.

"How to Find Western Water: Sell It." 1985. *New York Times.* Editorial, June 17.

Howe, Charles W., and Easter, William K. 1971. *Interbasin Transfers of Water: Economic Issues and Impacts.* Baltimore: Johns Hopkins Press for Resources for the Future.

Lohman, Loretta C., and Milliken, J. Gordon. 1985. *Informational/Educational Approaches to Public Attitudes on Potable Reuse of Wastewater.* Denver: Denver Research Institute.

MacDonnell, Lawrence J., et al. 1985. *Guidelines for Developing Areas-of-Origin Compensation.* Boulder: University of Colorado.

Marston, Ed. 1982. "Sweeping Court Decision Has Rollie Fischer Worried; River District May Have Won Too Big." *Western Colorado Report,* Mar. 29, pp. 8–10.

Martin, William E., et al. 1984. *Saving Water in a Desert City.* Washington, D.C.: Resources for the Future.

Martz, Clyde O. 1975. "Legal Principles Governing Water Allocation and Reuse." In J. Gordon Milliken et al., *Systems Analysis for Wastewater Reuse: A Methodology for Municipal Water Supply Planning in Water-Short Metropolitan Areas*. 2 vols. Denver: Denver Research Institute. (Updated in part to July 1977.)

McCormick, Senator Harold L. 1985. Interim Legislative Committee on Water and Natural Resources. 1985. Remarks at the Tenth Annual Colorado Water Workshop. Gunnison, Colorado.

"Metro Growth Rated Nation's Tenth-Fastest." 1985. *Denver Post*. July 9.

Metropolitan Water Development Agreement. July 13, 1982. (Copies may be available from the Denver Water Department.)

Metropolitan Water Roundtable, Water Conservation Committee. 1985. Report to Gov. Richard Lamm. Jan. 31. Quoted in *Metropolitan Water Providers Newsletter*, *2*(10).

Milliken, J. Gordon, et al. 1977. *Systems Analysis for Wastewater Reuse: A Methodology for Municipal Water Supply Planning in Water-Short Metropolitan Areas*. 2 vols. Denver: Denver Research Institute. (Updated in part to July 1977.)

Milliken, J. Gordon, et al. 1981. *Water and Energy in Colorado's Future*. Boulder: Westview Press.

Milliken, J. Gordon, Lohman, Loretta C., and Gougeon, Thomas A. 1985. *Economic and Social Impacts on the City and County of Denver of Alternative Metropolitan Water Supply Policies, 1985–2010*. Denver: Denver Research Institute.

Milliken, J. Gordon, and Taylor, Graham C.. 1981. *Metropolitan Water Management*. Water Resources Monograph 6. Washington, D.C.: American Geophysical Union.

Morris, John R., and Jones, Clive V. 1980. *Water for Denver: An Analysis of the Alternatives*. Denver: Environmental Defense Fund, Inc.

Morris, John R., Department of Economics, University of Colorado at Denver, and Luecke, Daniel, Environmental Defense Fund. 1985. Memorandum to the U.S. Army Corps of Engineers, Oct. 9.

Mosely, Earl L. 1957. "Western Slope Water Development for Denver." *AWWA Journal*, *49* (Mar.), 253.

National Water Commission. 1973. *Water Policies for the Future*. Washington, D.C.: U.S. Government Printing Office.

Pokorney, Edward E. 1982. *The Upper Colorado River Basin and Colorado's Water Interests*. Denver: The Colorado Forum.

"Prolonged Water Curbs Seen." 1977. *Denver Post*. June 4.

"'Reality' of Growth Accepted: Lamm Believes State Must Plan Properly." 1985. *Denver Post*. Jan. 17.

"Regional EPA Chief Urges Foothills' Demise." 1977. *Rocky Mountain News*. Sept. 27.

Rivkin, Malcolm D,. 1975. "Sewer Morartoria as a Growth Control Technique." In Randall W. Scott, ed., *Management and Control of Growth*. Vol. II. Washington, D.C.: The Urban Land Institute.

U.S. Army Corps of Engineers, Omaha District. 1985. *Metropolitan Denver Water Supply EIS*. Technical Appendix 5, *Development and Evaluation of Water Supply Scenarios*.

U.S. Department of Commerce, Bureau of the Census. 1973. *1970 Census of Population*. Vol. 1. *Characteristics of the Population*. Washington, D.C.

"Water Board Considers 9% Rate Increase." 1984. *Denver Post*. Sept. 18.

"Water Meter Work Could Begin in '87." 1986. *Denver Post*. June 11, p. 3B.

"Water War: Cities Draw Arkansas Valley's Lifeblood." 1984. *Rocky Mountain News*. Dec. 23, pp. 7, 40–41.

Weaver, Robert M. 1985. "Alternative Approaches to Transmountain Water Diversions." Delivered at the Tenth Annual Colorado Water Workshop, Gunnison, Colorado, Aug. 1.

9 New water policies for the West

MOHAMED T. EL-ASHRY
AND DIANA C. GIBBONS

From any close look at the major demands on western water resources emerges a sense of the inevitability of conflict. Evidence of pressure on water is everywhere. Most streams and rivers in the West have been fully appropriated. Salinity and toxic elements threaten traditional agricultural practices and impose high costs on subsequent water users. Groundwater tributary to a watercourse cannot usually be tapped unless equivalent surface flow rights are retired, and nontributary groundwater is being mined at high rates. In fact, use exceeds average streamflow in nearly every western subregion, and the deficits are being offset with groundwater and water imported from adjoining basins. (See Figure 9.1.)

Several major southwestern cities, including Tucson, are now mining groundwater aquifers to meet everyday demand. Reliable yields from these sources are falling, and economic exhaustion of the resource will soon be approached. Other cities, such as Los Angeles, face a reduction in long-held supplies as upstream states claim their compact-apportioned shares of Colorado River water. As cities attempt to increase supplies to meet growing demands, instream flows may be diminished and water use in other offstream sectors may be threatened politically, if not legally. Conflicts over water supplies and water-quality degradation are not limited to a particular state; they thrive in every state in the West and even extend across state borders.

Although the focus in this volume is on water demand in the municipal and agricultural sectors, overall competition for western water has been growing as strong demands emerge in other sectors as well. The energy sector is perceived as a new rival to traditional water users, although development of unsubsidized synfuels is not economically viable today. Recreational demand for water has grown along with population growth, and the value placed on instream flow for fishing, swimming, and boating has increased accordingly. When populations expand and urban areas sprawl, remaining wilderness areas and the water flowing through them also become more precious as legacies for future genera-

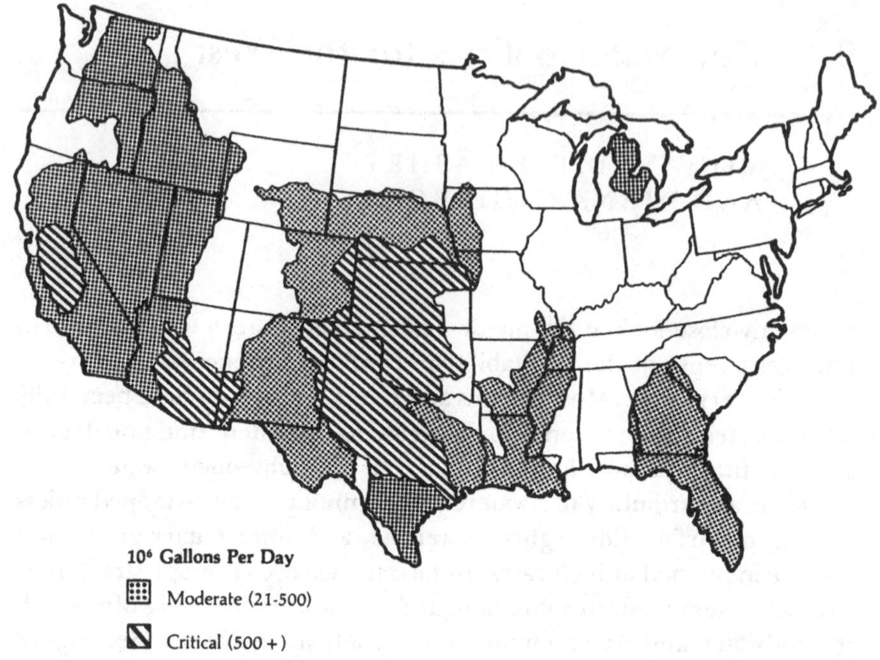

Figure 9.1. Groundwater overdraft in the United States (Engelbert and Scheuring, 1984, p. 28).

tions. Particularly on rivers developed to generate hydroelectric power, the economic value of water left flowing in the river can be significant – as much as $57 per acre-foot on the Colorado River (Gibbons, 1986). Similarly, as salinity levels reach new highs in the major rivers, the instream value of pristine water for dilution rises.

Competition for water supplies is perhaps most visible between cities and farms. Irrigated agriculture uses more water than any other activity in the West. Yet, this use is often inefficient, largely because policies and laws discourage conservation. Although irrigated agriculture has held a favored position in water allocation for decades and farms hold many of the senior appropriative water rights, urban water demands have grown rapidly, and citizens of the arid West intend to keep their oases green. When urban water managers eye the volumes of water used in irrigation, using words such as "waste," farmers' animosity toward the cities grows. Irrigators want to keep their supplies, especially if subsidized, or to realize an economic gain for forfeiting their water rights or investing in water use efficiency. Naturally, those holding the cards in uncertain times have a substantial advantage, so irrigation districts often prefer to retain their rights even when offered what appears to be a good deal – witness the Imperial Irrigation District's recent rejection of an initial water purchase offer from the Metropolitan Water District (Peterson, 1985). Nonetheless, thirsty cities, eager to secure supplies, can lobby for new definitions of reasonable use in agriculture or condemn agricultural water rights outright. Without active water markets, an increase in available water supplies, or some equitable adjustment in the rules governing the allocation and reallocation of resources, pressures on water resources will only mount (El-Ashry and Gibbons, 1986).

For many reasons, historical strategies for increasing surface supplies by capturing excess streamflows and constructing storage facilities are approaching obsolescence. First, the optimal reservoir locations have been developed, and marginal returns on the construction of new storage are diminishing. Remaining sites are suitable only for smaller dams, and they present engineering challenges that are inevitably costly. Costs for all types of capital construction are up, and federal funds for large water projects are drying up in an age of fiscal austerity. New formulas for cost sharing between the federal and state governments will probably shift more of the financial burden to the states. Further, a new alliance of budget cutters and environmentalists has focused attention on the full costs of new dams, including the loss of free-flowing river reaches and the inundation of fertile lands and the habitats of fish and wildlife. Even small-scale municipal reservoir projects that might pass the cost-benefit test face expensive legal battles over environmental impacts.

Proposals for interbasin transfers of water have met with the same resistance in recent years. Environmentalists point out the lost value of recreation sites and wildlife habitat as well as the potential for ecological disruption, especially where foreign aquatic species are introduced to a targeted watershed. Grandiose engineering feats still receive media attention, but most water managers now recognize that the costs are prohibitive, and they no longer take seriously such schemes as transferring water from the Great Lakes and Canada to Arizona and Texas. In addition, there are few river basins whose inhabitants would give up excess water supplies to another geographic area without compensation. Basins-of-origin protection clauses in state laws may be invoked to ensure that potential exporting basins are rewarded.

Another drawback of many proposed water projects is the fact that the marginal cost of developing new increments of supply usually exceeds the marginal value of water in the intended use. Federal benefit-cost estimates for irrigation projects invariably include the secondary indirect benefits of reservoir construction and increased farming activity so as to show a positive ratio and to justify a project. The marginal value of water in municipal use is not high enough to support many proposed projects for increasing urban supplies. Only when project costs are shifted – by drawing on general revenues rather than exposing costs directly through higher consumer water bills – does the public support such projects.

Acceptance of the limitation of further supply-side thinking in water resources management is not universal. Nonetheless, when citizens are allowed to choose among alternatives, they often reject large water projects resoundingly. A combination of regional self-preservation on the part of the water-rich and a universal distaste for heavy taxes helps explain why. Recently, California voters rejected plans for a north-south interbasin canal, the peripheral canal, suggesting that the historical pattern of moving more water from north to south in California will not continue (see Chapter 6).

When increasing competition limits choices for expanding usable supplies, the alternative is to reduce water demands. All evidence indicates that the number of people living in the arid West will continue to rise. Barring a highly unlikely mass movement to stop or slow growth, the region's carrying capacity will be pushed to its limit, or at least to that of its water supply. Arguably, the West's natural resources and environment are already endangered by current development.

If the western states are to meet the future with prosperous economies, satisfied citizens, and a healthy environment, policy must change. The future of western water use, as the case studies presented here have shown, lies in conservation, improved efficiency, and the reallocation of supplies among users and sectors. Unfortunately, the prior appropriation doctrine and the institutions that implement water law stress the initial allocation of water supplies. Their usefulness for an orderly, effective, efficient, and ongoing reallocation process is under scrutiny in every state in the West. Many observers believe that private property rights and free markets are just as necessary for water as for any other commodity (Anderson, 1983). But the extent to which the prior appropriation system can be modified and reinterpreted to allow water markets to operate effectively and fairly is not yet clear; nor has a consensus formed on the best ways to protect instream flows and water quality. Although improving water management appears simple enough, tremendous disagreement persists over the specific steps that should be taken.

Some change is already afoot. When economic pressures build and the marginal values of water use in different sectors diverge widely, institutions do evolve. Yet, too often, institutions change haphazardly to serve new powers – in the case of western water, urban interests. The greater play of competitive forces could help reallocate water to its most valuable uses, thereby increasing the total benefits of water use. But the challenge is to chart a path that does not ignore the consequences of transfers on third parties and to protect such public benefits as wildlife habitat, water quality, and instream flows for recreation.

Policy change aimed at increasing the efficiency of water use in all sectors, encouraging the movement of supplies from one use to another, and protecting the environment can take place at many levels. Managers of urban water agencies and irrigation districts must stretch their respective water budgets, and states must facilitate transfers among sectors while protecting the environment. Further, the obvious void in interstate institutions for basinwide water management challenges western states to cooperate and to maximize benefits from water resources throughout the West.

Urban water policies

Population growth in western cities has recently slowed somewhat, but forecasts call for continued increases in the absolute number of people living and working in them. Simply appropriating new supplies is not an option for most of these cities, and the costs, for both construction and legal fees, of importing new water supplies have become prohibitive. Given a limited water supply and a growing population, balancing water supply and demand will require new and creative approaches.

Temporarily tabling the question of buying or leasing existing water supplies from other sectors, the most viable option for cities with growing water demands and limited water budgets is to reduce and restructure that demand, thus lowering per capita water use. The most effective demand curtailment program would contain elements of pricing, regulation, and education. The carrot-and-stick approach to water management combines pricing water at its marginal cost to give consumers a clear idea of water scarcity with educating them on the need to reduce use and, perhaps, rewarding them for saving water (El-Ashry and Gibbons, 1986).

The key to making this policy package work is setting water prices so that the marginal cost of providing water is not obscured. Too often, the water service agency uses an average cost-pricing formula just to cover the total costs of water service. But in many cases, a city's water supply curve exhibits large discontinuities. For example, Tucson currently mines high-quality groundwater costing approximately $45 per acre-foot. But as groundwater mining is phased out, the city will have to turn to Central Arizona Project (CAP) water that costs approximately $250 per acre-foot (excluding subsidies) once it is treated to bring it up to comparable quality (see Chapter 7). Ideally, Tucson residents should pay the marginal cost of the new supplies, the CAP water, and thus base their consumption patterns on full knowledge of the cost of the *next* source of water. If water agencies are to maximize the benefits of resource use and allocate water efficiently, the marginal benefit of a consumer's water use should equal the marginal cost.

When average water costs are increasing, using average cost-pricing rules makes prices lower and consumption greater than they would be under marginal cost pricing. Clearly, imposing marginal cost pricing would dramatically affect total water demand in many cities. As shown in the Vaux case study (Chapter 6), a change to marginal cost pricing in the Metropolitan Water District of Southern California would reduce demand 260,000 acre-feet in 2000 and 340,000 acre-feet in 2020; per capita use reductions would be small, amounting to 9 percent in 2000 and 10 percent in 2020; and water prices would be 13.1 percent and 12.6 percent higher, respectively. These figures illustrate how small changes in daily household use can substantially change total water consumption within a basin.

Even with average cost pricing, the rate structure can make a difference. Traditionally, water is priced in declining block quantities: the more water a household uses, the less it costs for each successive increment. The odd result is that the heaviest users pay the lowest average water price. Many cities still using average cost pricing have inverted this rate structure so that they now charge more for each successive block – a sensible way to discourage profligate water use.

Closely related to water pricing is financing water-supply acquisitions and infrastructural improvements. Cities usually subsidize the development of urban water supplies with public funds. Back-door financing schemes that rely on general city taxes or special bonds in addition to water rate revenues keep water bills artificially low and thus encourage water consumption. Low water prices in western cities may already have aided population growth, and this expansion increases pressures on existing water supplies.

Of course, water pricing policy is linked to social concerns. Policy analysts, citizens, and politicians all want to know how increasing water rates affects the poor and those on fixed incomes. Also troubling is the inevitable clash with public utility law, which enjoins public water service agencies from profit making. These two related problems can both be avoided if either a rebate scheme or a lifeline block scheme is instituted. In the rebate program, profits made under marginal cost pricing could be redistributed to those for whom the higher water bills represent a hardship. Under a lifeline block program, an initial, necessary quantity of water for the average household is priced well below cost; additional blocks are priced above cost. On balance, the expense and red tape of administering rebates makes the lifeline scheme the more appealing approach.

From the standpoint of a region's overall water budget, separating household water use into indoor nonconsumptive use and outdoor consumptive use is crucial. In most arid cities, municipal wastewater is treated and used to irrigate public parks and golf courses or is appropriated farther downstream by other sectors. In some places, treated municipal effluent is traded for higher-quality water from the agricultural sector. Household consumptive uses, such as lawn irrigation and evaporative air conditioning, are thus the most important to curtail when supplies tighten. Because they are also the uses most responsive to water price, no water service agency can ignore the demand-management potential of rational pricing.

Directly regulating household water use is less effective alone than pricing, but it can be a valuable adjunct. Fortuitously, outdoor uses are also the easiest to regulate directly; imposing alternate-day lawn watering and banning ornamental fountains, for example, are not administrative nightmares. At one time or another during droughts, most cities have resorted to these restrictions. Another alternative is for the water utility to undertake or help finance conservation investments on the consumer side of the meter.

These potential gains notwithstanding, most managers are uncomfortable with the idea of controlling water demand or playing the role of water police. The mandate of most water service agencies is to develop and supply water to meet anticipated demand, whatever it might be. Typically, decisions on overall urban growth and urban planning are left to the city councils and other officials. But roles must change. As they find themselves in much the same position as the electric utilities in the mid-1970s, when the cost of new electric plants became prohibitive and the cost of reducing demand justified conservation investments, water service agencies will have to outgrow their traditional role as water purveyors. Urban water managers should note the success over the last ten years of energy conservation and education programs initiated by the electric utilities, institutions that once resisted demand management (Edison Electric Institute and Electric Power Research Institute, 1984).

Programs for encouraging conservation by appealing to the consumers' goodwill usually work best when drought brings out public camaraderie and civic responsibility. But education programs in less traumatic times can have a longer-lasting effect. Tucson's "Beat the Peak" campaign exemplifies a media blitz launched by a city to educate and enlist public cooperation in voluntary water conservation. When rebates, tax credits for purchasing water-saving appliances or landscaping, or other financial incentives are in place, results are likely to be impressive.

On the whole, it is far easier to propose broad policy changes than to implement these innovative programs. The public now views unlimited and inexpensive water almost as a basic right. Clearly, in most western cities, citizen awareness and involvement in the activities of their water service agency need improvement.

Unfortunately, the tradition of average cost pricing and the mandate for water provision (as opposed to demand management) are entrenched. One common argument against raising water prices is that municipal water use is not responsive to price. But volumes of evidence from economic and public policy studies say otherwise. Urban water managers also claim that most people balk at the lowering of lifestyle they think comes with reduced water use. But why not let households themselves decide whether to pay more for their lush lawns and golf courses or to cut back on their water consumption? Because taxes are higher in the end, residents of cities that finance supply-side developments rather than raise water prices end up paying one way or another.

Agricultural water policies

For urban water utilities, water demand management should be a high priority. But as long as water users in other sectors have plenty of water and are using it unproductively, the urban water manager trying to make ends meet should also try to obtain these local supplies through purchase, lease, or the exercise of eminent domain.

Without question, western agricultural interests feel the pressure to release water supplies for urban uses. Increases in irrigation efficiency could release water for other sectors without significantly decreasing agricultural output or accelerating land retirement. For example, if 90 percent of a state's water consumption is for irrigation, increasing efficiency by just 10 percent would double the amount of water available for urban residences and businesses.

Unfortunately, incentives for water conservation in agriculture are few, but disincentives are ubiquitous. Neither water pricing nor visible opportunity costs encourage the farmer to invest in the technology and management expertise needed to conserve water. Without remuneration or hope of later economic gain, a farmer has no reason to make expensive investments in irrigation efficiency.

A few states have recognized the need to provide conservation incentives. The groundwater mining problem on the Texas High Plains has spurred politicians and agricultural extension officials to ask the state to provide low-interest loans, education programs, and subsidies for investment in irrigation technology (see Chapter 4). But the Texas situation is unique because no other economic sectors are vying for agricultural water supplies, and the local indirect effects of aquifer depletion justify state intervention. In regions with diversified economies, state actions alone cannot offset the lack of economic incentives that would foster broad-scale conservation.

One of the most obvious disincentives to conservation is the low cost of water in agriculture. Congress and the Bureau of Reclamation have traditionally offered huge subsidies to agriculture. Farmers pay nowhere near the full costs of the federal water they use to irrigate crops, many of which have low water values. Congress struck at this problem when it passed the Reclamation Reform Act of 1982. The act increased from 160 to 960 acres the limit set in 1902 on acreage receiving federally subsidized water, but subsequent rules and regulations require individuals and districts wishing to take advantage of the higher acreage limit to amend their contracts with the Bureau of Reclamation and agree to pay full operation and maintenance costs on all water received. For the San Luis District in California, the current water price of $10.90 per acre-foot would rise fivefold to $49.89 per acre-foot (LeVeen and King, 1985). Obviously, political resistance to lowering subsidies in existing projects makes this a delicate subject. But the Bureau should find ways to renegotiate contracts and reduce federal subsidies to irrigators, and Congress should certainly require full-cost pricing in any new water projects.

The second disincentive to conserving water in irrigation, the lack of opportunity costs, is also difficult to remedy. Theoretically, a free market for water would produce the necessary price signals and incentives for conservation. If a farmer stood to gain more from selling or leasing water than from irrigating crops, that water would move from farming to another use of higher value. The operation of a water market could provide the mobility and price signals needed to channel water supplies to the most valuable uses and to maximize total benefits (El-Ashry and Gibbons, 1986).

Theory notwithstanding, a complete change to a water market system with absolute property rights could cause problems in some western states with unappropriated water. The prior appropriation doctrine holds that water belongs to all the people, with the ownership vested in the state as trustee. Allowing people to file for water rights that they intend to sell at a profit rather than use would be anathema to the intent of the law. In those few states with unappropriated supplies, the solution would be to let the states claim any remaining water and manage it for instream flow or resell it later when other users need the water.

Much more serious challenges to the efficiency of water markets are traditional market failures. With full markets, such public goods as clean water or instream flows for recreation could be undersupplied, and such common property resources as groundwater aquifers could be overutilized. Water quality protection and preservation of streams and rivers for recreation and wildlife habitat could fall between the cracks in a competitive marketplace for water, just as they have under the prior appropriation doctrine. Expanded water markets could also encourage groundwater extraction in some areas: people rushing to make profits on their surface rights will resort to pumping groundwater for their own use. A well-conceived program in which surface supplies and groundwater are co-managed could curb excess extraction.

Despite these potential problems, closing the door on water markets to improve resource management would be a mistake. By clarifying property rights in water and emphasizing and adjusting provisions in prior appropriation law, the western states could eradicate the obvious barriers to making water resources more flexible and productive. Limited water markets for trading or leasing would provide needed experience and standards for future operations. At the same time, states could help correct the classic market failures, not by replacing market forces with administrative decisions but by refining the rules and institutions to allow market forces to operate efficiently.

Unfortunately, legal and institutional barriers to water conservation and water markets are numerous. State laws generally do not give irrigators the property rights to conserved or salvaged water. If a farmer increased irrigation efficiency, the water saved could not be sold or applied to new lands. Instead, it would remain in the ditch or stream for use by the next senior appropriator. However, California has recently passed a law allowing the sale of conserved and salvaged water. In 1980 the Idaho Supreme Court also ruled that an appropriator retains his right for all water, including that salvaged by reducing seepage from transmission systems (Shupe, 1985, p. 20). On the other hand, Nevada, Arizona, and Wyoming still have counterproductive land appurtenancy requirements that prevent irrigators from using conserved water on other land.

The definition of property rights for water can also hinder water markets by making the cost of transfers too steep. In Colorado's adversarial and cumbersome system of water courts, where rights are defined by withdrawal quantities, it is up to would-be sellers to prove that the amounts for sale come from their traditional consumptive use and will not affect return flows. In short, legal fees and time-consuming proceedings remove much of the incentive to transfer water. Changing the basis of a water right from a withdrawal quantity for a specified use to a consumptive quantity has also been proposed as a way around the high costs of proving consumptive use for each intended transfer.

Another barrier to trade in water is the institutional structure for water use in agriculture (Lee, 1983). The irrigation district's power lies in its prerogative to approve or veto most transfers: individual irrigators do not hold the property rights to water. Obviously, the district management's objectives may not coincide with those of the individual members. The voting system used to elect district supervisors (one vote per acre or property owner or per dollar of assessed valuation) can skew political power within the district and affect its willingness to approve transfers. Nonetheless these special districts are creatures of the state legislatures and, as such, can be made subject to state water policies.

Some Bureau of Reclamation policies on the sale of water rights or change in type of use may be additional impediments to improved water management (see "Free Market Proposed," 1986). Experience in California indicates that water from a project chartered for multiple uses may be transferred from agricultural to urban users. (Transfers among irrigators are allowed in any case.) In projects whose original charter excludes multiple uses, Bureau policy is not clear, although the voluntary exchange of water has been allowed in the past. In the Emery County Project (agriculture only), Utah Power and Light Company was allowed to lease 6,000 acre-feet per year of irrigation water for 40 years (U.S. Department of the Interior, 1983, p. 5). Even though current policies prevent municipal and industrial users from enjoying the entire subsidy, Bureau contractors can still realize large profits on water sold.

Not surprisingly, many taxpayers disapprove of the potential windfall created if irrigation districts are allowed to sell federally subsidized water at market prices. But the resulting increases in irrigation efficiency, with concomitant decreases in salinity and other water quality problems, lower the taxpayers' water cleanup bills. And the overall gain in efficiency within the particular basin can offset the windfall further; urban dwellers are better off buying some agricultural water than developing new, expensive supply projects. In fact, both the irrigators, who can sell their water for more than its worth in growing crops, and cities, which can delay or even abandon plans for new projects, win (Willey, 1985). In California's South Coast Basin, the gains resulting from trade among the Imperial Irrigation District, the southern agricultural areas, and the cities, could reach $100 million per year by 2000 and almost $150 million per year by 2020, not including the benefits of water quality improvement (see Chapter 6).

A related problem is the oft-heard argument that expanded water markets could indirectly harm rural communities serving agricultural areas. Yet, because irrigation water use is so substantial in every western state, a great deal of water can be freed with quite achievable efficiency improvements – changes that alone would not eliminate agriculture. Where a substantial decline in agricultural activity is foreseen, the erosion of public revenues for such expenditures as schools and roads is a genuine concern (Engelbert and Sheuring, 1984, part III).

Fortunately, there is a compromise position on the questions of windfalls and the indirect effects of water sales. Some of the windfall can be recaptured by the federal, state, and local governments when water is transferred out of agriculture, and some can be reaped by the farmer. The keys here are making the farmer's share large enough to preserve the incentive to sell, repaying the federal government something for the national taxpayer who initially financed the projects, and making sure that the state and local governments collect enough to ameliorate any adverse local effects of reduced agricultural activity (El-Ashry and Gibbons, 1986).

Beyond the legal and political barriers to water markets, states need to consider the vehicles for water trading. Ideally, water rights could be sold or rented. Because leasing preserves the ownership of the right, it appeals to speculators and those who want a water hedge for times of drought. Conditional leasing also holds appeal, especially to cities that may need extra supplies only during droughts. One particularly flexible, but nonetheless formal, mechanism for handling trades is water banking. Although limited by the water-delivery system's physical constraints, water banks have worked before. During the 1976–1977 drought, the state of California and the Bureau of Reclamation arranged for water owned by the Metropolitan Water District but stored in the San Joaquin Valley to be sold to farmers in the valley and, through exchange agreements, to municipal water users in northern California (U.S. General Accounting Office, 1978, p. 22).

Environmental policies

Besides eliminating the barriers to effective water markets, state governments have a distinct role in redressing the inefficiencies that result when the private and social benefits and costs of water use diverge significantly from the social benefits and costs. Nowhere is this role more important than in the *protection* of instream flow and water quality.

Most western states recognize instream flow rights. But until the states systematically investigate their instream flow needs and exercise their legal authority to protect minimum flows, disputes over instream flows will chill the transfer of water rights. Ideally, state agencies, in close cooperation with federal land management agencies and Indian tribes, should investigate every river reach and decide on the appropriate level of flow. They should also provide legal protection by appropriating water under the name of state or federal rights, setting minimum streamflows beyond which water cannot be removed offstream, defining instream flow as a beneficial use open to individual or group appropriation, or legislating protection directly through a state-level Wild and Scenic Rivers Act. In addition, extending the Public Trust Doctrine to inland rivers, as it has been in California and Idaho, could help groups fighting reduction in streamflows and other negative environmental impacts of out-of-basin water transfers (Shupe, 1986, p. 2).

States should also protect the quality of instream flow. So far, progress in basinwide salinity control has been disappointing. The delay in implementing effective control measures does not reflect a lack of technically feasible and cost-effective measures: on-farm management practices that reduce leaching, deep percolation, and saline surface runoff are the most economical way to lessen the salt load from irrigation. But those most responsible for the problem see nothing to gain from reducing their saline outflows or, in the extreme case, retiring their lands. Thus, measures whose costs can be shared broadly are adopted even if they are not the most cost-effective means of control. For example, the projected capital cost for implementing the Bureau of Reclamation's water quality improvement package in Grand Valley, Colorado, is $280 million (in 1984 dollars), a sum that will come entirely from federal funds. The primary beneficiaries – to the tune of $15 million a year – are not the Grand Valley irrigators, but downstream water users in Arizona and southern California (El-Ashry, von Schlifgaarde, and Schiffman, 1985).

Traditionally, the Bureau of Reclamation's salinity control programs have been such capital-intensive structural projects as diverting river flows around natural sources of salinity, building desalination plants, or lining water-delivery canals with concrete – invariably expensive undertakings. Instead, the new emphasis of state and Bureau programs should be on controlling salinity from irrigated agriculture. The ideal encouragement for farm-level improvements would come from higher water prices and active water markets. Combined with numerical standards and damage assessment by the states, water markets would give irrigators the necessary incentive to improve efficiency and lower saline seepage and runoff.

Toxic elements in agricultural return flows also pose extensive environmental problems. The pollution of Kesterson Wildlife Refuge in California's Central Valley is the most celebrated example of selenium's fatal consequences for fish and wildlife, although the problem is widespread (Wahl, 1986, p. 11). Like salinity cleanup programs, selenium cleanup can be phenomenally expensive once the element has concentrated to lethal proportions in marshes and lakes. Once again, who pays? The taxpayer is not likely to subsidize ongoing cleanup and disposal programs for the toxic by-product of an already heavily subsidized irrigation operation. As with salinity, the optimal solution is to encourage farmers with drainage problems to sell their water at market prices and retire their lands from production.

Interstate water policies

The prior appropriation doctrine that governs water rights within states colors the way water is viewed among states. The Colorado River Compact assigns shares of Colorado River flows to the basin states, with the stipulation that if an Upper Basin state cannot use its entire share, the excess water flows downstream to satisfy water demands in the Lower Basin states. Although the Upper Basin thus holds paper rights to develop its water, this division impedes efficient basinwide water use. So long as an Upper Basin state cannot benefit from unused releases to the Lower Basin and fears the loss of its water rights in the future, it will continue to support uneconomic water development projects and inefficient water use within its borders. The Central Utah project and the Dolores and Animas–La Plata projects in Colorado exemplify this preemptive development: Utah and Colorado citizens want to know for certain that Colorado River supplies will be there when they need them, whether for agricultural or municipal use (El-Ashry, 1977).

In a sense, under Colorado River Compact provisions, the states do not even own their waters. Colorado cannot charge Los Angeles for the unused outflows that eventually provide that city with part of its municipal water supply. However, the Lower Basin users could collectively pay the Upper Basin states not to divert and develop their entitlements further if such interstate water agreements and water markets were allowed. The problem with a specific irrigation district's negotiating with a Lower Basin city to sell unused water is that, legally, those released waters should go automatically to the next senior appropriator. The Colorado River flows that reach California are allocated on a strict priority basis, with large irrigation districts, not cities, at the top of the list. The Galloway Group's plan to sell surface water from Colorado to municipal users in San Diego is a case in point (Gross, 1985). To date, no compact explicitly allows water sales across state lines. Some legal experts suggest that interstate transactions with water that has been "propertyized," or allocated by the prior appropriation system within an Upper Basin state, will be possible under certain interpretations of compact provisions. But difficult questions surround water that has not yet been given a vested property right by an Upper Basin state (Meyers, 1986).

The dilemma surrounding management of the Colorado River is that the Upper Basin states continue to use water inefficiently and to invest in uneconomic projects despite the sizable economic value of water (both instream and offstream) derived from letting water flow downstream to the Lower Basin instead. The crux of the matter is that the Upper Basin states cannot share in those returns. Additionally, while the Bureau of Reclamation, the Environmental Protection Agency, and the states continue to search for cost-effective solutions, salinity remains the major water quality problem in the Colorado River, with the Upper Basin states contributing most of the salt load and the Lower Basin states shouldering the costs of lowered water quality.

Today's emerging conflicts in the Colorado River Basin highlight problems that will only worsen if current policies are continued. The states of the Colorado River Compact now need to reorder their water development priorities, facilitate fair and equitable interstate water markets, and improve the current fragmented and inefficient management of a valuable and scarce resource. This tall order will require considerable cooperation and discussion among the states – which need to devise a forum for resolving disputes and anticipating and heading off new problems – and will be furthered if the states become serious about data gathering, water quality monitoring, and developing appropriate compensatory arrangements. Clearly, western states must act in concert.

Unless the rules and customs for using the Colorado River can be adjusted to the changing demands being placed upon it, intrusion by federal agencies, the courts, and Congress will escalate.

In addition to resolving some of the externalities of salinity and instream flow benefits, reinterpreting and revising the Colorado River Compact and state agreements to include interstate water markets could alleviate some of the dilemmas that rapidly growing western cities face. Many of the urban areas in the arid Southwest, such as Phoenix and Tucson, face stark limitations on their water supplies. But because the physical components of the delivery system are in place – the river itself and the various aqueducts that carry water to central Arizona and southern California – water markets could enable cities in the Lower Basin to keep up with their growing water demands. Still, reallocating irrigation water through an interstate water market, helpful as it may be in the short term, cannot solve all the problems of unchecked urban growth.

Conclusions

The arid West has entered a new era. What was once an endless frontier is now a vital, populated region fully integrated with the rest of the country. What were once seemingly endless stores of natural wealth – water, land, forests, blue skies – are now resources under stress. Water, one of the West's most essential resources, is no less physically abundant today than it was decades ago, but it is now oversubscribed and, in many cases, polluted.

In light of emerging conflicts over water allocation and water quality degradation, western states must now reevaluate their water management priorities. As the demand for water for cities and for recreation and instream flow protection grows, water supplies are being stretched to their limits. The laws and institutions governing natural resources in the old West have survived the transition from plenty to scarcity, but they may not be providing the best, most efficient resource management under today's conditions of rapid population growth and urbanization. Although irrigation remains by far the largest western water user, urban thirst threatens traditional water use patterns.

Growing urban demand is not the only threat to agricultural water use, either. Soil and water salinity and toxic return flows highlight the negative environmental impacts of traditional irrigation practices. The usual structural solutions to water quality problems, such as cleaning up rivers downstream from the saline sources, are phenomenally expensive and economically unwarranted when compared to cheaper on-farm alternatives. But just as water quantity problems are not easily solved given existing rules and customs for water allocation, neither are water quality problems.

The time-honored strategy of increasing water supplies and correcting water quality degradation through capital-intensive projects has reached its limits; financial, environmental, and legal obstacles are overwhelming. For the most part, alternative water management schemes should now replace structural water development. At the heart of decisions on western water management should be least-cost accounting, incorporating all the environmental and third-party costs of any proposed path.

Growing cities, for example, that have already instituted marginal cost pricing but still need more water should follow least-cost rules in choosing between new water supply projects and such alternatives as purchase or lease of water from agriculture (including irrigation efficiency investments in exchange for water supplies) or demand-reduction programs. In many cases, water service agencies will undoubtedly discover that "new" supplies found through water conservation programs are cheaper than new supplies obtained by expanding the system or purchasing irrigation water rights in a water market.

As a natural response to water markets, higher water use efficiency in irrigated agriculture makes more water available and has many positive environmental consequences – reduced saline and toxic return flows from irrigation and augmented supplies for recreation or waste dilution. In essence, these new policies constitute a positive-sum game. Everyone would be better off: cities could buy water more cheaply than they could build new projects, national taxpayers would not be asked to foot ever-higher salinity control and water development bills, and farmers could sell water for more than it could return in irrigated crop production.

Overall, the new water policies presented in this volume are rooted in the commonsense notion that arid and semiarid regions are going to have to live within their limited water budgets. Even full mobility of water supplies and much more efficient water use cannot change the fact that much of the West is a desert, or close to it. Yet, increasing the productivity and substitutability of water through water markets will ease competition and conflict, enabling western states not only to survive, but also to thrive in this era of water scarcity.

These recommendations also stem from the realization that many of the traditional water policies set when the West had fewer people and a simpler economy no longer serve the public interest and do not meet the new challenges of urbanization and environmental degradation. Resolving emerging conflicts over water will require creative and sometimes controversial steps – but the West needs to act now, *before* crises govern the management of this vital resource.

References

Anderson, Terry L., ed. 1983. *Water Rights: Scarce Resource Allocation, Bureaucracy, and the Environment.* Pacific Institute for Public Policy Research, San Francisco.

Edison Electric Institute and Electric Power Research Institute. 1984. *Demand-Side Management: Overview of Key Issues.* Washington, D.C.

El-Ashry, Mohamed T. 1977. Statement at Department of the Interior Water Projects Review Team. Hearings on Fruitland Mesa, Dolores, and Savery-Pot Hook Water Projects, Mar. 21, Grand Junction, Colorado.

El-Ashry, Mohamed T., and Gibbons, Diana C. 1986. *Troubled Waters: New Policies for Managing Water in the American West.* Washington, D.C.: World Resources Institute.

El-Ashry, Mohamed T., van Schilfgaarde, Jan, and Schiffman, Susan. 1985. "Salinity Pollution from Irrigated Agriculture." *Journal of Soil and Water Conservation, 40*(1), 48–52.

Engelbert, Ernest A., and Scheuring, Ann Foley, eds. 1984. *Water Scarcity: Impacts on Western Agriculture.* Berkeley: University of California Press.

"Free Market Proposed." 1986. *U.S. Water News, 2*(11), 1.

Gibbons, Diana C. 1986. *The Economic Value of Water.* Washington, D.C.: Resources for the Future.

Gross, Sharon P. 1985. "The Galloway Project and the Colorado River Compacts: Will the Compacts Bar Transbasin Water Diversions?" *Natural Resources Journal, 25,* 935–960.

Lee, Dwight R. 1983. "Political Provision of Water: An Economic/Public Choice Perspective." In James N. Corbridge, Jr., ed., *Special Water Districts: Challenge for the Future.* Proceedings of the Workshop on Special Water Districts held at the University of Colorado, Sept. 12–13. Boulder: University of Colorado School of Law. 51–70.

LeVeen, E.P., and King, L.B. 1985. *Turning Off the Tap on Federal Subsidies.* San Francisco: Natural Resources Defense Council.

Meyers, Charles J. 1986. Personal communication, May 19.

Peterson, Cass. 1985. "California's Liquid Asset." *Washington Post,* Nov. 3.

Shupe, Steven J. 1985. "Wasted Water: The Problems and Promise of Improving Efficiency under Western Water Law." Paper presented at a conference on Colorado Water Issues and Options: The 90's and Beyond. University of Colorado School of Law, Natural Resources Law Center, Oct. 8–9.

Shupe, Steven J. 1986. "Emerging Forces in Western Water Law." *Resource Law Notes.* Boulder: University of Colorado, Natural Resources Law Center.

U.S. Department of the Interior, Office of Policy Analysis. 1983. "Meeting Water Needs through the Voluntary Exchange of Water – Continuation of Past Policy." Staff paper. Washington, D.C.

U.S. General Accounting Office. 1978. "Better Water Management and Conservation Possible – But Constraints Need to Be Overcome." Washington, D.C.

Wahl, Richard W. 1986. "Cleaning Up Kesterson." *Resources Newsletter*. Washington, D.C.: Resources for the Future.

Willey, Zach. 1985. *Economic Development and Environmental Quality in California's Water System*. Berkeley: University of California, Institute of Governmental Studies.

Appendix: Advisory Panel, Arid Lands Project

Paul Bloom	Attorney; former General Counsel, New Mexico State Water Engineer's Office
John Bryson	Senior Vice President, Southern California Edison Company
Harold E. Dregne	Director, International Center for Arid and Semi-Arid Lands Studies, Texas Tech University
Robert M. Hagan	Professor, Department of Land, Air and Water Resources, University of California at Davis
Jim Hightower	Commissioner of Agriculture, State of Texas
Carl N. Hodges	Director, Environmental Research Laboratory, University of Arizona
Helen Ingram	Professor of Political Science, University of Arizona
H.O. Kunkel	Dean, College of Agriculture, Texas A&M University
Dean E. Mann	Professor of Political Science, University of California at Santa Barbara
Don Maughan	Deputy Director, Water Management, State of Arizona (now Chairman of the California State Water Resources Control Board)
Charles Meyers	Attorney, Gibson, Dunn, and Crutcher, Denver; former Dean, Stanford Law School
Gilbert White	Gustavson Distinguished Professor Emeritus of Geography, Institute of Behavioral Sciences, University of Colorado
Zach Willey	Senior Economist, Environmental Defense Fund, Berkeley

Index